WITHDRAWN

Benchmark Papers in Systematic and Evolutionary Biology

Editor: Carl Jay Bajema—Grand Valley State Colleges

MULTIVARIATE STATISTICAL METHODS: Among-Groups Covariation / *William R. Atchley and Edwin H. Bryant*
MULTIVARIATE STATISTICAL METHODS: Within-Groups Covariation / *Edwin H. Bryant and William R. Atchley*
CONCEPTS OF SPECIES / *C. N. Slobodchikoff*
ARTIFICIAL SELECTION AND THE DEVELOPMENT OF EVOLUTIONARY THEORY / *Carl Jay Bajema*
NATURAL SELECTION THEORY: From the Speculations of the Greeks to the Quantitative Measurements of the Biometricians / *Carl Jay Bajema*
SEXUAL SELECTION THEORY PRIOR TO 1900 / *Carl Jay Bajema*
PALEOBOTANY, PART I: Precambrian Through Permian / *Thomas N. Taylor and Edith L. Smoot*
PALEOBOTANY, PART II: Triassic Through Pliocene / *Thomas N. Taylor and Edith L. Smoot*
CLADISTIC THEORY AND METHODOLOGY / *Thomas Duncan and Tod F. Stuessy*

THE ORIGIN OF EUKARYOTIC CELLS

Edited by

BETSEY DEXTER DYER
Wheaton College

and

ROBERT OBAR
Worcester Foundation
for Experimental Biology

A Hutchinson Ross Publication

VNR VAN NOSTRAND REINHOLD COMPANY
——————————————— New York

Copyright © 1985 by **Van Nostrand Reinhold Company Inc.**
Benchmark Papers in Systematic and Evolutionary Biology, Volume 9
Library of Congress Catalog Card Number: 85-20261
ISBN: 0-442-21952-0

All rights reserved. No part of this work covered by the copyrights
hereon may be reproduced or used in any form or by any means—
graphic, electronic, or mechanical, including photocopying,
recording, taping, or information storage and retrieval systems—
without permission of the publisher.

Manufactured in the United States of America.

Published by Van Nostrand Reinhold Company Inc.
115 Fifth Avenue
New York, New York 10003

Van Nostrand Reinhold Company Limited
Molly Millars Lane
Wokingham, Berkshire RG11 2PY, England

Van Nostrand Reinhold
480 Latrobe Street
Melbourne, Victoria 3000, Australia

Macmillan of Canada
Division of Gage Publishing Limited
164 Commander Boulevard
Agincourt, Ontario MIS 3C7, Canada

15 14 13 12 11 10 9 8 7 6 5 4 3 2 1

Library of Congress Cataloging in Publication Data
Main entry under title:
The origin of eukaryotic cells.
  (Benchmark papers in systematic and evolutionary
biology; 9)
  "A Hutchinson Ross publication."
  Includes indexes.
  1. Eukaryotic cells—Evolution—Addresses, essays,
lectures.   2. Symbiosis—Addresses, essays, lectures.
I. Dyer, Betsey Dexter.   II. Obar, Robert.   III. Series.
QH371.O73   1985   575   85-20261
ISBN 0-442-21952-0

To Lynn Margulis

In non-living habitats, an organism either exists or it does not. In the cell habitat an invading organism can progressively lose pieces of itself, slowly blending into the general background, its former existence betrayed only by some relic. Indeed one is reminded of Alice in Wonderland's encounter with the Cheshire cat. As she watched it, "it vanished quite slowly, beginning with the tail, and ending with the grin, which remained some time after the rest of it had gone." There are a number of objects in a cell like the grin of a Cheshire cat. For those who try to trace their origins, the grin is challenging and truly enigmatic (D. C. Smith, Paper 9).

CONTENTS

Series Editor's Foreword	xi
Preface	xiii
Contents by Author	xv
Introduction	1

PART I: HISTORICAL OVERVIEW

Editors' Comments on Papers 1 through 6   8

1   **WALLIN, I. E.:** The Mitochondria Problem   10
Am. Nat. **57:**255-261 (1923)

2   **GOKSØYR, J.:** Evolution of Eucaryotic Cells   17
Nature **214:**1161 (1967)

3   **SAGAN, L.:** On the Origin of Mitosing Cells   18
J. Theor. Biol. **14:**225-274 (1967)

4   **SCHWARTZ, R. M., and M. O. DAYHOFF:** Origins of Prokaryotes, Eukaryotes, Mitochondria, and Chloroplasts   68
Science **199:**395-403 (1978)

5   **RAVEN, P. H.:** A Multiple Origin for Plastids and Mitochondria   77
Science **169:**641-646 (1970)

6   **DOOLITTLE, W. F.:** Revolutionary Concepts in Evolutionary Cell Biology   83
Trends Biochem. Sci., June, pp. 146-149 (1980)

PART II: MECHANISMS NECESSARY FOR SERIAL SYMBIOSIS

Editors' Comments on Papers 7 through 12   88

7   **CUTTING, J. A., and H. M. SCHULMAN:** The Biogenesis of Leghemoglobin. The Determinant in the Rhizobium-Legume Symbiosis for Leghemoglobin Specificity   90
Biochim. Biophys. Acta **229:**58-62 (1971)

8   **LORCH, I. J., and K. W. JEON:** Rapid Induction of Cellular Strain Specificity by Newly Acquired Cytoplasmic Components in Amoebas   95
Science **211:**949-951 (1981)

9   **SMITH, D. C.:** From Extracellular to Intracellular: The Establishment of a Symbiosis   97
R. Soc. London Proc., ser. B, **204:**115-130 (1979)

Contents

10	SONEA, S.: Bacterial Plasmids Instrumental in the Origin of Eukaryotes? *Rev. Can. Biol.* **31**:61-63 (1972)	113
11	MARTIN, J. P., JR., and I. FRIDOVICH: Evidence for a Natural Gene Transfer from the Ponyfish to Its Bioluminescent Bacterial Symbiont *Photobacter leiognathi*. The Close Relationship Between Bacteriocuprein and the Copper-Zinc Superoxide Dismutase of Teleost Fishes *J. Biol. Chem.* **256**:6080, 6082-6089 (1981)	116
12	FAIRFIELD, A. S., S. R. MESHNICK, and J. W. EATON: Malaria Parasites Adopt Host Cell Superoxide Dismutase *Science* **221**:764-766 (1983)	125

PART III: ORIGIN OF MITOCHONDRIA

Editors' Comments on Papers 13 through 18		130
13	THOMAS, D. Y., and D. WILKIE: Recombination of Mitochondrial Drug-Resistance Factors in *Saccharomyces Cerevisiae* *Biochem. Biophys. Res. Commun.* **30**:368-372 (1968)	132
14	BARATH, Z., and H. KÜNTZEL: Cooperation of Mitochondrial and Nuclear Genes Specifying the Mitochondrial Genetic Apparatus in *Neurospora crassa* *Natl. Acad. Sci. (USA) Proc.* **69**:1371-1374 (1972)	137
15	JOHN, P., and F. R. WHATLEY: *Paracoccus denitrificans* and the Evolutionary Origin of the Mitochondrion *Nature* **254**:495-498 (1975)	141
16	GRAY, M. W., T. Y. HUH, M. N. SCHNARE, D. F. SPENCER, and D. FALCONET: Organization and Evolution of Ribosomal RNA Genes in Wheat Mitochondria *Structure and Function of Plant Genomes*, O. Ciferri and L. Dure III, eds., Plenum Publishing Corporation, New York, 1983, pp. 373-380	145
17	VAN DEN BOOGAART, P., J. SAMALLO, and E. AGSTERIBBE: Similar Genes for a Mitochondrial ATPase Subunit in the Nuclear and Mitochondrial Genomes of *Neurospora crassa* *Nature* **298**:187-189 (1982)	153
18	OBAR, R., and J. GREEN: Molecular Archaeology of the Mitochondrial Genome *J. Mol. Evol.* (1985)	156

PART IV: ORIGIN OF PLASTIDS

Editors' Comments on Papers 19 through 24		162
19	LEWIN, R. A., and N. W. WITHERS: Extraordinary Pigment Composition of a Prokaryotic Alga *Nature* **256**:735-737 (1975)	164

20	BOYNTON, J. E., N. W. GILLHAM, E. H. HARRIS, C. L. TINGLE, K. VAN WINKLE-SWIFT, and G. M. W. ADAMS: Transmission, Segregation and Recombination of Chloroplast Genes in *Chlamydomonas* *Genetics and Biogenesis of Chloroplasts and Mitochondria,* Th. Bücher, W. Neupert, W. Sebald, and S. Werner, eds., Elsevier/North-Holland Biomedical Press, Amsterdam, The Netherlands, 1976, pp. 313-322	167
21	GIBBS, S. P.: The Chloroplasts of *Euglena* May Have Evolved from Symbiotic Green Algae *Can. J. Bot.* **56:**2883-2889 (1978)	177
22	BONEN, L., and W. F. DOOLITTLE: Ribosomal RNA Homologies and the Evolution of the Filamentous Blue-Green Bacteria *J. Mol. Evol.* **10:**283-291 (1978)	184
23	STERN, D. B., and J. D. PALMER: Extensive and Widespread Homologies Between Mitochondrial DNA and Chloroplast DNA in Plants *Natl. Acad. Sci. (USA) Proc.* **81:**1946-1950 (1984)	193
24	SCHIFF, J. A.: Origin and Evolution of the Plastid and Its Function *N.Y. Acad. Sci. Ann.* **361:**166, 167, 172-174, 186, 187-188 (1981)	198

PART V: ORIGIN OF NUCLEOCYTOPLASM

Editors' Comments on Papers 25 and 26		206
25	STARR, M. P.: *Bdellovibrio* as Symbiont: The Associations of Bdellovibrios with other Bacteria Interpreted in Terms of a Generalized Scheme for Classifying Organismic Associations *Symbiosis,* Symposia of the Society for Experimental Biology, No. 29, Society for Experimental Biology, London, England, 1975, pp. 95-104, 121-124	208
26	MARGULIS, L.: Mitochondria: Acquisition by Whom? *Symbiosis in Cell Evolution,* W. H. Freeman and Company Publishers, San Francisco, 1981, pp. 205-213	223

PART VI: ORIGIN OF MOTILITY ORGANELLES

Editors' Comments on Papers 27 through 32		236
27	MAY, H. G., and K. GOODNER: Cilia as Pseudo-Spirochaetes *Am. Microsc. Soc. Trans.* **45:**302-305 (1926)	240
28	BEISSON, J., and T. M. SONNEBORN: Cytoplasmic Inheritance of the Organization of the Cell Cortex in *Paramecium Aurelia* *Natl. Acad. Sci. (USA) Proc.* **53:**275-282 (1965)	244
29	HEIDEMANN, S. R., G. SANDER, and M. W. KIRSCHNER: Evidence for a Functional Role of RNA in Centrioles *Cell* **10:**337-350 (1977)	252

Contents

30	MARGULIS, L., and D. CHASE: Microtubules in Prokaryotes *Science* **200**:1118-1124 (1978)	266
31	GRIMES, G. W., M. E. MCKENNA, C. M. GOLDSMITH-SPOEGLER, and E. A. KNAUPP: Patterning and Assembly of Ciliature are Independent Processes in Hypotrich Ciliates *Science* **209**:281-283 (1980)	273
32	HUANG, B., Z. RAMANIS, S. K. DUTCHER, and D. J. L. LUCK: Uniflagellar Mutants of Chlamydomonas: Evidence for the Role of Basal Bodies in Transmission of Positional Information *Cell* **29**:745-753 (1982)	276

PART VII: ORIGIN OF MITOSIS, MEIOSIS, AND SEX

Editors' Comments on Papers 33 and 34 — 286

33	CLEVELAND, L.: Sex Produced in the Protozoa of *Cryptocercus* by Molting *Science* **105**:16-17 (1947)	288
34	MARGULIS, L., AND D. SAGAN: Evolutionary Origins of Sex *Oxford Surveys in Evolutionary Biology*, R. Dawkins and M. Ridley, eds., Oxford University Press, Oxford, England, 1985, pp. 30-47	290

PART VIII: CLASSIFICATION OF EUKARYOTES

Editors' Comments on Papers 35A through 35E — 310

35A	MARGULIS, L., and K. SCHWARTZ: Monera *Five Kingdoms*, W. H. Freeman and Company Publishers, San Francisco, 1982, pp. 24-29	311
35B	MARGULIS, L., and K. SCHWARTZ: Protoctista *Five Kingdoms*, W. H. Freeman and Company Publishers, San Francisco, 1982, pp. 68-71	317
35C	MARGULIS, L., and K. SCHWARTZ: Fungi *Five Kingdoms*, W. H. Freeman and Company Publishers, San Francisco, 1982, pp. 144-147	321
35D	MARGULIS, L., and K. SCHWARTZ: Animalia *Five Kingdoms*, W. H. Freeman and Company Publishers, San Francisco, 1982, pp. 160-165	325
35E	MARGULIS, L., and K. SCHWARTZ: Plantae *Five Kingdoms*, W. H. Freeman and Company Publishers, San Francisco, 1982, pp. 248-251	331

Author Citation Index — 335
Subject Index — 343
About the Editors — 347

SERIES EDITOR'S FOREWORD

The Systematic and Evolutionary Biology series reprints classic scientific papers on the evolution and systematics of organisms. The volumes in this series do more than just provide scholars with facsimile reproductions or English translations of classic papers on a particular topic. The interpretative commentaries and extensive bibliographies prepared by each editor provide busy scholars with a review of the primary and secondary literature of the field from a historical perspective and a summary of the current state of the art.

Biologists employ comparative methodology to scientifically reconstruct phylogenies—the evolutionary history of life on this planet. Scientists studying the fossil record have identified microfossils of prokaryotes (bacteria) in rocks known to be 3.7 billion years old. Microfossils of eukaryotes have been found in rocks as old as 1.4 billion years. Most of what we know now of the fossil records of these organisms has been discovered within the last 20 years. Most of the classic papers comparing living prokaryotes with living eukaryotes have also been published within the last 20 years. Drs. Betsey Dexter Dyer and Robert Obar have chosen the classic comparative studies of living as well as fossil prokaryotes and eukaryotes for inclusion in *The Origin of Eukaryotic Cells*.

The philosopher of science, Alfred North Whitehead, is reported to have said "We give credit for an idea not to the first man to have it, but to the first one who takes it seriously." It is for this reason that we give a woman scientist credit for developing the symbiotic theory of the origin of eukaryotes from prokaryotes. Lynn Margulis first championed her theory in 1967, only 18 years ago. She has successfully employed hypothetico-deductive-testing scientific methodology in her research. Many of the predictions she has deduced from her theory that eukaryotes originated as the result of symbiotic interactions among prokaryotes have been tested and verified by her as well as others.

Evolutionary biology is a successful science because its theories are testable and have been scientifically tested against empirical evidence; its theories have unified many diverse problems in biology and solved them with the same problem solving strategy; and its theories have been fruitful in opening up new dimensions of scientific investigation. Lynn Margulis's symbiotic theory of the origin of eukaryotic cells from prokaryotic cells is an example of such a successful evolutionary theory.

CARL JAY BAJEMA

PREFACE

The present understanding of the origin of eukaryotic cells is based on the development of two new fields of research in the past 20 years: micropaleontology (the study of microfossils of prokaryotes and early eukaryotes) and biochemical and genetic research on the origin and nature of eukaryotic organelles (mitochondria, plastids, motility organelles). The second of these two fields is the topic of our book.

The publication of the discovery of microfossils in 2 billion-year old rocks by S. Tyler and E. Barghoorn in 1965 exploded the notion that the fossil record began abruptly and inexplicably with large, hard-bodied eukaryotes about 500 million years ago. The field of micropaleontology continues to grow. Microfossils of prokaryotes are now known in 3.7 billion-year old rocks while presumptive eukaryotic fossils have been found in rocks as old as 1.4 billion years. Micropaleontology is an integral part of any understanding of early life and the initial stages of eukaryote evolution.

The first full length treatment of the theory of a symbiotic origin for eukaryotic cells was presented by L. Margulis in 1967 (Paper 3). Opposition to this paper was so strong that it nearly remained unpublished. The theory stated that the mitochondria, plastids, nucleocytoplasm, and motility organelles (that is, most of the major components of eukaryotes) originated as separate prokaryotes (bacteria) that came together in symbiotic associations, forming the first eukaryotes. In her 1967 paper, Margulis unified data and observations on eukaryotic cells into one theory and made specific predictions, many of which were tested and verified in subsequent years. As a result, the theory of a symbiotic origin of eukaryotes is no longer controversial at this writing (1984) and is even included in most discussions of cell evolution in new elementary biology textbooks.

In this book we have assembled classic and in some cases difficult-to-obtain papers that deal with the symbiotic origin of eukaryotes, as well as several early papers of historical interest. A collection of this sort has never before been published in one volume. We believe that it will become a valuable and convenient reference for college libraries and for educators and advanced students in most fields of biology and biochemistry. Many of the experiments described are elegant and inspirational, and many of the papers are exemplary for their literary style as well.

We hope that our introductory list of research criteria for determining whether an organelle or other intracellular structure is of symbiotic origin

Preface

will be particularly useful to theorists and researchers. Also included is a section on the still relatively controversial symbiotic origin of motility organelles, with a collection of the classic papers on this topic. We hope that this volume will encourage potential researchers in the field, as well as provide a collection of background references.

We thank Carl Bajema for critical reading of the manuscript. We thank Lynn Margulis with love, for years of generosity, support, inspiration, tolerance, and enthusiasm.

BETSEY DEXTER DYER
ROBERT OBAR

CONTENTS BY AUTHOR

Adams, G. M. W., 167
Agsteribbe, E., 153
Barath, Z., 137
Beisson, J., 244
Bonen, L., 184
Boynton, J. E., 167
Chase, D., 266
Cleveland, L., 288
Cutting, J. A., 90
Dayhoff, M. O., 68
Doolittle, W. F., 83, 184
Dutcher, S. K., 276
Eaton, J. W., 125
Fairfield, A. S., 125
Falconet, D., 145
Fridovich, I., 116
Gibbs, S. P., 177
Gillham, N. W., 167
Goksøyr, J., 17
Goldsmith-Spoegler, C. M., 273
Goodner, K., 240
Gray, M. W., 145
Green, J., 156
Grimes, G. W., 273
Harris, E. H., 167
Heidemann, S. R., 252
Huang, B., 276
Huh, T. Y., 145
Jeon, K. W., 95
John, P., 141
Kirschner, M. W., 252
Knaupp, E. A., 273
Küntzel, H., 137
Lewin, R. A., 164
Lorch, I. J., 95
Luck, D. J. L., 276
McKenna, M. E., 273
Martin, J. P., Jr., 116
May, H. G., 240
Meshnick, S. R., 125
Obar, R., 156
Palmer, J. D., 193
Ramanis, Z., 276
Raven, P. H., 77
Sagan, D., 290
Sagan-Margulis, L., 18, 223, 266, 290, 311, 317, 321, 325, 331
Samallo, J., 153
Sander, G., 252
Schiff, J. A., 198
Schnare, M. N., 145
Schulman, H. M., 90
Schwartz, K., 311, 317, 321, 325, 331
Schwartz, R. M., 68
Smith, D. C., 97
Sonea, S., 113
Sonneborn, T. M., 244
Spencer, D. F., 145
Starr, M. P., 208
Stern, D. B., 193
Thomas, D. Y., 132
Tingle, C. L., 167
Van den Boogaart, P., 153
Van Winkle-Swift, K., 167
Wallin, I. E., 10
Whatley, F. R., 141
Wilkie, D., 132
Withers, N. W., 164

THE ORIGIN OF EUKARYOTIC CELLS

INTRODUCTION

The category *eukaryotic organisms* includes animals, plants, fungi, and protoctists. The cells of eukaryotes contain nuclei (double membrane-bound packages containing a DNA genome) and, in most cases, mitochondria (double membrane-bound packages in which oxidative respiration occurs). Some eukaryotes (plants and some protoctists) have plastids, which are double membrane-bound packages containing the light-capturing pigments and enzymes for photosynthesis. Eukaryotes undergo mitotic cell division in which the chromosomes (DNA genome plus associated proteins) are moved by microtubule spindle fibers made primarily of tubulin protein. The eukaryotic motility organelle (undulipodium) is a bundle of tubulin microtubules usually with a 9 + 2 cross-sectional arrangement of the microtubules (e.g., cilia and sperm tails) (Paper 30). In addition to tubulin, the motility organelles (undulipodia) are made of about 200 other proteins and the structure is intrinsically motile if provided with an energy source.

Prokaryotes (i.e., the bacteria) differ from eukaryotes in that the DNA of prokaryotes is not contained within a nucleus, and metabolic processes such as photosynthesis are not contained in separate packages or organelles. Prokaryotes also have no tubulin-based motility systems, although at least three species reportedly have tubulin-like proteins (Part IV), the functions of which are unknown. Prokaryotes do not undergo mitosis with tubulin spindle fibers. The prokaryotic flagellar motility system is based on a unique set of proteins, *flagellins,* which are not intrinsically motile (Paper 30).

Eukaryotes are divided into four distinct kingdoms: protoctists, fungi, animals, and plants (Papers 35A, B, C, D, and E). The protoctists make up an extremely diverse group of uni- and multi-cellular organisms that are mostly aquatic. The group lacks extensive tissue differentiation and embryonic stages and includes members that are probably direct descendants of the first eukaryotes, which began to evolve about 2 billion years ago. There are about 30 distinct phylum-level groups of protoctists. Examples of some photosynthetic protoctists are *Chlamydomonas, Euglena, Volvox,* diatoms, dinomastigotes, red algae, and brown algae. These are distinct from the heterotrophic protoctists, organisms that either ingest or absorb their food such as amoebae, sporozoans (a parasitic group), mastigotes, and myxomycetes (slime molds).

Fungi form thread-like cells called *hyphae* that absorb nutrients from the environment. The group has no tubulin-based motility organelles (undulipodia), but it does have mitosis, with tubulin spindle fibers, and mitochondria. The fungal kingdom includes the mushrooms and molds.

The plants and animals both have extensive tissue differentiation and embryonic stages. The animals are heterotrophic, ingesting or absorbing their food. They have mitochondria and, in at least one stage of their life cycles, eukaryotic motility organelles (undulipodia). Plants are oxygenic green photosynthesizers. Descendants of the earliest plants (mosses, ferns, ginkgos) have motile sperm. Motile forms were entirely lost in later plant groups (gymnosperms, angiosperms). Plant cells have mitochondria and chloroplasts.

All organisms interact with other organisms in nature; there are no solitary individuals. The environment of every organism is filled with other organisms, thus the selection pressure on a community of organisms is not a result of merely the physical environment but of other organisms as well. The limitation of resources on Earth causes part of the selection pressure by which interactions between organisms evolve. Metabolic reactions are not perfectly efficient; there are extra products and reactants and usually waste products. Organismal interactions evolve to use these products. The problem of limited space on Earth is also solved in part by interactions in which organisms live upon and inside of each other. Interactions range from situations where one organism benefits from the relationship and the other does not (as in predator/prey relationships and host/parasite relationships) to interactions in which both organisms benefit (symbioses). The boundaries are blurred between the definitions of organismal interactions. Many parasites are benign or are harmful only in a certain life-cycle stage of the host. Most symbioses probably originate as less equitable relationships. The ongoing tendency to interact is probably one of the major mechanisms by which the first eukaryotes began to evolve 2 billion years ago.

The first evidence for life on Earth is 3.7 billion years old and includes microfossils of bacteria-like organisms and laminated sedimentary structures (stromatolites) formed by communities of bacteria-like organisms (Fig. 1). It may be extrapolated that life originated at least 4 billion years ago in order for a diversity of organisms to be present 3.7 billion years ago. The "age of bacteria," in which prokaryotes

Introduction

Figure 1. Proterozoic stromatolite (bacterial skyscraper) (*photo:* Geological Survey of Canada).

diversified and covered every available part of the Earth, lasted for about 2 billion years. During that period almost all of the metabolic processes of prokaryotes and eukaryotes were evolved by the prokaryotes. Fieldwork indicates that modern prokaryotes tend to live in complex communities, and there is evidence for such communities in the fossil record. Modern prokaryotes also have many interactions (ranging from predator/prey to symbioses) with each other and it is presumed (although evidence has not been preserved in the fossil record) that ancient prokaryotes also interacted extensively.

About 2 billion years ago a crisis occurred. Oxygen began to accumulate in an atmosphere which, up to that point, had been

Introduction

anaerobic. Almost all prokaryotes living previous to that time were anaerobic and would have required mechanisms to deal with the accumulating free oxygen. The oxygen, a waste product of photosynthesizing blue-green bacteria, had actually been produced by prokaryotes for almost 2 billion years, but it only began to accumulate in the atmosphere after all reduced rocks on the surface of the Earth had been oxidized. Prokaryotes adapted to the bane of oxygen in several ways: some evolved enzyme systems to bind free oxygen; some evolved metabolic processes that could utilize the oxygen (e.g., respiration, chemoautotrophy); some restricted themselves to deep anaerobic muds where their descendants still live today; and, most importantly for the evolution of eukaryotes, some began interactions with organisms that had found ways to live with the accumulating oxygen. Some of the interactions just mentioned led to the existence of pre-eukaryotes. The evolution of such interactions forms the basic framework for the endosymbiotic theory for the evolution of eukaryotes, and was the major focus in our selection of the papers collected in this volume.

It is difficult to establish with precision just when the first eukaryotes evolved because of difficulties with interpretation of the fossil record. First, not all organisms form recognizable fossils and the best candidates for the first eukaryotes, amoeba-like protoctists, fall into this category. Second, the distinguishing characteristics of eukaryotes—the nucleus, mitochondria, plastids, and motility organelles (undulipodia)—do not preserve well and interpretations of such structures may be ambiguous. The oldest, fairly definite pieces of evidence for eukaryotes are fossils dated at about 1,400 milion years ago (Fig. 2). These fossils are presumably protoctist-like, but how they fit into the evolution of the modern protoctist kingdom is not easily determined. Protoctists (except for those with mineralized parts) do not have a well preserved fossil record.

The first animals are seen in the fossil record at 670 million years ago, the first fungi at 400 million years ago, and the first plants at 425 million years ago (Papers 35A, B, C, D, and E). While there are more fossilizable parts (shells, hard cell walls, skeleton) in members of these three kingdoms, the early record is still unclear. Soft animals, very delicate plants, and unicellular fungi are generally not present in the record.

The endosymbiotic theory is the major theory that explains how the distinguishing characteristics of eukaryotes, nuclei, mitochondria, plastids, and motility organelles (undulipodia) evolved. The theory maintains that the nucleus and cytoplasm of eukaryotes originally comprised a prokaryote, which entered into a symbiotic relationship with an aerobic bacterium (the mitochondrion), a photosynthetic

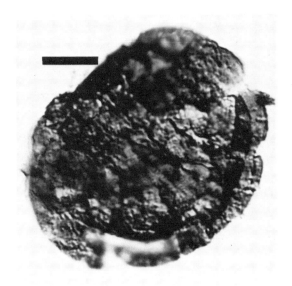

Figure 2. Fossilized eukaryote from Chamberline Shale Formation, Montana, 1.4 billion years old (10 micron bar) (*photo:* R. J. Horodyski).

bacterium (the plastid), and a motile spirochete bacterium (the motility organelle or undulipodium). The participants in this first series of symbiotic relationships gradually evolved such a strong dependence on the association that the component parts are almost indistinguishable as having once been separate organisms. There are, however, many important biochemical, genetic, and morphological clues supporting the serial endosymbiotic theory.

The other major theory that attempts to explain the origin of eukaryotic cells is the direct filiation theory, which maintains that the eukaryotic organelles all evolved within one prokaryotic lineage (proto-eukaryote) as invaginations of the inner cell membrane and as differentiations of the internal membrane system of the sort present in some modern prokaryotes (especially photosynthesizers). However, the basic biochemical and genetic characteristics of the mitochondria and plastids have been shown to be quite different in may ways from the characteristics of the nucleocytoplasm. At the same time, these organelles have been shown to be more similar to prokaryotic counterparts. Both plastids and mitochondria are somewhat autonomous in that they contain their own genomes and have a genetics separate from that of the nucleus. Advances in the methodology and instrumentation of macromolecular sequence analysis in the 1970s has led to the accumulation of large quantities of data for systematic evaluation.

The theory of direct filiation does not adequately explain these

Introduction

observations, and biologists concerned with the origin of eukaryotes have, over the past 15 years, gradually replaced this theory with the endosymbiotic theory. As more biochemical and genetic data have become available on eukaryotic organelles (including entire sequences of some mitochondrial genomes), the endosymbiotic interpretation of organellar origins becomes stronger.

The theory of the endosymbiotic origin of eukaryotic cells makes the predictions about eukaryotic organelles listed below, including mitochondria, plastids, motility organelles (undulipodia), and other less common organelles such as the anaerobic hydrogenosomes and possibly the extrusomes (secreting organelles of protoctists). For each organelle in question, if the organelle is acquired as a symbiont, then one expects to find evidence of the following:

1. a genome, with characteristics more similar to that of a prokaryotic than a eukaryotic genome
2. ribosomal RNA, transfer RNA, and messenger RNA more similar to that of a prokaryote than a eukaryotic cytoplasm
3. enzymes and protein complexes more similar to prokaryotic than eukaryotic cytoplasmic analogs
4. an ability to replicate and a genetics that is separate from the nuclear genetics
5. a free-living prokaryotic counterpart with strong genetic, biochemical, and morphological resemblances: an example of a similar symbiotic relationship of the counterpart organism with a host
6. an *all or nothing* phenomenon in which one either finds the organelle as a whole or does not find it at all (if it has not been acquired or was secondarily lost); one does not expect to find intermediate stages of the organelle if it was acquired all at once as a symbiont.

In *The Origin of Eukaryotic Cells* the above-listed criteria will be considered for each of the major eukaryotic organelles: the mitochondrion, the plastid, the motility organelle (undulipodium), and the nucleocytoplasm. The origin of mitotic and meiotic division, as they are related to the motility organelles (undulipodia), will also be discussed.

We have chosen only representative papers by authors such as W. F. Doolittle and his collaborators, M. Gray and L. Bonen, and D. C. Smith, who have published a number of outstanding articles in the fields of symbiosis and molecular evolution. We heartily encourage interested readers to make use of the various lists of references within the reprinted papers of this book to gain a fuller appreciation of research in cell evolution.

Part I

HISTORICAL OVERVIEW

Editors' Comments
on Papers 1 through 6

1 **WALLIN**
   The Mitochondria Problem

2 **GOKSØYR**
   Evolution of Eucaryotic Cells

3 **SAGAN**
   On the Origin of Mitosing Cells

4 **SCHWARTZ and DAYHOFF**
   Origins of Prokaryotes, Eukaryotes, Mitochondria, and Chloroplasts

5 **RAVEN**
   A Multiple Origin for Plastids and Mitochondria

6 **DOOLITTLE**
   Revolutionary Concepts in Evolutionary Cell Biology

The idea of a symbiotic origin for eukaryotic cells is not a new one. In 1910 Mereschkowsky published a remarkable evolutionary tree that included specific symbiotic events as mechanisms in eukaryotic evolution (see Fig. 3 in Mereschkowsky, 1910), although his work was virtually ignored by the scientific community. Wallin (Paper 1), an imaginative researcher with unpopular ideas, presented several papers in the 1920s suggesting that mitochondria were symbiotic bacteria and that symbiosis may be an important mechanism in evolution.

It was not until 1967 that the idea of symbiosis playing an important role in evolution was publicly revived. Almost simultaneously, and completely independently, Goksøyr (Paper 2) and Sagan* (Paper 3) presented their hypotheses for the symbiotic origin of eukaryotic organelles. Goksøyr's short paper, included here for historical interest, was not followed up with further work by him. Sagan, however, almost single-handedly reintroduced the subject to the Western scientific

*Sagan is also referred to by her married name, Margulis.

community. At first the opposition was substantial. Paper 3 was rejected and even lost by several journals before its 1967 publication.

Now, almost twenty years later, the endosymbiotic theory, particularly in regard to the origin of mitochondria and plastids, is accepted by most evolutionists, biochemists, and molecular biologists. Papers by Schwartz and Dayhoff (Paper 4), Raven (Paper 5), and Doolittle (Paper 6) are examples of work by some of the major proponents of the endosymbiotic theory.

REFERENCE

Mereschkowsky, C., 1910, Theorie der zwei Plasmaarten als Grundlage der Symbiogenesis, einer neuen Lehre von der Entstehung der Organismen, *Biol. Centralbl.* **30**:366.

1

Copyright © 1923 by The University of Chicago
Reprinted from *Am. Nat.* **57**:255-261 (1923), by permission of The University of Chicago Press

THE MITOCHONDRIA PROBLEM
I. E. Wallin

The small bacteria-like structures which may be demonstrated in the cytoplasm of all organized cells and which have come to be known as mitochondria have been recognized by biologists although their nature remains obscure. The discovery of these structures may justly be credited to Altmann (1) who believed that they are microorganisms and represent the ultimate units of life embedded in a lifeless cytoplasm of the host cell. Among others, Bütschli and Safftigen upheld a view that these structures are microorganisms having a symbiotic relationship to the host cell. A large number of investigators, however, have assumed a cytoplasmic origin of mitochondria and have preferred to consider them in the rôle of cell organs. That this latter conception has not met with universal acceptance is attested to by the brief consideration that is given to mitochondria in modern textbooks and to the varied opinions that have been advanced by numerous biologists regarding the nature of mitochondria.

In two recent papers (Wallin, 2 and 3), the writer has challenged the validity of the "microchemical" and staining methods that have been used in an attempt to establish a cytoplasmic origin for mitochondria. From the results of the experiments recorded in these papers, together with evidence by analogy therein submitted, the writer concluded that there is more evidence in favor of a microorganismal nature of mitochondria than there has been submitted in the literature to support a cytoplasmic origin of these structures. Since the publication of these papers by the writer, Cowdry and Olitsky

(4) have taken issue with the conclusions arrived at by the writer and assert: "that the discussion provoked may easily lead investigators from the main problem, which is, we take it, the elaboration of new methods for the determination of the chemical constitution of mitochondria as a prerequisite to the study of their rôle in cellular physiology and pathology." The article by Cowdry and Olitsky was followed by an unsigned editorial in the *Journal* of the American Medical Association for November 25, 1922, in which the editor exhibits an apparent unfamiliarity with the evidence in the "controversy."

It is apparent from a perusal of Cowdry and Olitsky's article and the above-mentioned editorial that certain fundamental conceptions must be clarified. These conceptions are primarily concerned with the nature of bacteria. In a previous paper, the writer (Wallin, 2) called attention to the lack of a definition of bacteria. Cowdry and Olitsky offer the following from Park and Williams' textbook on pathogenic microorganisms as a definition for bacteria: "The properties of bacteria ... which are fairly constant under uniform conditions and which have been more or less used in systems of classification, are those of spore and capsule formation, motility (flagella formation), reactions to staining reagents; relation to temperature, to oxygen and to other food material, and, finally, their relation to fermentation and disease." To which Cowdry and Olitsky add: "the property of forming smaller or larger aggregates (colonies), the individuals of which are, however, physiologically independent (Zinsser), and the development of characteristic modes of growth on suitable artificial media." It is obvious, from the above-quoted statement, that this does not constitute a definition of bacteria nor was it, apparently, so intended by Park and Williams. A definition would include a statement of those properties that are common to *all* bacteria. It would appear unnecessary to emphasize the fact that all bacteria do not exhibit motility, spore formation and capsules, nevertheless, Cowdry and Olitsky, apparently, assume that *all* bacteria possess *all* the characteristics that Park and Williams include in their system of classification, for, to quote from their recent article: "It must be admitted in all these respects that a great gulf remains between mitochondria and bacteria. In respect to mitochondria definite spores and capsules have not been noted. The mitochondrial blebs which Wallin suggests may be due to fixation bear in our judgment no resemblance to spores. . . . Motility due to flagellar action has not been observed." It appears superfluous to emphasize that biologists in general and bacteriologists in particular have not even assumed that *all* bacteria possess capsules and exhibit motility and spore formation. The absence or presence of these properties serve as a basis for a system of classification of bacteria which in itself implies that these properties are not common to all strains of bacteria. It may appear, in the above quotation, that the writer suggested that mitochondrial blebs may bear a relationship to spores. Such a suggestion was not made by the writer, but to the contrary, he suggested that the mitochondrial blebs described by Cowdry (5) may be due to the action of chemicals just as such bleb-like appearances were produced in bacteria by the action of chemicals (Wallin, 2).

In the absence of a clear and specific definition of bacteria it is important to mention some of the properties of bacteria, particularly those properties that have a bearing on a possible relationship of mitochondria and bacteria. The probable rôle of bacteria in organic evolution has been noted by many biologists and has been clearly indicated particularly by Osborn (6). Their relationship to natural processes in the "economy of nature" is significantly illustrated in the "nitrogen cycle." Space would not permit in this article to give even an abbreviated catalogue of the more important "natural" activities that are due to bacteria. Suffice to mention such activities as the production of flavor in many of our foods, nitrogen fixation, acetic acid production, etc. The relationship of bacteria to other forms of life is varied and has been investigated chiefly from the standpoint of pathogenicity. It is well known that there are many strains of bacteria which have an intimate relationship to other forms of life, but are not necessarily pathogenic in their reactions. A suggestive and significant example of such a relationship is to be found in the *Bacillus coli* which is nourished by the food in the intestines and in turn renders the food more easily digestible by the host organism (Marshall, 7). It is only recently that this type of relationship of bacteria to higher forms of life has come to occupy the attention of investigators. Aside from a few forms like the root nodule bacteria, the symbiotic bacteria in *Psychotria bacteriophila,* and a few others, very little study has been directed toward the nature of these important life relationships.

The reactions of bacteria to stains and staining methods have also been shown to be quite varied. While the majority of free-living bacteria respond in a fairly uniform manner to the ordinary bacteriological technique, there are many strains which demand a special technique for their demonstration. Apparently, the majority of the strains which require a special technique for their demonstration are those strains that have an intimate relationship with the tissues of higher forms of life.

The cultural properties of bacteria also vary within wide limits. Here again, it appears that those forms which require very special culture media are such strains which have a symbiotic or partial symbiotic relationship to some higher form of life. In some instances an artificial culture medium has not as yet been devised which will encourage the growth of the symbiotic organism. The most significant of such organisms are the microorganisms associated with Rocky Mountain spotted fever. These organisms, whose pathogenicity and life cycle have been established by Wolbach (8) and a number of co-workers, furnish us with an undisputable example of a known microorganism which exhibits the essential characteristics of mitochondria. Briefly stated, these characteristics are: fragility, as represented by sensitiveness to acetic acid, alcohol, and mechanical manipulation; non-pathogenic in the wood tick and may be transmitted from one generation to another in the egg cell; an intracellular position in the host (apparently, also extracellular); the necessity of a special staining technique that respects their fragility. Concerning these organisms, or more particularly the organisms of typhus fever which have somewhat the same characteristics, Cowdry and Olitsky state: "In pathology these views (*i.e.*, a bacterial conception of mitochondria), if verified, would have a distinct bearing upon our conception of the significance of Rickettsia bodies in typhus fever, trench fever, and other conditions and would leave us in some doubt regarding the criteria to be employed in the identification of microorganisms the nature of which remains obscure." It may be true that a bacterial conception of mitochondria may have a bearing on the ultimate solution of the nature of Rickettsia bodies and it appears to the writer that this is not only possible, but probable. The possibility that the establishment of such a view might "leave us in some doubt regarding the criteria to be employed in the identification of microorganisms the nature of which remains obscure" is far-fetched and is quite irrelevant to the problem as Wolbach (9) has recently called attention to. The known nature, relationships and reactions of Rickettsia bodies leave no basis for any possible confusion with mitochondria, but if there were occasion for the suggested confusion the writer is convinced that *no criterion* is preferable to a *false criterion* in the search for truth.

Cowdry and Olitsky, apparently, demonstrated that

mitochondria are colored in an exceedingly dilute solution (1 : 500,000) of Janus green while some six or eight different strains of bacteria which they subjected to a like dilution of the dye were unstained. Granted that Janus green in a highly diluted solution were a specific stain for mitochondria, would such a staining reaction be convincing evidence against their bacterial nature? On the basis of our knowledge concerning stains and staining reactions in general, the writer is convinced that such evidence is inconclusive. Further, Cowdry and Olitsky failed to stain mitochondria with the Giemsa stain and conclude: "The failure of this so-called nuclear dye to color mitochondria specifically points to a difference which we must consider to be fundamental between mitochondria and bacteria." The writer has definitely stained mitochondria with the Giemsa stain in smear preparations of dog fœtus tissue dried in the air and followed by absolute alcohol fixation.

It is obvious, after a broad survey of the nature of bacterial organisms, that a wide range of variations exists in the bacteria. These variations are concerned with physical and chemical properties and, of greatest significance in a bacterial conception of mitochondria, there are pronounced variations of the relationships of bacteria to other forms of life. This latter variation ranges all the way from a negative, through an indifferent and a partial symbiotic relationship, to a possible absolute symbiosis in which the bacterial symbiont is incapable of development in any natural medium with the exception of a living medium. It is apparent that, if mitochondria are of a bacterial nature, we would not expect them to exhibit the properties of hardy, free-living microorganisms, but rather the properties of such forms that have developed an absolute symbiotic relationship. Again, there are no reasons why one should assume that a bacterial mitochondrium should exhibit pathogenic properties, at least under normal conditions.

It must be borne in mind that only a small fraction of one per cent. of all bacteria are pathogenic. It must, further, be emphasized that some influence in the host cell produces a limitation of growth of the symbiotic organism. Such an influence is, apparently, present in the root nodules of leguminous plants.

The mitochondria problem is primarily concerned with a determination of the nature of these structures and is still an open problem. Further, we believe that the fundamental bearing its solution will have on the future trend of cytological investigation warrants the most thorough and exhaustive researches into the nature of mitochondria. We can not agree with Cowdry and Olitsky that biological investigators may easily be led from the main problem.

LITERATURE CITED.

1. Altmann, R. 1890. Die Elementarorganismen und ihre Beziehungen zu den Zellen. Leipzig.
2. Wallin, Ivan E. 1922. On the Nature of Mitochondria. I. Observations on Mitochondria Staining Methods applied to Bacteria. II. Reactions of Bacteria to Chemical Treatment. *Am. Jour. Anat.*, Vol. 30, No. 2.
3. Wallin, Ivan E. 1922. On the Nature of Mitochondria. III. The Demonstration of Mitochondria by Bacteriological Methods. IV. A Comparative Study of the Morphogenesis of Root-nodule Bacteria and Chloroplasts. *Am. Jour. Anat.*, Vol. 30, No. 4.
4. Cowdry, E. V., and Olitsky, P. K. 1922. Differences between Mitochondria and Bacteria. *J. Exp. Med.*, Vol. 36, No. 5.
5. Cowdry, E. V. 1918. The Mitochondrial Constituents of Protoplasm. Carnegie Inst. Pub., Contr. to Embr., No. 25.
6. Osborn, H. F. 1917. The Origin and Evolution of Life. New York.
7. Marshall, C. E. 1912. Microbiology. Phila.
8. Wolbach, S. B. 1919. Studies on Rocky Mountain Spotted Fever. *Jour. Med. Research*, Vol. XLI.
9. Wolbach, S. B. 1922. Correspondence in *Jour. of A. M. A.*, Vol. 79, No. 24, p. 2022.

EVOLUTION OF EUCARYOTIC CELLS

J. Goksøyr

IT seems generally accepted that the eucaryotic cellular form (that is, cells with developed nuclei) has developed from procaryotic forms. Stanier[1] suggested that this transition may have taken place in the blue–green algae, because these have the same photosynthetic mechanism as do other algae and higher plants.

It often seems to be taken for granted that the transition took place in such a way that one procaryotic cell has developed into a eucaryotic one. It may, however, be just as fruitful to discuss the possibility that one eucaryotic cell has evolved from a number of procaryotic cells, for example, that it originates from something like a coenocytic relationship (that is, a relationship in which the cells are brought into contact without intervening cell walls) between procaryotic cells.

Fig. 1 gives an outline of this idea. The first step would be the establishment of a coenocytic relationship between anaerobic procaryotes, most probably of a single species. The DNA from the individual cells must be expected to accumulate in the centre of the compound cell, and may also form concatenates. The success of the cell at this stage would depend to a large extent on how efficiently the DNA can be distributed to the daughter cells when the cell is dividing. One can imagine the development of a primitive mitotic mechanism, perhaps based on the same principles that apply to the distribution of DNA in connexion with the division of the procaryotic cells. The development of a nuclear membrane from the endoplasmatic reticulum would seem to be a logical next step, as the mitotic process develops further. The resulting cell would be an anaerobic eucaryote.

With oxygen in the atmosphere (presumably from photosynthetic blue–green algae), aerobic procaryotes must have developed. We can then assume that some of the anaerobic eucaryotes established an endocellular symbiotic relationship with aerobic procaryotes. Such a symbiosis would certainly give them an evolutionary advantage over the anaerobic forms. During further evolution, the aerobic partner must necessarily have lost a great part of its autonomy. On a molecular basis, the loss of autonomy must mean a loss of DNA. This DNA may have become incorporated into the nuclear DNA, giving the eucaryotic cell a still better control over its aerobic partner. The final step in this evolutionary process would be the development of mitochondria as we know them from eucaryotic cells today.

The aerobic eucaryotic cell produced in this way could enter a new symbiotic relationship, this time with blue–green algae of a primitive kind. The symbiotic relationship with blue–green algae must have developed along the same lines as the previous relationship, so that the photosynthetic partner must have lost a great part of its autonomy and DNA, to appear as the chloroplasts we know from eucaryotic algae and higher plants today.

An evolutionary line such as is suggested here can hardly be proved, although DNA/RNA hybridization experiments between blue–green algae and the chloroplasts of other algae, perhaps primitive Rhodophyceae, might give some information. The evolutionary advantage of the development of a eucaryotic cell like the one indicated here would be that it contained more DNA. In the beginning this would probably only mean the same information repeated several times, but mutations could give the basis for more information.

A further logical conclusion is that the eucaryotic cell which developed would take its genetic material mainly from the procaryotic forms making up the coenocytic system. Such coenocytic systems may have developed a number of times, from different procaryotic forms. Present-day eucaryotic organisms do not necessarily, therefore, have to be developed from one original species. This might even explain some of the rather puzzling parallels that exist between groups of procaryotic and eucaryotic organisms.

[1] Stanier, R. Y., in *The Bacteria* (edit. by Gunsalus, I. C., and Stanier, R. Y.), **5**, 445 (London, 1964).

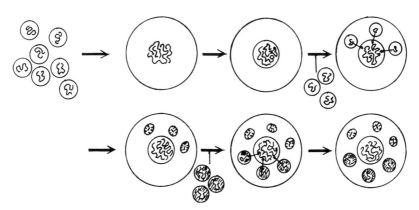

Fig. 1. Suggested evolutionary development of a eucaryotic photosynthetic cell from procaryotic forms. The small arrows within the cells indicate transfer of genetic material to the nucleus. See text for further details.

3

Copyright © 1967 by Academic Press Inc. (London) Limited
Reprinted from *J. Theor. Biol.* **14**:225-274 (1967)

ON THE ORIGIN OF MITOSING CELLS

L. Sagan

[*Editors' Note:* Figure 1 and Table 1 are foldout pages that could not be reproduced here. See figures in Papers 35A through 35E for similar material.]

A theory of the origin of eukaryotic cells ("higher" cells which divide by classical mitosis) is presented. By hypothesis, three fundamental organelles: the mitochondria, the photosynthetic plastids and the (9+2) basal bodies of flagella were themselves once free-living (prokaryotic) cells. The evolution of photosynthesis under the anaerobic conditions of the early atmosphere to form anaerobic bacteria, photosynthetic bacteria and eventually blue-green algae (and protoplastids) is described. The subsequent evolution of aerobic metabolism in prokaryotes to form aerobic bacteria (protoflagella and protomitochondria) presumably occurred during the transition to the oxidizing atmosphere. Classical mitosis evolved in protozoan-type cells millions of years after the evolution of photosynthesis. A plausible scheme for the origin of classical mitosis in primitive amoeboflagellates is presented. During the course of the evolution of mitosis, photosynthetic plastids (themselves derived from prokaryotes) were symbiotically acquired by some of these protozoans to form the eukaryotic algae and the green plants.

The cytological, biochemical and paleontological evidence for this theory is presented, along with suggestions for further possible experimental verification. The implications of this scheme for the systematics of the lower organisms is discussed.

1. Introduction

All free-living organisms are cells or are made of cells. There are two basic cell types: *prokaryotic* and *eukaryotic*. Prokaryotic cells include the eubacteria, the blue-green algae, the gliding bacteria, the budding bacteria, the pleuropneumonia-like organisms, the spirochaetes and rickettsias, etc. Eukaryotic cells, of course, are the familiar components of plants and animals, molds and protozoans, and all other "higher" organisms. They contain subcellular organelles such as mitochondria and membrane-bounded nuclei and have many other features in common.

> "The numerous and fundamental differences between the eukaryotic and prokaryotic cell which have been described in this chapter have been fully recognized only in the past few years. In fact, this basic divergence in cellular structure which separates the bacteria and blue-green algae

from all other cellular organisms, probably represents the greatest single evolutionary discontinuity to be found in the present-day living world" (Stanier, Douderoff & Adelberg, 1963).

This paper presents a theory of the origin of this discontinuity between eukaryotic (mitosing or "higher") and prokaryotic cells. Specifically, the mitochondria, the (9+2) basal bodies of the flagella, and the photosynthetic plastids can all be considered to have derived from free-living cells, and the eukaryotic cell is the result of the evolution of ancient symbioses. Although these ideas are not new [Merechowsky (1910) & Minchin (1915) in Wilson (1925), Wallin (1927), Lederberg (1952), Haldane (1954), Ris & Plaut (1962)],† in this paper they have been synthesized in such a way as to be consistent with recent data on the biochemistry and cytology of subcellular organelles. In accord with both the fossil record and this theory, the lower eukaryotes (protozoans, eukaryotic algae and fungi) can now be included on a single phylogenetic tree (Fig. 1). In contrast to previous thought on the subject (Cronquist, 1960; Dougherty & Allen, 1960; Fritsch, 1935), many aspects of this theory are verifiable by modern techniques of molecular biology.

In defending the idea that the eukaryotic cell arose by a specific series of endosymbioses, a plausible scheme for the origin of mitosis itself emerges. ("Mitosis" is meant only in the classical sense; the analogous equal distribution of genes to daughter cells in prokaryotes is not relevant here.)

The paper is organized into three parts. The first presents the theory of the origin of the eukaryotic cell. The second part is a compilation of the scientific literature in support of the sequences presented first. The last section suggests some experimental results which can be predicted from the theory.

2. Hypothetical Origin of Eukaryotic Cells

2.1. THE EVOLUTION OF PROKARYOTIC CELLS IN THE REDUCING ATMOSPHERE

Prokaryotic cells containing DNA, synthesizing protein on ribosomes, and using messenger RNA as intermediate between DNA and protein, are ancestral to all extant cellular life. Such cells arose under reducing conditions of the primitive terrestrial atmosphere ($4 \cdot 5$–$2 \cdot 7 \times 10^9$ years ago). All earlier events leading to the origin of a population of free-living entities upon which natural selection could act (i.e. cells) are outside the province of this

---

† "More recently, Wallin (1922) has maintained that chondriosomes may be regarded as symbiotic bacteria whose associations with other cytoplasmic components may have arisen in the earliest stages of evolution ... to many, no doubt, such speculations may appear too fantastic for present mention in polite biological society; nevertheless, it is in the range of possibility that they may some day call for more serious consideration" (Wilson, 1925, see p. 378).

discussion, in other words they occurred prior to the events described herein (Bernal, 1957).

2.2. THE EVOLUTION OF PORPHYRIN SYNTHETIC PATHWAYS, PHOTOSYNTHESIS, AND RESPIRATION IN PROKARYOTIC CELLS

Early in the history of prokaryotic cells, nucleotide sequences (genes) which coded for porphyrin syntheses, evolved. The following is considered to be a plausible historical sequence: photodisassociation of water vapor in the upper atmosphere resulted in the escape of free hydrogen which led to the production of molecular oxygen. This threatened the highly reduced nucleic acid of early self-replicating cellular systems. Associated metal-chelated porphyrins protected these systems from oxidation. Genes coding for the pathways involved in the syntheses of such chelated porphyrins (e.g. the coenzymes of peroxidase and catalase) were selected for and retained under this continuing threat of naturally produced oxidizing agents. The incidental fact that such antimutagenic compounds preferentially absorb visible light was put to evolutionary advantage: cells eventually evolved containing mechanisms to produce ATP (and the other nucleotides) utilizing visible solar energy absorbed by chlorophyll-like porphyrins. These cells, now photosynthetic, replaced earlier ones in which ATP was synthesized by direct ultraviolet absorption. With energy released from photoproduced ATP, H atoms from atmospheric hydrogen and hydrogen sulfide were used in the reduction of CO_2 for cell material.

In other populations of heterotrophic microbes dependent upon fermentation of sugars and amino acids for ATP production, cells which used their porphyrins for more efficient oxidation of carbohydrates were eventually selected for. This resulted in microbes capable of cytochrome-mediated production of ATP *via* electron transport systems (e.g. anaerobic respirers: nitrate and sulfate reducers, etc.).

Eventually, among the primitive phototrophs, a population of cells arose using photoproduced ATP with water as the source of hydrogen atoms in the reduction of CO_2 for the production of cell material. This led to the formation of gaseous oxygen as a by-product of photosynthesis. Such oxygen elimination by microbial photosynthesizers increased the partial pressure of oxygen in the atmosphere. In the anaerobic respirers, the abundance, at least in some locations, of free oxygen led to the evolution of the final and aerobic step in respiration, i.e. the complete oxidation of carbohydrate *via* the cytochrome system with the elimination of CO_2 and H_2O.

Among autotrophic microbes which used porphyrins in anaerobic photosynthesis in the light, mutants which could use these same porphyrins for aerobic respiration in the absence of light were eventually selected for. This resulted in the evolution of prokaryotic algae ancestral to extant blue-greens —with both photosynthetic and respiratory mechanisms for ATP production. These versatile prokaryotes represented a large step in cellular efficiency. They continued to eliminate gaseous oxygen in photosynthesis and further accelerated the transition to the oxidizing atmosphere. Thus, some time after the origin of the earth and before the deposition of oxidized rocks and micro-fossils—a period of about 2400 million years—populations of prokaryotes with the major photosynthetic and respiratory metabolic capabilities evolved ($4 \cdot 5$–$2 \cdot 1 \times 10^9$ years ago).

2.3. THE EVOLUTION OF EUKARYOTIC CELLS FROM PROKARYOTIC CELLS BY SYMBIOSIS

The continued production of free oxygen as a by-product of photosynthesis resulted in a crisis: all cells had to adapt to an atmosphere containing oxygen, or they had to survive in a specialized anaerobic environment. The geological evidence indicates oxygen was present in the atmosphere as early as $2 \cdot 7 \times 10^9$ years ago and became relatively abundant $1 \cdot 2 \times 10^9$ years ago (Cloud, 1965). By this time, then, all production of abiogenic organic matter must have come to a halt; not only was its source of energy removed by ozone absorption of ultraviolet light in the upper atmosphere, but such organic matter, if it were produced, would have been rapidly oxidized (Abelson, 1963). Therefore all terrestrial life became dependent, either directly or indirectly, upon cellular photosynthesis before $1 \cdot 2 \times 10^9$ years ago. To insure replication of their nucleic acids, heterotrophs were forced to eat organic matter produced by photosynthetic or chemoautotrophic processes.

It is suggested that the first step in the origin of eukaryotes from prokaryotes was related to survival in the new oxygen-containing atmosphere: an aerobic prokaryotic microbe (i.e. the protomitochondrion) was ingested into the cytoplasm of a heterotrophic anaerobe. This endosymbiosis became obligate and resulted in the evolution of the first aerobic amitotic amoeboid organisms.

By hypothesis, some of these amoeboids ingested certain motile prokaryotes. Eventually these, too, became symbiotic in their hosts. The association of the motile prokaryote with the amoeboid formed primitive amoeboflagellates. In these heterotrophic amoeboflagellates classical mitosis evolved.

The evolution of mitosis, insuring an even distribution of large amounts of nucleic acid (i.e. host chromosomes containing host genes) at each cell division, must have taken millions of years. It most likely occurred after the transition to the oxidizing atmosphere, since all eukaryotic organisms contain mitochondria and are fundamentally aerobic. Based on the abundance of fossil eukaryotes, it must have occurred before the dawn of the Cambrian. Hence, the most likely period for the evolution of mitosis is between $1 \cdot 2 - 0 \cdot 6 \times 10^9$ years ago—a period of about 600 million years.

The first symbiotic acquisition, that of the protomitochondrion, produced cells with the typical eukaryotic carbohydrate oxidative pathways. The anaerobic breakdown of glucose to pyruvate along the Embden-Myerhof pathway occurred in the soluble cytoplasm under the direction of the host genome. Further oxidation of glucose using molecular oxygen *via* the Krebs cycle (H atoms from organic acids combine with DPN, FAD, and cytochromes; ATP is generated; and water is eliminated) occurred only in the symbiotic mitochondrion under the direction of its own genes.

This mitochondrial-host symbiosis may have resulted in the typical eukaryotic phospholipid membrane and steroid synthesis and, in particular, the formation of a nuclear membrane and endoplasmic reticulum. The greater amounts of energy available after the incorporation of the mitochondrion resulted in large cells with amoeboid and cyclotic movement. However, the diversity in types and amounts of proteins such cells could make would have been limited by the amount of DNA available to administer protein synthesis. Hence, in the absence of any efficient mechanism to insure the equal distribution of newly synthesized DNA, the size of the amoeboids host cell must have been limited. Multinuclearity and duplicated cistrons or entire genomes (polyploidy) may have been early mechanisms to distribute greater amounts of DNA evenly to daughter cells, since multiple copies greatly increase the probability that each daughter will get at least one copy of the genome. However, the inefficiency of polyploidy and the selective advantage of linkage groups over polyploidy have been demonstrated (Gabriel, 1960). The problem of equal segregation of the daughter linkage groups remained.

2.4. THE EVOLUTION OF MITOSIS IN AMOEBOFLAGELLATES

How did the efficient mitotic mechanism which equally segregates daughter linkage groups evolve? The dividing nuclei of the lower eukaryotes themselves suggest a plausible evolutionary sequence for mitosis.

It is likely that the large size and heterotrophic habits of the amoeboids led to the ingestion of all kinds of small prokaryotes (Trager, 1964). Among these, it is hypothesized that some ancestral amoeboid acquired a motile parasite, perhaps spirochaete-like, by ingestion. The genes of the parasite, of course, coded for its characteristic morphology, $(9+2)$ fibrils in cross section. Amoeboid host cells which retained their endosymbionts had the

I *Tetramitus* (Wilson, 1925, p. 82)

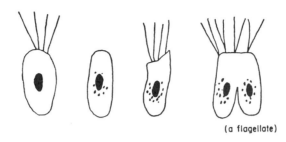

(a flagellate)

II.A. (a) *Amoeba tachypodia* (Wilson, 1925, p. 206)

II.A. (b) *Amoeba polypodia* (Wenyon, 1926, p. 63)

II.A. (c) *Hartmanella hyalina* (Wenyon, 1926, p. 63)

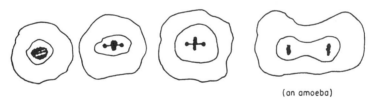

(an amoeba)

FIG. 2. Mitotic figures of some representative eukaryotes. These figures were adapted from the illustrations which may be found in the references cited. The names of the organisms given are generic. The name in parenthesis indicates the larger group in which the organism can be found in Fig. 1, the phylogenetic tree. A common description is given for some of the more obscure genera. The Roman numeral groups correspond to branches on the phylogenetic tree as discussed in sections 2.5.1 to 2.5.6 of the text. The organisms in groups I, III and V are thought to be premitotic. In groups II, IV and VI the subgroupings A and B refer to "premitotic" and "eumitotic", respectively. [The lower case letters (a), (b) . . . are simply included for reference.]

immense selective advantage of motility long before mitosis evolved; they could actively pursue their food.

By hypothesis, the motile prokaryotic endosymbiont itself is the ancestor of the complex flagellum of eukaryotic cells. The replicating nucleic acid of the endosymbiont genes (which determines its characteristic (9+2) substructure) was eventually utilized to form the chromosomal centromeres and centrioles of eukaryotic eumitosis and to distribute newly synthesized host nuclear chromatin to host daughters.

How did endosymbiont nucleic acid differentiate into chromosomal centromeres and centrioles? At least two series of mutational steps were required: one leading to the development of some attraction between nucleic acid of the host and that of the symbiont, and eventually, to permanent connections of daughter endosymbionts precisely to daughter chromosomes of the hosts in the formation of chromosomal centromeres; a second leading to the segregation of the replicated daughter endosymbionts to opposite poles of the host cell. In any case, in each generation only those amoeboid daughter cells containing a full euploid genome—that is, at least one copy of each gene—were selected. This insured continuing selection pressure for improved mechanisms of host nucleic acid segregation. Judging from present-day mitotic figures among lower eukaryotes, it probably led to the development of eumitosis at various times in various primitive amoeboflagellates. This might be considered analogous to the development of different mechanisms determining sexual polymorphism in higher organisms. The biological goal of all of these is the same: insurance that genetic recombination systematically accompanies reproduction. Just as the various sex determining mechanisms are not directly homologous (e.g. ovaries in insects, mammals, and flowering plants), it is likely, too, that some eumitoses in lower eukaryotes are analogous rather than directly homologous.

A corollary of this is the assertion that primitive eukaryotes can be classified according to the various lines they represent in the evolution of eumitosis. An attempt to reconstruct these lines on the basis of cytological and genetic information accumulated from the literature is presented in sections 2.5.1 to 2.5.6 below. The plausible steps in the differentiation of the chromosomal centromeres and centrioles from "(9+2) homologue" flagellar basal bodies have been reconstructed from the available data and the immediate selective advantages of each step is discussed in sections 2.5.1 to 2.5.6. The large groups of organisms which presumably derived from these steps, are presented on branches I to VI of the phylogenetic tree (Fig. 1). The groups were formed to include as much data as possible on the eumitosis of lower eukaryotes in a way which is considered to be consistent with the general classification of these organisms, *exclusive of their photosynthetic plastids.*

232 L. SAGAN

II.B. (a) *Porphyra lacinata* (Copeland, 1956, p 42)

II.B. (b) *Euglypha* (Minchin, 1912, p.113)

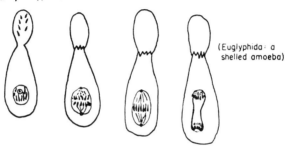

II.B. (c) *Amoeba limax* (Goldschmidt & Popoff, 1907)

FIG. 2. Mitotic figures of some representative eukaryotes (*cont.*).

2.5 STEPS IN THE EVOLUTION OF EUMITOSIS

The corresponding figures, together comprising "Fig. 2", were adapted from the literature as examples of actual mitotic figures. The term "(9+2) homologue" refers to the flagella and cilia, as well as to the more specialized cirri (fused cilia), axopodia, and other cellular organelles with the characteristic electron microscope cross section. "Basal bodies" (found at the base of these motile organelles, maybe self-reproducing), centrioles, or some other submicroscopic homologues are considered to be the repository of the nucleic acid of the replicating system and, thus, the descendants of the genes of the original symbiont. The "division center" is considered to be homologous also (Wilson, 1925, see p. 204). Therefore, the term "(9+2) homologue" is used here in its most general sense; it implies the genetic system that codes for the development of flagella, cilia, "division centers", centrioles, or any other of these homologous organelles. Diagrams I to VI, which illustrate the hypothesized relationships between host genome and (9+2) homologue genome, include in each generation the ratios of symbiont to host genomes suggested by the theory. Although at least one dividing (9+2) homologue on each host linkage group (i.e. a chromosomal centromere) is required to insure euploidy, multiple numbers of chromosomes in the schematic diagrams have been omitted for clarity; their omission changes in no way the general argument. We know that very closely related species can differ in their total chromosome number and, in general, the number *per se* of any fundamentally replicative structure itself is a very poor criterion in taxonomy (Dillon, 1962). In the diagrams, the numerator in the ratio "symbiont (9+2) genomes per host nuclear genome" refers only to *functionally* distinct (9+2) homologues.

"Premitosis", in Diagrams I to VI, explicitly illustrates the phenomena presumably at the basis of the lack of "sex" (meiosis and fertilization) in many lower eukaryotes. These organisms have the general features of eukaryotic cells (mitochondria, nuclear membranes, etc.), but their "aberrant" division figures often lack the spindles and centrioles of typical eumitosis. (*Amoeba, Euglena, Tetramitus*, etc.: see Fig. 2). They are probably not degenerate phytoflagellates, but eukaryotic organisms which are premitotic in the sense that they branched off the main lines of higher cell evolution before eumitosis evolved. The diploid-haploid cycle (i.e. "sex") is the surest indication of the evolution of eumitosis in any group of organisms. Eumitotic organisms, therefore, are those which demonstrate the typical genetic patterns described as Mendelian.

2.5.1. *(9+2) Homologue as basal body for flagella only*

In the first group of primitive eukaryotic flagellates, the nucleic acid of the motile symbiont is used only for its own replication and synthesis of

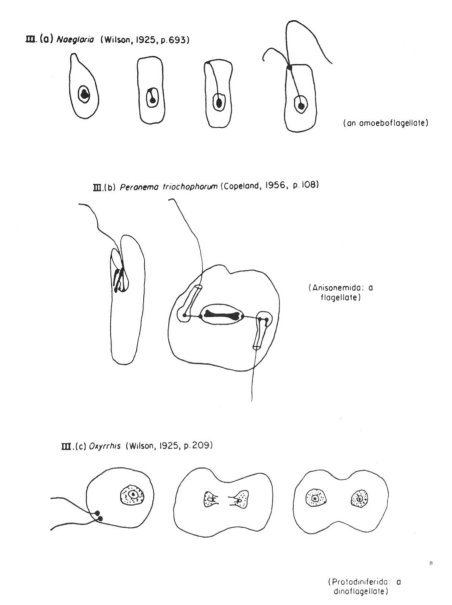

FIG. 2. Mitotic figures of some representative eukaryotes (*cont.*).

its own protein. The immediate selective advantage of the acquisition of the (9+2) endosymbiont was motility. Relic of this event may persist in some isolated groups of small flagellates, e.g. *Tetramitus* (Wilson, 1925, see p. 82.) "Mitosis" in these organisms is by no means standard and, of course, sexuality in the meiotic sense is unknown and, by hypothesis, will never be found (Fig. 2. I, Diagram I).

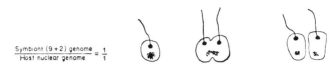

DIAGRAM I

2.5.2. (9+2) *Homologue incorporated into the nucleus for division center, no flagella*

Mutations occurring in populations of organisms described in section 2.5.1, led to attractions between host and (9+2) endosymbiont nucleic acid, which resulted in the incorporation of the replicating genome of the (9+2) symbiont into the nucleus of the host. When the (9+2) symbiont divided, it was utilized by the host as an intranuclear "division center". (The term "division center" is taken from the classical literature, see Wilson, 1925, p. 204.) The body stains deeply with nuclear dyes. Its homologies to basal bodies of flagella, etc., were recognized (and disputed) in the elegant light-microscopic studies on protozoan mitosis early in this century (Wilson, 1925, p. 206; Wenyon, 1926, see pp. 62 and 102).

These events first produced premitotic amoebae, which never have flagella at any stage of their life cycle but contain intranuclear division centers. Possibly multinuclear and other asexual amoebae are relics of this premitotic stage. For example, Dobell says of division in *Amoeba lacertae*: "No equatorial plate is formed and the 'chromosomes', or chromatic granules wander irregularly toward the poles, the whole karyosome meanwhile drawing out into a spindle shape and finally dividing. It is doubtful whether we can here speak of chromosomes or even mitosis, but such a type of division might well form the point of departure for the evolution of a true mitotic process" (Wilson, 1925, p. 213; Wenyon, 1926, p. 101) [Fig. 2. II A. (a) to (c)].

Eventually different groups of eumitotic organisms evolved from these various amoebae, namely, some eumitotic amoebae; the zygomycetes, ascomycetes, and basidiomycetes; the cellular slime molds; the conjugating green algae; the two great classes of red algae and their green counterparts

IV. A. (a) *Collodictyon* (Wilson, 1925, p.677)

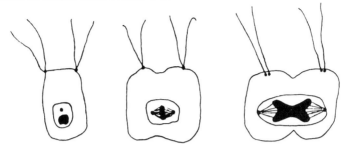

(a flagellate)

IV. A. (b) *Chilomonas* (Belar, 1915)

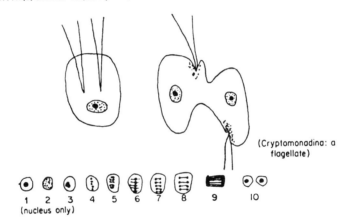

(Cryptomonadina: a flagellate)

FIG. 2. Mitotic figures of some representative eukaryotes (*cont.*).

—the *Schizogoniacea*. All of these organisms lack flagella at all stages in their life cycles, even though many of them are aquatic [Fig. 2. IIB. (a) to (c)] (Copeland, 1956, p. 42; Minchin, 1912, p. 113; Goldschmidt & Popoff, 1907).

How did eumitosis begin in this line which permanently lacked flagella? The first step may have involved replications of the (9+2) homologue (after it had become the intranuclear division center) in the absence of nuclear division. This made available *pairs* of intranuclear (9+2) homologues. Some mutation then occurred causing one member of the pair to be attracted to host chromatin and eventually to function as chromosomal centromere. The other, not attracted to host chromatin, functioned as an intranuclear centriole. The members of these pairs attracted one another. The attraction of the (9+2) homologue (chromosomal centromere) dragging with it host chromatin as it moved toward its sister (9+2) homologue (intranuclear centriole) may be analogous to whatever mechanisms are generally involved in attractions of prokaryotic cells to each other prior to mating.

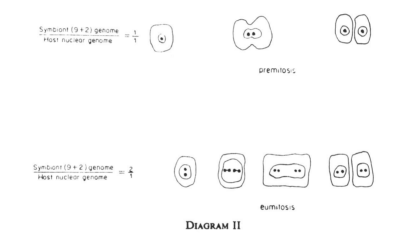

DIAGRAM II

2.5.3. (9+2) *Homologue used both as intranuclear division center and as flagellar basal body*

Off of this ancestral animal flagellate line, cells evolved which were motile at certain stages in the life cycle. At other stages the (9+2) flagellar basal body entered the nucleus and divided. While the (9+2) homologue divided, it was "borrowed" as an intranuclear division center for segregation of host chromatin. After division, the (9+2) homologue resumed its function at the base of the flagellum. This line may have led to a number of amoebo-flagellates which do not divide during their motile stages; in some a morphological relationship between flagella and nucleus may be seen [Fig. 2. III (a) to (c)].

IV. B. (a) *Allium* (Wilson, 1925, p. 697)

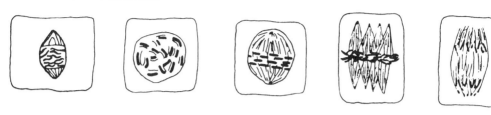

(Plantae: onion)

IV. B. (b) *Paramecium caudatum* (Wilson, 1925, p. 610)

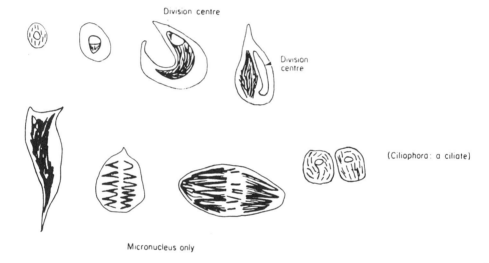

(Ciliophora: a ciliate)

FIG. 2. Mitotic figures of some representative eukaryotes (*cont.*).

Relics of this line are, for example: *Naeglaria* (Wilson, 1925, p. 693), *Anisonemids* (peranemids) (Copeland, 1956, p. 108), and dinoflagellates (Wilson, 1925, p. 209); *Dimastigamoeba, Mastigella* and *Mastigina* (Goldschmidt, 1907). The first three groups are considered to be premitotic in that sexual phenomena are not known, or its occurrence seriously disputed (Diagram III).

premitosis

DIAGRAM III

2.5.4. $(9+2)$ *Homologues used as basal bodies to flagella and other* $(9+2)$ *homologues permanently differentiated as intranuclear division centers*

Mutations occurring off the general line of amoeboflagellates essentially produced two separate $(9+2)$ homologue "clones". One mutant $(9+2)$ homologue produced "offspring" which served permanently as intranuclear division centers. The other $(9+2)$ homologue produced offspring which remained as basal bodies to the flagella only. An intranuclear division center and eventually eumitosis evolved in a line in which $(9+2)$ homologue-basal bodies to independent flagella were retained. This series of mutations (which may have occurred more than once) presumably led to groups of protozoans, green algae, golden algae, etc. Some of these groups are clearly premitotic (e.g. *Euglenids*) [Fig. 2. IV (a), (b)], (Wilson, 1925, p. 697; Belar, 1915).

One, or perhaps several, of these lines led to the eumitotic green plants with anteriorly flagellated motile stages and (by hypothesis) intranuclear division centers (Wilson, 1925, p. 151); another may have evolved some sporozoans (Copeland, 1956, p. 17).

The selective advantage of these mutations is clear: they insure the organism both equal distribution of its chromatin and retention of its flagellar motility.

The relatively homogenous groups of protozoans, the ciliates with their dimorphic nuclei, presumably evolved from an analogous series of mutations off the primitive amoeboflagellate line. They have a eumitotic micronucleus (with an intranuclear spindle) reserved for genetic continuity (Weinrich, 1954). They also maintain replicating basal bodies to the $(9+2)$ cilia of the cortex independent of the mitotic apparatus. However, the mitotic "germ line" and premitotic "soma" are strangely differentiated in the two types of nuclei in these organisms. The endopolyploid macro-

V. (a) *Dimorpha mutans* (Picken, 1962, p. 259)

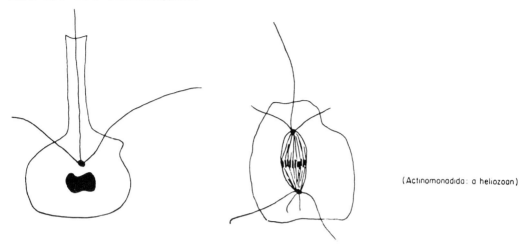

(Actinomonadida: a heliozoan)

V. (b) *Ochromonas* (Doflein & Reichenow, 1929)

(Ochromonadalea: a golden or chrysophysean flagellate alga)

FIG. 2. Mitotic figures of some representative eukaryotes (*cont.*).

nucleus contains no chromosomal centromeres or individual chromosomes and divides by amitosis. Presumably, the series of mutations which were selected for in the evolution of eumitosis in the micronucleus began in a binucleate cell, one in which many copies of the genome were regularly produced and reserved for the active administration of protein synthesis. The evolution of eumitosis never occurred in the macronucleus [Fig. 2. IV B. (b)] (Wilson, 1925, p. 610).

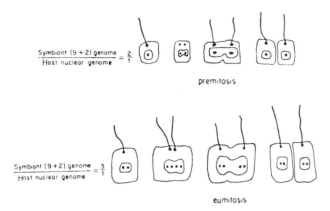

DIAGRAM IV

2.5.5. *(9+2) Homologue used both as basal body to flagella and as extranuclear "division center"*

In members of the early amoeboflagellate populations already containing intranuclear (9+2) homologues mutations must have occurred leading to cells in which the basal bodies of the flagella, when they divided, functioned as extranuclear division centers. Such striking examples of the basal body functioning as an extranuclear division center during mitotic stages exist in quite different forms among many extant organisms [*Dimorpha mutans* (Picken, 1962), *Clathrina* (Wilson, 1925, p. 42), *Ochromonas* and *Centropyxis* (Doflein & Reichenow, 1929)]. Some of these organisms [e.g. *Acanthocystis* (Calkins, 1909), *Wagnerella* (Wilson, 1925, p. 677)] have animal-type division figures, but ones in which centrioles are indistinguishable from the flagellar basal bodies. During division the flagella of many of these organisms are still attached to the centrioles at the poles of the mitotic figure. In some cases when the (9+2) homologue divides, one product of the division functions as centriole, and the other gives rise to the flagellar basal body which differentiates and is no longer capable of continued replication (Renaud & Swift, 1964).

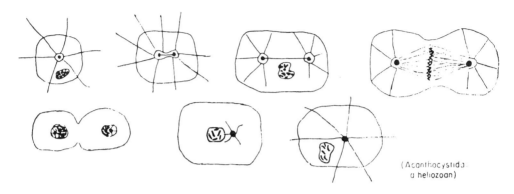

Fig. 2. Mitotic figures of some representative eukaryotes (*cont.*).

On this evolutionary line, in which dividing extranuclear basal bodies functioned as centrioles attracted to intranuclear chromosomal centromeres, it is likely that a series of specialized flagellates arose, which maintain throughout the life cycle some morphological connection between the flagellar apparatus and the mitotic figure, e.g. *Trichomonas* (Wilson, 1925, p. 205); *Polymastix* (Wilson, 1925, p. 694); *Heteromita* (Wenyon, 1926, p. 118); *Prowazekia* (Belar, 1915) and *Eudorina* (Hartmann, 1921). In some organisms this connection was probably secondarily lost, e.g. *Cryptobia* (Copeland, 1956, p. 160); *Herpetomonas* (Wenyon, 1926, p. 118); *Vaucheria* (Fritsch, 1935, p. 70); *Dictyota* (Wilson, 1925, p. 200); *Paramoeba* (Goldschmidt & Popoff, 1907); *Gurleya* (Wenyon, 1926, p. 742). Hypermastigote protozoans, whose mitotic centrioles form from the flagellar band in the living cell, are especially illustrative of this group (Copeland, 1956, p. 171; Cleveland, 1956, 1963). For example, after much study of the life cycle of the centrioles in the hypermastigote, *Barbulanympha*, Cleveland concludes:

". . . the reorganization process (of the centrioles) just described shows several things clearly; a definite relationship between the hypermastigote centrioles and those of higher forms of life; the ability of the centriole at certain times to function more than once in the formation of flagella axostyles, and parabasals, just as it is able to do at all times in the formation of the achromatic figure (i.e. mitotic apparatus); the ability of the centrioles to function in the production of extranuclear organelles without reproducing themselves, and also without accompanying nuclear or cytoplasmic reproduction; the inability of flagella, parabasals and axostyles to reproduce themselves; and most important of all, the fact that the anterior tip of these unusually large centrioles of flagellates is their reproducing portion" (Cleveland, 1956).

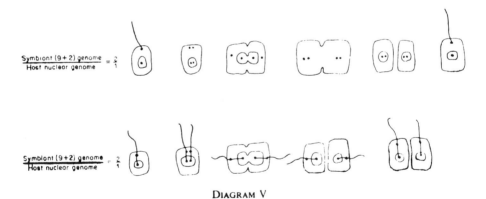

DIAGRAM V

2.5.6. (9+2) *Homologues used as basal bodies to flagella and other* (9+2) *homologues permanently differentiated as extranuclear division centers (centrioles)*

This series of mutations leading to the typical centrioles and amphiasters of animal-type mitosis involve steps analogous to those in section 2.5.4. Two "clones" of (9+2) homologues were produced: in one, the (9+2) homologue permanently gives rise to basal bodies which only gives rise to other basal bodies which produce flagella. In the other, the (9+2) homologue gives rise to extranuclear centrioles. This may have involved mutations leading to the loss of the two central fibers of the (9+2) homologue leading to the non-motile (9+0) centriole structure. [It may be the central fibers confer motility on the flagella and are somewhat analogous to the axial fiber of spirochaetes (Stanier *et al.*, 1963, p. 158).] The selective advantage of a permanent functional differentiation of the two organelles is clear, as in section 2.5.4, flagellar motility is completely independent of the mitotic apparatus: flagellated cells, in eumitotic organisms, too, can divide. Examples of this type of mitosis with amphiasters surrounding the extranuclear centrioles are found in some protozoans, and most eumetazoan groups (Wilson, 1925, p. 124). The conspicuous extranuclear centriole of the diatoms may be functionally homologous, too. This is consistent with the interesting late arrival of this relatively homogeneous group of advanced algae on the fossil scene, during the Cretaceous [Fig. 2. VI (b)] (Fritsch, 1935, p. 660).

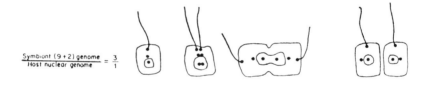

DIAGRAM VI

2.6. THE EVOLUTION OF EUKARYOTIC PLANTS FROM VARIOUS LINES OF PROTOZOANS WHICH ACQUIRED SYMBIOTIC PROKARYOTIC ALGAE

As outlined above, the evolution of mitosis is assumed to have occurred millions of years after the evolution of photosynthesis. This hypothetical origin of eukaryotes is completely incompatible with the notion that a simple phytoflagellate is the ancestor to extant higher cells—eukaryotic plant cells did not evolve oxygen-eliminating photosynthesis which later "packaged" into membrane-bounded plastids; they acquired it by sym-

V.(e) *Eudorina* (Hartmann, 1921)

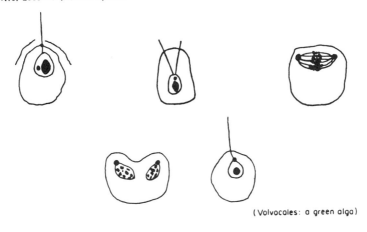

(Volvocales: a green alga)

V.(f) *Herpetomonas* (Wenyon, 1926, p. 118)

(Oikomonadacea: an an animal flagellate)

V.(g) *Paramoeba eilhardi* (Goldschmidt & Popoff, 1907)

(Paramoebida: an animal flagellate)

FIG. 2. Mitotic figures of some representative eukaryotes (*cont.*).

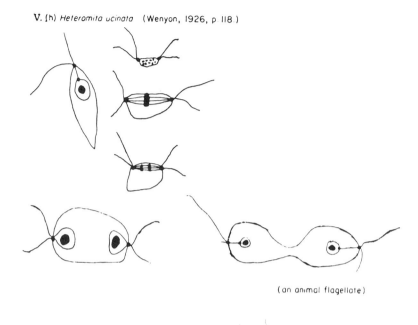

Fig. 2. Mitotic figures of some representative eukaryotes (*cont.*).

biosis. The diversity of cell structure and the life cycle in lower eukaryotic algae imply that different photosynthetic prokaryotes (protoplastids) were ingested by heterotrophic protozoans at various times during the evolution of eumitosis. The protoplastids themselves evolved from oxygen-eliminating prokaryotes, homologous to blue-green algae. To their own selective advantage they remained in the protozoan hosts which had ingested them and eventually became the obligately symbiotic plastids, retaining their characteristic photosynthetic pigments and pathways. In many eukaryotic algae and all higher plants, of course, the symbiosis continued to evolve a great deal, and vestiges of heterotrophy were eventually lost. This implies that in eukaryotic plants the plastids are homologous to blue-green algae; the non-chloroplast part of plants is directly homologous to eukaryotic heterotrophs.

The Phylogenetic Tree—Fig. 1. Based on the above hypothesis, Fig. 1 presents the major groups of eukaryotic lower organisms on a phylogenetic tree. In an extremely original taxonomic treatise, H. F. Copeland classified many of these lower eukaryotic organisms into relatively isolated natural groups. Except for accepting genetic autonomy of photosynthetic plastids, Copeland's independent work is remarkably consistent with the evolutionary theory presented here. His book (Copeland, 1956) was invaluable to the development of this phylogeny.

Thus, there are two major innovations presented in the phylogenetic tree. One involves the separation of lower eukaryotic algae into groups based on the supposed homologies of the "host" or protozoan part of the organism. The lettered circles representing plastids on the lines indicate the homologies of the prokaryotic photosynthetic endosymbiont; with less accuracy, the positions of these circles represent the stage in evolution at which the symbiotic acquisition occurred. The second is an attempt to reclassify admittedly heterogeneous groups of protozoans (e.g. sporozoans, sarkodina, etc.) according to their mitotic cytologies and, hence, the stages they presumably represent in the evolution of eumitosis. Aside from these major modifications, standard taxonomy has been followed. Copeland's terminology is adapted on the phylogenetic tree for all organisms except the green algae. The terminology of Fritsch is used for these. Except for organisms explicitly mentioned elsewhere on the tree, the groups are considered to be equivalent to those of Copeland.

The acquisition of photosynthesis by eukaryotic heterotrophs then, may be thought of as quite analogous to some recognized symbioses, which on the basis of the identification of the photosynthetic symbiont, must have originated much more recently, e.g. *Paramecium bursaria*, chrysophytes in Radiolarians, *Hydra viridis*, the algal-fungal associations in lichens. Occur-

V. (j) *Polymastix* (Wilson, 1925, p. 205)

(Polymastigida an animal flagellate)

VI (a) *Ascaris* (Wilson, 1925, p 124)

   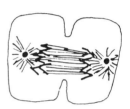

(Animalia: a roundworm)

VI. (b) *Surirella* (Fritsch, 1935, p. 660)

centriole

(Bacillaracea: a diatomaceous alga)

Fig. 2. Mitotic figures of some representative eukaryotes (*cont.*).

rences of green, blue-green and cryptomonad algal symbionts in the tissues have been reported in most groups of lower animals (Buchner, 1930, 1953).

The entire chronology presented here is summarized in Table 1.

In order to document this theory, the following, at the very least, are required: (1), the theory must be consistent with the geological and fossil records; (2), each of the three cytoplasmic organelles (the mitochondria, the (9+2) homologues, and the plastids) must demonstrate general features characteristic of cells originating in hosts as symbionts. None may have features that conflict with such an origin; (3), predictions based on this account of the origin of eukaryotes must be verified. The rest of this paper discusses the evidence for this theory in terms of assumptions based on molecular biology which can be made concerning evolutionary mechanisms.

3. Evidence from the Literature

3.1. ON CRITERIA FOR A NATURAL MICROBIAL PHYLOGENY

Throughout the systematic literature there exists the underlying assumption that the more traits two organisms possess in common, the more closely related they are. Higher organisms, such as vertebrates and flowering plants, show consistent morphological patterns which change in steps that can be related directly to adaptive value. Observation of living organisms, comparative physiology and anatomy, and the fossil record all provide much evidence for this assumption.

However, the lack of a comparable series of consistent morphological criteria has thwarted attempts to trace the evolutionary history of the "lower organisms", e.g. the "phycomycetes", the "mastigophora", the "sarkodina", etc., as any investigation of the literature on the subject will show.

On what, then, can the reconstruction of the evolutionary history of microbes be based? Recent advances in molecular biology provide some idea of the relative significance of various criteria in the development of microbial phylogeny, even if a great deal of fundamental data are still not available. For example, it is clear that we may *not* simply collect an arbitrarily large number of equal valued "traits" and try to group microorganisms on the basis of the largest number of such common "traits". Instead, as the genetic basis of many "traits" becomes known, we can rank them on the basis of the total number of single step mutations required to evolve them. In determining the relationship of two microbes—that is, the amount of time elapsed since they diverged from a common ancestor—we may ask: how many homologous base pair sequences in DNA do they share? The number of mutational steps which occurred to produce one from the other is related to the number of generations elapsed since the two populations diverged.

Thus, although two organisms may differ in a single measurable "trait", it must be clearly recognized that the genetic basis of such a trait may vary from a single (or a very few) DNA base pair changes to thousands or tens of thousands of such changes.†

Clearly, to construct a taxonomy reflecting natural phylogeny and not evolutionary analogy, microbes must be grouped on the basis of the total amount of genetic homology they share. A single mutation in a microorganism, resulting in a small chemical change with a profound phenotypic effect with respect to selection could easily mislead the taxonomist of lower organisms. On the other hand, organisms may share phenotypic traits and still be distantly related. For example, the fact that two microbes both metabolize glucose, but along entirely different pathways, implies a large number of different cistrons and, therefore, a long time since the two organisms diverged from a common ancestor.

Because of the lack of information, it is impossible of course to determine the number and order of DNA base pairs coding for a particular advantageous cistron, however certain criteria for ascertaining the degree of relationship between two microbes can be ranked in order of general validity (Table 2). For example, the homology of an entire metabolic pathway is a much more significant taxonomic criterion than the presence or absence of a single enzyme or pigment. In fact, the point at which metabolic pathways diverge in two otherwise similar microbes may help determine the elapsed time since the two microbes themselves diverged from a common ancestor.

The necessity for comprehending entire metabolic patterns, rather than individual biochemical traits, is especially relevant in botanical evolutionary

---

† This can be illustrated by an example drawn from two species of single-celled algae. Streptomycin resistant and streptomycin sensitive *Chlamydomonas* may differ only in a single trait—so may "bleached" and green strains of *Euglena gracilis* differ only in the single trait of "green color". However, it is likely that the difference in *Chlamydomonas* is due only to a single muton (Sager & Tsubo, 1961), whereas in clones of *Euglena* the one trait of green color (e.g. chlorophyll) in permanently "bleached" cells has been related to the presence of the entire chloroplast, and the potentiality for its formation (Lyman, Epstein & Schiff, 1961). Thus "green" and "bleached" *Euglena*, differing in the single obvious trait of color, betray a genetic difference of thousands of mutons. In the *Euglena* chloroplast there are at least 15 different kinds of enzymes (Smillie, 1963) and each one can roughly be estimated to be about 100 amino acid residues long. With a coding ratio of three nucleotides to each amino acid, the presence of a chloroplast implies enough genetic material, for structural genes alone, to code for $15 \times 100 \times 3 = 4500$ independent mutons. On the assumption that one cell in about a million contains a random mutation which turns out to be favorable for the evolution of the structural genes in the chloroplast, it would have taken about ($2^n \approx 10^6$, $n \approx 20$) twenty generations of *Euglena* to derive each of the 4500 favorable mutations in the pathway. Therefore, the trait of green color in *Euglena*, which may be permanently lost by exposure of the cells for a few minutes to ultraviolet light (Lyman, Epstein & Schiff, 1961), must have taken (very roughly but probably an extreme lower limit), 20×4500, or 90,000 generations to evolve.

TABLE 2

Taxonomic criteria in the formation of a natural phylogeny for microbes
(listed roughly in order of relative importance)

Criterion	Techniques by which measured
Total homology of DNA base pairs	Direct DNA nucleotide sequence data Agar-gel technique for DNA homologies Ability to genetically recombine (i.e. classical genetic techniques) DNA base ratios on CsCl density gradient DNA denaturation (melting point) determinations
Homologous metabolic pathways	Classical biochemistry
Homologous cistrons, same "genetic code letters"	Homologous messenger RNA's (DNA-RNA homologies). Identity of individual transfer RNA's for specific amino acids
Ultrastructural morphology	Electron microscopy
Morphology and life cycle	Light microscopy, classical cytology
Single biochemical pigments, enzymes, etc., in common	Spectroscopy, classical biochemistry
Molecular structure of single pigment, or enzyme	Classical chemistry
Common phenotypic traits	Ability to grow on same carbohydrate, production of same end product, motility, etc.

work on "lower plants", where common photosynthetic pigments and metabolic products have been considered primary phylogenetic markers. Some groups of organisms have been considered closely related on the basis of their pigments [i.e. *Schizogoniacea*, *Euglenids*, *Conjugacea*, and other chlorophytes (Fritsch, 1935, p. 70), *Myxochrysis* and *Crysothylakion* (Copeland, 1956, p. 63]. Even red algae are thought to have been derived from blue-greens in some botanical literature:

"Fucoxanthin, which is shared by chrysomonads, diatoms and brown algae supports a common evolutionary origin for these members of the 'brown' line of metaprotists. The common possession of diadinoxanthin by diatoms and dinoflagellates is suggestive of an affinity between these groups. . . . The common possession of this class of chromoproteins—the bili-proteins—by the blue-green and red algae fits very well with derivation of the red algae photosynthetic apparatus from that of the blue-green algae. . ." (Dougherty & Allen, 1960, p. 129).

Classically photosynthetic organisms have been segregated into separate Kingdoms (or Classes) from their plastid-lacking counterparts, regardless of their cellular morphologies which imply (at least to zoologists) that they should be grouped together [e.g. "*Chrysamoebida* (order): Rhizochrysidaeae, Chrysarachiaceae, Myxochrysidacea (families) . . . organisms in which only plastids distinguish them from various different groups of Rhizopods, Heliozoans and Sarkodina"] (Copeland, 1956, p. 63). Since zoologists assign relatively less importance to plastid characteristics, and tend to place such pigmented organisms in their respective protozoan groups, inconsistencies are rampant in the taxonomic literature of lower eukaryotes.

3.2. SOLUTION OF SOME TAXONOMIC PROBLEMS

According to the fundamental thesis of this paper, these problems are resolved if the validity of both views are seen simultaneously. For example, in the *Chrysamoebida*, it is unnecessary for zoologists to hypothesize that chrysophysean-type photosynthesis evolved separately, yet exactly analogously, in several diverse heliozoan, rhizopodal and sarkodinal lines; it is equally unnecessary for botanists to believe that the various rhizopods, heliozoans, and sarkodinae evolved from immediate chrysophysean ancestors by loss of photosynthetic plastids. Consistent with the above-mentioned hypothesis is the theory that algae, with chrysophysean photosynthetic characteristics, evolved millions of years before from photosynthetic prokaryotes and were acquired symbiotically in heliozoan, rhizopodal, and sarkodinal species, and that these prokaryotic algae became the obligately symbiotic plastids in the various protozoans. In other cases (e.g. *Chloramoeba* and *Chrysamoebae*), both chrysophysean and typical "green" prokaryotic algae were ingested by very similar amoeboflagellates whose descendants subsequently evolved "algal" ways of life. Presumably, too, typical green prokaryotic algae were acquired by euglenids, chloromonads, *Schizogoniacea*, etc., explaining the remarkable similarity in the chloroplast-related features in these different "host" organisms. There is really no other reasonable explanation for the evolution of certain natural groups of highly specialized flagellates which are found to contain distinct types of plastids with distinct pigments and storage products [e.g. *Cryptomonadina* (family): *Rhodomonas* (red), *Cryptochrysis* (chrysophysean), *Chilomonas* (colorless)]. As is now recognized in lichen classification, the host and plastid components of photosynthetic eukaryotes, like the component fungi and algae of lichens, ought to be classified separately. Perhaps, then, a natural phylogeny of the lower eukaryotes could be developed which is satisfactory to botanists, zoologists and mycologists.

In some cases, no doubt, the secondary loss of plastids from various protozoans have resulted in their return to the heterotrophic habit. The probability that the host can tolerate the loss of the symbiont must be related to the extent to which the symbiosis itself has evolved. For example, *Paramecium bursaria* and various other animal cells harboring symbiotic algae can survive induced loss of their photosynthetic capabilities (Siegel & Karakashian, 1959). *Euglena gracilis* can survive loss of its chloroplasts. In both of these organisms, "curing the cells" of their photosynthetic algae or chloroplasts involves some loss of viability in most media, however (Karakashian, 1963). For example, *Paramecium bursaria* grown in the dark will lose its algae. Although this ciliate can still divide asexually, it cannot go through the meiotic events prior to conjugation in the absence of its symbiotic zoochlorellae (Siegel, personal communication). The symbiont-host relationship can be restored readily in *P. bursaria* by feeding the specific zoochorellae (which can be grown *in vitro*) to the "cured" ciliate (Siegel, 1960). However, growth of the chloroplast *in vitro* and its reintroduction into *Euglena* have posed insurmountable technical problems to date. The dependence of the "host" on the symbiotic plastid is, of course, still more pronounced in most higher plants unable to survive a loss of photosynthetic capacity at all except, perhaps, under very special conditions. (For example, people may keep albino corn alive by feeding the plants sugar directly through their leaves).

Thus, in general, a greater mutual dependency probably reflects a longer host-symbiont association. This is because new syntheses, made possible by the symbiosis, will be selected for and gradually will become necessary for survival. It is highly probable that many of the pathways particular to plant metabolism—the formation of the plant cell wall, alkaloid syntheses, formation of certain storage products, etc.—originate from the ancient protozoan-prokaryotic algal symbioses, analogous to starch synthesis in *Peliaina*.

"*Peliaina cyanea* is a flagellate (chrysomonad or cryptomonad in its affinities) which harbors from one larger to six smaller cyanella (blue-green algae). Rare assymetric fissions yield colorless monads (host flagellates) which produce oily reserves rather than starch like the *Peliaina* complex. Since free cyanophytes do not produce starch either, the complex has achieved a new function. The syncyanom must be of some antiquity as there is no way the flagellate could have ingested the cyanella. Phagotrophic, amoeboid phases have been described for some primitive flagellates, and the ingestion of a free living cyanophyte by an amoeboid ancestor presumably initiated the symbiosis . . ." (Lederberg, 1952).

On the basis of a very general biological argument too, it can be said that photosynthesis evolved in the prokaryotes and mitosis evolved in the protozoans. That is, from our knowledge of evolution in higher organisms, it seems that to identify the population in which a multicistronically determined trait evolved, related organisms must demonstrate a large range of small variations in the given trait that can be correlated to specific selective environmental factors.

Lack of variation in a multicistronically determined trait in a natural group of organisms suggests the trait evolved earlier in some other ancestral population. (For example, in mammals: the metabolism of glucose; the histology of bone tissue; lungs developed from an outpocketing in the gut; the closed circulatory system containing red blood corpuscles; the vertebral column; the dorsal, hollow nervous system, etc.; in angiosperms: Krebs cycle intermediates; green plant photosynthesis; meiosis and fertilization; vascular tissue; seeds; etc.)

On the other hand, that the variations can be understood in terms of adaptive value to the population is considered evidence that, for example, these traits evolved in the following natural groups: triploblastic development in metazoans; the five-digit tetrapod in amphibious vertebrates; development through the total life cycle in dry environments in amphibians and reptiles; wide variety of beaks and feet forms in Galapagos Island finches; the mammary glands in mammals; and the flower in angiosperms.

We have already discussed some of the variations on the theme of mitosis in protozoans, sections 2.4 to 2.5.6; for the discussion of the variations in photosynthetic metabolism in contemporary prokaryotes, see sections 3.4 and 3.5.

3.3. GENERAL PROPERTIES OF SYMBIOSIS

The argument presented for the origin of the eukaryotic cell unequivocally points to the acquisition by symbiosis of mitochondria; the genome of the $(9+2)$ complex flagellum; and the plastid. What are the general criteria of organelles derived by symbioses?

(1) A symbiont originated as a free-living cell and therefore must have once been able to replicate its own DNA on its own protein synthesizing machinery. As outlined above, these subcellular organelles are hypothesized to have evolved along the main line of terrestrial cellular evolution and, therefore, must have contained DNA, messenger and ribosomal RNA, etc., that is, all minimal requirements for cell reproduction common to terrestrial cellular life. This is not meant to preclude the possibility that alternative modes of cellular replication antedated the ones with which we are concerned (Pirie, 1959). Thus at the very least, a symbiont must have had: (a) DNA; (b) messenger RNA (mRNA) complimentary to that DNA; (c) a functioning protein synthesizing system; (d) a source of ATP and other nucleotides; (e) a source of small molecules from which to make proteins and nucleic acids; and (f) a cell membrane synthesizing system. Upon entry into a host, such a symbiont may lose from none to all of its

synthetic capabilities *except the ability to replicate its own DNA* and synthesize complementary mRNA from that DNA—the *sine qua non* of any organism. Hence, if any of these organelles originated as symbionts, their characteristic specific DNA must be present in the host at every stage of the host life cycle. It, of course, follows, too, that the ratio of host genome DNA synthesis to symbiont genome DNA synthesis must approximate 1 : 1. If the ratio is greater than 1 : 1 the host will outgrow the symbiont and give rise to daughters which lack it; if the ratio of host DNA synthesis to symbiont DNA synthesis is less than 1 : 1 the symbiont will outgrow the host and lysis will ensue.

By analogy with parasitism and mutualism in higher organisms, it is highly likely that after long association the redundancy intrinsic in symbiotic relationships will be selected against; the symbiont will tend to relegate all dispensable metabolic functions to the host. This tendency results ultimately in symbioses which become progressively more obligate.

(2) If an organism or organelle has been acquired by symbiosis, it will be retained intracellularly in its host if, and only if, there exists some mechanism to insure that each daughter cell of the host receives at least one copy of the symbiont genome at each division. Any mutation that insures the symbiont's distribution to both daughters in each division will be of high selective value for the complex (e.g. a common effective but expensive mechanism would be the presence of many copies of the symbiont within the host cells, increasing the probability that each daughter receives at least one symbiont). Indeed, when viewed in this way, many classical cytological observations on the behavior of centrioles, mitochondria and chloroplasts in cells of their "hosts" can easily be interpreted, as they indeed were, to be mechanisms that insure the genetic continuity of the organelle genome.

(3) If a cellular organelle is acquired by symbiosis, there should be no organism containing intermediate intracellular stages of the organelle. The entire series of metabolic capabilities conferred on the host by the symbiont must be acquired together, i.e. "packeted" as a unit.

(4) If the symbiont is lost, all metabolic characteristics coded for on the symbiont genome must be lost together. Once lost, a symbiont can never be regained unless it is reacquired by ingestion. Indeed, unless the reingestion quickly succeeds the loss, it is unlikely that precisely the same symbiont will be reacquired. For example, such reingestion has been suggested for the origin of *Glaucocystis nostoc*, an organism with the "host" features of an *Oocystaceae* containing distinctly blue-green algae-like, rather than "green", plastids (Fritsch, 1935, p. 186) and for *Gloeochaete*, "long referred as an anomalous genus of the *Myxophyceae*, but now known to represent a

colorless Tetraspraceous form in which the blue-green chromatophores are symbiotic blue-green algae" (Fritsch, 1935, p. 125).

Of course, a clear distinction must be made between the loss and the dedifferentiation of the symbiont. In dedifferentiation, the genetic potential for symbiont formation is retained implying a complete retention of the permanent symbiont genome and, therefore, of its DNA. Loss of the symbiont implies permanent loss of symbiont DNA.

Metabolic pathways reflected in some morphology (e.g. ribosomes, lysosomes, endoplasmic reticulum, nuclear membranes, etc.) which are coded for on the nuclear genome if lost from the cytoplasm, may be replaceable in the proper environment by the action of the nuclear genes. This can never be true of organelles which originated by symbiosis.

(5) Since any intracellular symbiont must have its own genes, a correlation can be made between genetic traits conferred on the host by the symbiont and the morphological presence of the symbiont. For example, in all eukaryotes the number of mitochondria and plastids will not necessarily be halved at meiosis and doubled at fertilization to establish their constancy. Hence, they will not necessarily display the Mendelian distribution of characteristics from generation to generation. The observations that in some cases mitochondria and plastids are inherited only with the larger female gamete in anisogametic fertilizations, led to the early hypotheses that these organelles were bearers of "cytoplasmic heredity". Thus, non-mendelian genetics should be found in those organisms in which mitochondria or chloroplasts are inherited uniparentally. The transmission of the trait should be associated with the donor parent [e.g. sea urchins where all paternal mitochondria are found in only one blastomere at the 32-cell stage (Wilson, 1925, p. 713). In exceptional cases this might apply to (9+2) homologues as well (e.g. patrilinear inheritance of the centrosome, e.g. in *Culex* (Darlington, 1958, p. 174)]. In general, of course, genetic traits carried by the (9+2) chromosomal centromeres will show strict Mendelian inheritance patterns.

(6) If an organelle originated as a free-living cell, it is possible that naturally occurring counterparts still can be found among extant organisms. Even if precise extant morphological and physiological codescendants cannot be found, the organelle must have genetic and physiological characteristics known to be consistent with those generally present in terrestrial cells.

Applying the above criteria, the nucleus of the eukaryotic cell could not have originated by symbiosis. The great body of literature on the genetics and cytology of higher organisms defends the thesis that nuclear genes control cytoplasmic syntheses which make possible the duplication of nuclear DNA and growth of the cell prior to the next division (Brachet, 1957). Recent evidence indicates a large number of sites spread over many nuclear chromosomes control the production of the cytoplasmic RNA (Prescott, 1964). The nucleus and cytoplasm of the eukaryotic system, are clearly part of a highly integrated continuous system; it seems there is little evidence that can be cited for their independent origins.

To document this theory, the rest of the discussion is devoted to the origin of prokaryotic cells hypothetically destined to become organelles of eukaryotes, and to the present status of cytoplasmic organelles in terms of the general criteria for symbiotic origin developed above.

3.4. ANCIENT ANAEROBIOSIS AND MICROBIAL PHOTOSYNTHESIS

The earth's atmosphere, both in its absence of hydrogen and presence of large quantities of oxygen is cosmically atypical. Although there is no consensus on details astronomers and geologists today believe that the original terrestrial atmosphere, which was subsequently lost, was composed primarily of hydrogen. Many different laboratory attempts to simulate the origin of life have been made; a variety of molecules (e.g. amino acids, pyrimidines, ATP) found in contemporary organisms can be produced if the net chemical conditions are reducing. However, if conditions are oxidizing, such as they are in the present atmosphere, the production of common organic compounds is exceedingly inefficient. All of the evidence is consistent with the widely accepted hypothesis that life arose under reducing conditions of the primitive atmosphere (Sagan, 1965b).

Photosynthesis, in the general sense of the utilization of solar energy in the production of organic compounds, probably antedates the origin of life itself. The strong ultraviolet light absorption capabilities of the major nucleic acid components, of ATP, and of amino acids and peptides may have led to their local accumulation in the absence of replicating cellular systems upon which natural selection acted, as early as 4 to 5 billion years ago (Sagan, 1961).

When did the atmosphere become oxidizing? Geochemical studies independent of biological speculation provide evidence on the question of dating the transition to the oxidizing atmosphere. As recently as 2 to 3 billion years ago, sediments in Canadian, Brazilian and South African shields were deposited containing uraninite (UO_2) that had not been oxidized to pitchblends, suggesting the absence of an appreciable oxygen partial pressure at the time of deposition. However, oxidized rocks as old as 2.5×10^9 years have been found, suggesting that free oxygen was abundant enough to oxidize iron to form limonite red-beds at this time (Rutton, 1962).

Recently, Cloud has reviewed this literature and amassed a good deal of evidence resolving the apparent discrepancies in the data. He has found microstructures, indisputably fossils, which he considers to greatly resemble extant prokaryotes (blue-green algae and iron bacteria) in rocks definitely dating $2 \cdot 1 \times 10^9$ years ago. [In fact, microstructures of fossil prokaryotes—(e.g. *Eobacterium isolatum*) have now been reported in rocks as old as $3 \cdot 1 \times 10^9$ years (Barghoorn & Schopf, 1966).] In concluding his review, Cloud points out,

"The combined evidence of paleontology and stratigraphy, therefore, indicates that the potentiality for the evolution of oxygen by green plant photosynthesis existed at least $2 \cdot 7$ to $2 \cdot 1 \times 10^9$ years ago, and that atmospheric oxygen first began to be available in relatively large quantities probably about $1 \cdot 2 \times 10^9$ years ago (a conclusion independently reached on a different line of reasoning by Lepp and Goldich)" (Cloud, 1965).

(The "green plant" photosynthesis referred to by Cloud includes prokaryotic blue-green algal photosynthesis; it is distinguished from bacterial photosynthesis by the elimination of gaseous oxygen in the process.)

Correct identification of the fossils implies the origin of photosynthetic metabolism which uses the visible portions of the spectrum, occurred in cellular systems earlier than $2 \cdot 1$ billion years ago or even $3 \cdot 1 \times 10^9$ years ago. This is consistent with independent astronomical arguments which indicates that uv absorbing reduced compounds were probably in the atmosphere contemporaneous with these prokaryotic photosynthetic algae (Sagan, 1965*b*).

In any case, the utilization of visible light in microbial photosynthesis must have evolved more than $1 \cdot 2 \times 10^9$ years ago (the time at which free gaseous oxygen was clearly present) because it is likely that, at that time, the ultraviolet light that penetrated the ozone layer of the upper atmosphere was insufficient to provide energy for the formation of organic compounds upon which all life is based. Even if the production of new organic matter continued, it would have been destroyed by oxidation. Thus, within the confines of these dates, this outline for the evolution of cellular photosynthesis was reconstructed.

Porphyrins (metal-chelated tetrapyrroles such as those found in catalase and peroxidase) of some kind are of universal occurrence in all extant cells with the exception of some obligate anaerobes (Lascelles, 1964). Their ubiquity has been related to their ability to reduce mutagenic oxidizing agents. Molecular oxygen is continually being produced by the photolysis of water in the upper atmosphere and the escape of hydrogen (Urey, 1959). This oxygen must have been lethal to the early self-replicating systems. Atmospheric photoproduction of oxygen throughout the ages presumably provided selection pressures for the retention of these ancient cistrons involved in the synthesis of porphyrin antimutagens. By hypothesis, the incidental fact that these antimutagenic tetrapyrole compounds were strong visible light absorbers was later put to advantage in the evolution of chlorophyll-mediated microbial photosynthesis (Sagan, 1961).

The CO_2 fixation (the dark reactions) of photosynthesis is known to be present for the production of cell material in many different non-photosynthetic organisms (e.g. chemolithotrophs: bacteria capable of oxidizing hydrogen gas, sulfide, sulfur, ammonia and nitrite). It is possible that mutations causing the porphyrins to be used in cellular photoproduction of ATP originally occurred in microbes which already contained the CO_2 fixing "dark reactions". This could have resulted in the evolution of the fundamental photosynthetic mechanism, providing a new source of ATP, produced by visible light absorbed by porphyrins. In some organisms, chemotrophic or direct heterotrophic methods of obtaining ATP presumably were then replaced by the visible light-chlorophyll mediated methods characteristic of all microbial photosynthesizers.

The relationship of porphyrin synthesis to intermediary metabolism, i.e. the synthesis *via* glutamate, α-ketoglutaric acid, and succinyl-CoA, suggests that the porphyrins originally selected for because of their antimutagenic properties—for example, in the reduction of hydrogen peroxide—may have eventually been used in the more efficient oxidation of carbohydrate. Organisms unable to form these porphyrins were doomed to eternal obligate anaerobiosis (Stanier *et al.*, 1963, p. 85). Thus, heterotrophic, anaerobic respirers (microbes which oxidize carbohydrates in the absence of molecular oxygen to make ATP *via* iron-chelated porphyrins such as cytochromes) could also have evolved from primitive porphyrin synthesizers.

Eventually in prokaryotic anaerobic photosynthesizers, mutations must have occurred that led to the development of organisms capable of using H atoms from (the much more abundant) water instead of from (the less abundant) hydrogen gas or hydrogen sulfide (e.g. the anaerobic photosynthesizer, *Thiorhodacea*) as electron donors to chlorophyll. The natural consequence of these mutations was the elimination of oxygen derived from that water in the photosynthetic process.

The fact that oxidized limonite red-beds are as old as $2 \cdot 5 \times 10^9$ years is strong evidence that mutations for oxygen eliminating photosynthesis occurred in some microbial autotrophs well before that time.

3.5. ATMOSPHERE OXYGEN AND THE ORIGIN OF AEROBES

The presence of increasing amounts of oxygen, produced by the new photosynthesis must have been extremely deleterious: there was no longer any abiogenic synthesis of organic compounds; the obligate anaerobes found progressively fewer niches; and all life became ultimately dependent on visible light photosynthesis to produce nucleotides and oxidizable carbohydrates. Under such conditions, constant selection pressure must have been placed on any microbes capable of tolerating and utilizing oxygen. In the case of the cytochrome-containing anaerobic respirers—cells which already had genes to code for earlier steps in the respiratory pathways—mutants were selected that could transfer hydrogen atoms to oxygen (instead of to nitrogen or sulfate). This could have resulted in the evolution of ATP pathways involving the complete oxidation of production *via* the aerobic carbohydrate. Such mutations leading to the aerobic breakdown of carbohydrates presumably occurred in some photoautotrophs too, evolving the ancestors of blue-green algae.

It is consistent with both microbial metabolism and the geological evidence to assume, then, from $2 \cdot 1$ to $0 \cdot 6 \times 10^9$ years ago the prokaryotic photosynthesizers eliminated more and more oxygen into the atmosphere. In response to this, different types of microbes evolved (including the proto-mitochondrion and prokaryotic aerobic algae), capable of coping with the increasing abundance of oxygen. The goal of the long metabolic sequences coded for on microbial genomes was then, as it is today, maintenance of the highly reduced surroundings required for reproduction. Ironically, the change from reducing to oxidizing conditions to which microbial cells were forced to adapt had been caused by the cells themselves.

Interestingly enough for our argument, although micro-fossil forms quite analogous to extant prokaryotes are found from $3 \cdot 1 \times 10^9$ years ago, no eukaryotic algae appear in dated rocks until the dawn of the Paleozoic. *Dasycladaceans* are known from early Ordovician, about $0 \cdot 4 \times 10^9$ years ago, and algal filaments perhaps reds or brown (*Epiphyton*) are known as old as early Cambrian $0 \cdot 50$ to $0 \cdot 55 \times 10^9$ years ago (Cloud, 1965, personal communication).† Thus, not only are "missing links" between prokaryotic and eukaryotic algae conspicuously absent from all present day flora, but they are missing from the fossil record during those 2700 million years (i.e. from $3 \cdot 1 \times 10^9$ to $0 \cdot 4 \times 10^9$ years ago), in which first prokaryotic cells (Schopf & Barghoorn, personal communication) and later atmospheric oxygen were known to be present!

Besides the direct evidence of the fossil record, there are biological reasons for regarding photosynthesis as a fundamentally anaerobic process which evolved in prokaryotes. Both blue-green algae and bacterial photosynthesizers are clearly related to other prokaryotes.

† *Note added in proof:* It is now likely that Australian (Bitter Springs) PreCambrian fossils about $0 \cdot 8 \times 10^9$ years old contain primitive eukaryotic coccoid algae (Schopf & Barghoorn, 1966, personal communication).

All bacterial photosynthesizers are morphologically typical gram negative bacteria. The exception to this is *Rhodomicrobium*, photosynthetic budding bacteria, which also have typically prokaryotic nonphotosynthetic counterparts (*Hyphomicrobium*). In all these bacterial photosynthesizers, the photosynthetic process itself is anaerobic. It is only in the dark that any bacterial photosynthesizers are aerobic and respire. Even in blue-green algae, oxygen uptake is inhibited by light (Marsh, Galmiche & Gibbs, 1964).

Blue-green algae also have nonphotosynthetic morphological counterparts among prokaryotes (the filamentous gliding bacteria). Some of these, like *Beggiota*, metabolize H_2S to form sulfur. These organisms grow using CO_2 as their only source of carbon; the H atoms from H_2S are transferred to "fix" CO_2 for reduction to cell material. In the blue-greens H atoms from H_2O are transferred to "fix" CO_2 for reduction to cell material using photosynthetically produced ATP. Thus these gliding bacteria are physiologically as well as morphologically related to blue-greens.

The fact that some blue-greens metabolize along the pentose phosphate (rather than the Embden-Meyerhof pathway so ubiquitous in mitochondria-containing eukaryotes) also reflects their origin and relationship to prokaryotes (Fewson, Al-Hafidh & Gibbs, 1962). This difference in the first steps of carbohydrate metabolism in these two great groups of oxygen eliminating algae indicates their ancient evolutionary divergence† (Haldane, 1954).

The monophyletic evolution of eukaryotic algae from blue-greens is hard to reconcile with many other facts. Blue-greens are typical prokaryotes: they lack a membrane bounded nucleus, mitochondria, flagella, mitosis, sex, simultaneity of respiration and photosynthesis, and, in general, they are smaller than other algae. Furthermore, no intermediate organisms between blue-green and green algae have ever been found, even though they have been extensively sought.

For a comprehensive review of recent literature demonstrating the striking similarities in cell structure of the blue-green algae and the bacteria, see Echlin & Morris, 1965.

Why then are the photosynthetic mechanisms in prokaryotic blue-green algae and eukaryotic green algae so similar? If the prokaryotic blue-greens were not ancestral to the eukaryotic algae and the higher plants, what organisms were?

† This type of reasoning was first used by Haldane in 1929. He pointed to the uniformity of the first steps in anaerobic metabolism and the diversity in the molecular oxygen utilizing —or avoiding—later steps to indicate the greater age of the anaerobic processes. From this, he inferred something of the nature of the reducing atmosphere—in the absence of the laboratory simulation experiments and the geological data that is available to us today!

This apparent paradox can be resolved by recognizing the validity of the thesis presented in this paper, namely, that the evolution of photosynthesis preceded the evolution of the eukaryotic cell by millions of years, and that green plant oxygen-eliminating photosynthesis—so characteristic of both blue-green algae and "chlorophytes"—evolved in prokaryotes and was later acquired symbiotically by various eukaryotes.

This view is valid only if the evolution of the eukaryotic cell itself from an aerobic, heterotrophic ancestor can be understood. What evidence have we that mitosing eukaryotic cells evolved monophyletically and independent of the evolution of photosynthesis? What is the basis for the claim that mitochondria and (9+2) homologues originated as endosymbionts, and that our hypothesis concerning the evolution of the eukaryotic cell is, at least in essential outline, correct?

3.6. THE REPRODUCING MITOCHONDRION

A plethora of recent studies elegantly reviewed by Gibor & Granick have presented inexorable testimony for the following: mitochondria contain specific DNA and RNA; they are self-duplicating bodies that do not arise *de novo*; the multigenic system of the organelle is responsible, in part, for the specific biochemical properties of the organelle; and mitochondrial development (in yeast cells, at least) are controlled by an adaptive mechanism which is responsive to oxygen. There have been some reports that mitochondria are capable of limited incorporation of amino acids into proteins and they contain their own protein synthesizing mechanisms, as well. For a detailed description and bibliography of the pertinent literature, the reader is referred to this excellent review (Gibor & Granick, 1964). That mitochondria have sources of ATP and small molecules, is indisputable. Whether the mitochondrial membranes are coded for by the mitochondrial genes themselves is not clear. In any case, the mitochondria satisfy the first criteria for cells originating as endosymbionts as discussed earlier (section 3.3).

In 1927, Wallin argued that mitochondria originated as endosymbionts in higher cells. His evidence was based on the size, shape, staining properties and general cytological behavior of the organelles which he claimed were comparable to bacteria. Of course, at that time he could have had little concept of their physiology. The most convincing evidence for the genetic autonomy of mitochondria involved studies of their continuity. These observations reviewed by Wilson in 1925, indicated that mitochondria are invariably included in sperm cells and are, in general, present at every stage of the life cycle in eukaryotic cells. Some of the mechanisms by which mitochondria are retained in daughters during cell divisions are listed in Table 3. Full discussions appear in the classical literature (Wilson, 1925).

TABLE 3

Cytological mechanisms for retention of organelles throughout life cycle

Organisms	Mechanism
A. *Mitochondria*	
Many plants and animal cells; dividing germ cells of vertebrates; cleavage stages	Many mitochondria randomly distributed throughout cell (Wilson, 1925, p. 712)
Spermatocytes of some scorpions, such as *Opisthancanthus, Hadrurus*, etc.	Primary spermatocytes: small and numerous mitochondria join together to form 24 spheroids. Secondary spermatocytes: 12 spheroids are segregated to each daughter cell. Spermatids: six spheroids are segregated to each pole; thus six mitochondrial spheroids are present in each sperm cell (Wilson, 1925, p. 163)
Spermatocytes of scorpion, such as *Centrurus*	Primary spermatocytes: mitochondria aggregate into ring-shaped body, oriented on spindle, is cut transversely by division of cell into two half rings; each half ring forms a rod. Secondary spermocyte: each rod is carried on spindle and cut by cell division into a half rod (Wilson, 1925, p. 365)
Spermatocytes of worm such as *Ascaris*	Mitochondria are oriented on spindles toward centrioles. They do not divide but are segregated to daughter cells by virtue of their orientation (Wilson, 1925, p. 163)
Some insects, such as *Hydrometra*	Elongate rods of mitochondria are oriented on spindles; some of these mitochondrial rods are cut by cell equator (Wilson, 1925, p. 164)
Some ciliates	Mitochondrial divisions of numerous mitochondria synchronous with nuclear division (Wilson, 1925, p. 13)
Vicia, a bean	Two groups of mitochondria orient at opposite poles at division (Wilson, 1925, p. 163)
Micromonas, a photosynthetic flagellate	One mitochondrion divides synchronously with nucleus (Gibor & Granick, 1964)
B. *Chloroplasts*	
Diatoms	Small and constant number evenly distributed in mitosis (Darlington, 1958)
Micromonas, a photosynthetic flagellate	One chloroplast divides (Gibor & Granick, 1964) synchronously with nucleus
Chlamydomonas, a photosynthetic flagellate	One large chloroplast cut by cleavage plane (Gibor & Granick, 1964)
Most higher plants	Many chloroplasts randomly distributed at mitosis (Wilson, 1925, p. 162)

Recently, the absence of visible mitochondria in the cytoplasm of anaerobic yeast (as seen through the electron microscope) has again been taken as evidence for the lack of genetic continuity of the organelle. Since we now know the fundamental replication event takes place on the molecular level, conclusive experimental evidence for the contrary could come from simply showing that yeast "satellite band DNA" (e.g. mitochondrial DNA) is still present in those organisms which lack visible mitochondria but retain the potential to form them.

The entries in Table 3 can be interpreted as mechanisms which insure that each daughter cell of the host receives at least one copy of the symbiont genome at each division. In the formation of sperm, mitochondria are often wrapped around the axial fiber, a $(9+2)$ homologue, insuring continuity of the organelle through fertilization as well (Wilson, 1925, p. 373) (criterion (2), section 3.3).

Do mitochondria fulfil other requirements for organelles originating in symbionts? There seem to be no extant examples of organisms that totally lack mitochondria and contain other characteristics of eukaryotes. This is probably because the mitochondrial symbiosis is so ancient it is obligate in all eukaryotes: total loss of the organelle is invariably lethal (criteria (3), section 3.3). However, neither are there nucleated or plastid-containing organisms which contain mitochondrial enzymes "unpackaged" (criteria (4), section 3.3).

Examples of mitochondrial cytoplasmic heredity are well known (Gibor & Granick, 1964; Jinks, 1964) (criterion (5), section 3.3). Many aerobic bacteria provide free-living extant counterparts to the protomitochondrion, since aerobic cytochrome-mediated glucose metabolism is well known among bacteria (criterion (6), section 3.3).

Regardless of the history, it must have taken thousands of mutational steps to evolve mitochondria (i.e. approximately 100 enzymes $\times$ 100 amino acids per enzyme $\times$ 3 nucleotide pairs per amino acid $\approx 3 \times 10^4$ nucleotide pairs). The evolution of the chloroplast in an aerobic cell containing mitochondria resulting in green plant photosynthesis precisely analogous to that of blue-green algae is highly improbable. Such a monophyletic origin of chloroplasts would require thousands of further highly specific mutations to evolve photosynthesis. The absence of both fossil evidence and extant intermediate organisms in the evolution of this "primitive phytoflagellate" makes it unlikely that plastid-contained aerobic photosynthesis itself evolved in a cell containing mitochondria. Since we have already presented evidence that photosynthesis is fundamentally an anaerobic process which evolved in prokaryotes (section 3.5), this is especially unlikely.

On the other hand, if the evolution of mitochondria followed the evolution of plastids, there is no obvious reason why photosynthetic eukaryotes should show remarkable uniformity in mitochondrial structure and metabolism. Plastid-containing organisms with alternative carbohydrate metabolic patterns should have evolved, especially since such alternative pathways are known to be present in blue-green algae.

3.7. THE REPRODUCING (9+2) HOMOLOGUES AND THEIR RELATIONSHIP TO THE NUCLEUS

"... that a fundamental dualism exists in the phenomenon of mitosis, the origin and transformation of the achromatic figure being in large measure independent of those occurring in the chromatic elements. Mitosis consists, in fact, of two closely correlated but separable series of events" (Wilson, 1925).

The homology of the "achromatic figure" (i.e. mitotic apparatus, basal bodies, centrioles, etc.) and its relative independence of nuclear chromatin is at the base of the above statement which summarizes the results of some 40 years of cytological observation, by perhaps the finest biologists of the time (c. 1885-1925).

Classical cytogenetics and more recent experimental work involving modern techniques have confirmed and extended the observations of the early workers. Especially relevant are the careful studies of Cleveland which experimentally disassociate chromosome replication and cell division.

"Oxygen concentrations of 70-80% destroy all the chromosomes of the hypermastigote flagellate, *Trichonympha*, provided the oxygen treatment was carried out during early stages of gametogenesis when chromosomes are in process of duplicating themselves. This treatment does no damage to the cytoplasm and its organelles, following the loss of chromosomes. The centrioles function in the production of the achromatic figure (e.g. the mitotic apparatus), the flagella and the parabasal bodies. Then the cytoplasm divides, thus producing two anucleatic gametes which make some progress in the cytoplasmic differentiation characteristic of the normal male and female gametes of *Trichonympha*" (Cleveland, 1956).

On the other hand, Cleveland's observations of binucleate 5-centriole cell showed that "... without centrioles no achromatic figure is formed, there is no poleward movement of the chromosomes to form daughter nuclei. The chromosomes reproduce themselves but the nucleus does not. However, two or more centrioles must be present and must be fairly close to the nucleus if the nucleus is to reproduce itself" (Cleveland, 1963). When there is a choice in these multicentrioled cells, chromosomes will move along other than the central spindle.

The fundamental difference between eukaryotic mitosis and the equal distribution of genes in prokaryotic cell division is the total amount of DNA which can be distributed to the daughter cells. If newly synthesized cellular DNA could be attached to some intracellular self-replicating body which at the time of host division segregated from *its* sister offspring, then a mechanism for equal distribution of genetic material could result. This mechanism, operating quite independent of the "messages" carried by the host DNA, would simply insure the segregation of the newly synthesized host DNA associated with the self-replicating intracellular body. In such a scheme, it is unnecessary to explain how the self-replicating body differentiated from the nucleus *before* it was selected as a division center. The alternative assumption that a pre-existing replicating endosymbiont, the basal body of the flagellum, was utilized in a new role, is consistent with traditional belief—that evolution is opportunistic and not foresighted. The role of host chromatin segregation is the recognized function of the centromere; cells lacking centromeres simply do not arrive at the poles of the mitotic spindle to be incorporated into daughter nuclei, and chromosomes always proceed to the poles centromere first. Chromosomes may contain two centromeres that travel to opposite poles of the dividing cell; often when these dicentric chromosomes break, each fragment attached to its centromere is incorporated into the resulting daughter cell.

The function of the flagellar basal apparatus in mitosis and its use in distribution of cellular organelles was widely recognized in the classical cytological literature (Table 3). A striking example of this is to be found in *Trypanoplasma* (Belar, 1915). The division of the blepharoplast at the base of the flagellum—called the *blephoplasteilung*—is quite as conspicuous as the nuclear division, forming an apparent second "mitotic apparatus" related to the flagella [for example, Fig. 2. V (d); [criterion (2), section 3.3].

In *Leishmania* the kinetoplast attached to the flagellar apparatus, for example, is Fuelgen positive and known to divide. Very clear evidence for a specialized DNA satellite band associated with the organelle has been presented (Du Buy, Mattern & Riley, 1964) consistent with evidence that the organelle incorporates thymidine into DNA (Gibor & Granick, 1964). Electron microscopy indicates that the kinetoplast contains the single mitochondrion which differentiates during the part of the life cycle in which oxidative metabolism is required. It is possible that a specialized association between (9+2) homologue and the one mitochondrion has evolved to insure distribution of the mitochondrion to the daughter cells in a way analogous to the distribution of daughter nuclear genomes. The homology of the kinetoplast DNA may be ascertained by DNA hybridization experiments with the mitochondrial and flagellar DNA's.

The homology of the centriole and the flagellar basal body, first suggested in 1898 (Wilson, 1925, p. 697), is now widely accepted especially as their structure (as revealed by the electron microscope) has been elucidated (Sleigh, 1962). The excellent evidence for the "genetic autonomy" of these cytoplasmic self-replicating (9+2) organelles has been well reviewed (Jinks, 1964). There are reports in the literature of irreversible losses of (9+2) homologues (Lederberg, 1952; Jinks, 1964) (criterion (4), section 3.3). A specific (9+2) homologue DNA which presumably codes for the characteristic proteins, has been suggested by some studies (Seaman, 1960) but has not been definitely shown (Hoffman, 1965) (criterion (1), section 3.3). No intermediate organisms between those containing the (9+2) complex flagella and the simple eubacterial flagella are known (criterion (3), section 3.3). Hence, the origin of the (9+2) homologues as endosymbionts is not inconsistent with the evidence; but the argument that these homologues were of exogenous origin and did not pinch off from the cell (i.e. did not originate as episomes) must be justified.

For various reasons, spirochaete or spirochaete-like organisms have been suggested as likely candidates for free-living counterparts of the motile (9+2) endosymbiont which later differentiated into the flagellum, centrioles, and chromosomal centromeres. Spirochaetes are known to be associated with protozoans; for example, in *Dienympha* and *Pyrsonympha* (eukaryotic animal flagellates found in insects) the family (Dienymphida) to which they belong has been characterized by the following description: ". . . elongate flagellates, the four or eight anterior flagella adherent to the body and spirally twisted with it, free at their distal ends. Often they are beset with spirochaetes which have been mistaken for additional flagella; the family has been misplaced in the order *Hypermastigina*" (Copeland, 1956, p. 166). The same is true of another family of eukaryotic flagellates, the *Devescovinida*: "Spirochaetes which share the habitat of these organisms are commonly found adhering to their cell membranes, and were mistaken for additional flagella in the original descriptions of some of the genera" (Copeland, 1956, p. 167).†

Spirochaetes are approximately the same size as flagella, they are made of subunit strands (Stanier *et al.*, 1963, p. 158) varying in number depending on the species. They are usually found in micro-aerophilic environments; they are always motile; their motility is sensitive to ATP.

Indeed, it would be most interesting if the flagellar ATPase, other characteristic proteins (Gibbons, 1963), and (9+2) sub-

---

† *Note added in proof:* Recent elegant studies of *myxotrixa paradoxa* indicate symbiotic spirochaetes on the surface of these cells are responsible for the hosts' movements! Moreover these flagellates typically have at least three different types of symbiotic prokaryotes associated with them. (Grimstone, A. V. & Cleveland, L. R. *Proc. R. Soc. Lon.* **159**, 668, 1964.)

structure could be identified with the proteins and axial fiber of some free-living spirochaete (criterion (6), section 3.3).

If the flagellar basal body had escaped from the host genome for the purpose of distributing nuclear DNA, it is likely that such episomal escape would have occurred in many different lines of microbes and that not all extant examples would be homologous. There is no immediate selective advantage to the escape of such a piece of host nucleic acid. If episomal in origin, we might expect examples of organelles which lyzed the host for their own continued replication. If (9+2) homologue centrioles pinched off and were selected for because they distributed cellular DNA efficiently, they should not necessarily be related to the flagella at all. If they escaped from the nucleus, these (9+2) homologues should be more sensitive to treatments which affect the nucleus than to those which are related to the destruction of the flagella. In fact, the opposite is true (Weinrich, 1954, p. 135; Yow, 1961).

3.8. THE REPRODUCING CHLOROPLAST

Presumably because it was the least ancient of the symbioses, the exogenous origin of the photosynthetic plastid is the easiest to defend (Ris & Plaut, 1962; Echlin, 1966). Chloroplast DNA has been found in many different photosynthetic eukaryotes. [In *Euglena*, direct autoradiographic evidence for nuclear and plastid DNA has recently been presented. The *in situ* DNA has been correlated with the presence of labeled main band (nuclear) and satellite band (chloroplast) DNA on a cesium chloride density gradient (Sagan, Ben-Shaul, Schiff & Epstein, 1965).] This satellite DNA band is totally absent in permanently "bleached" cells, those which permanently lack the potential for chloroplast formation (Leff, Mandel, Epstein & Schiff, 1963) (criterion (4), section 3.3). The disappearance of the "satellite band" DNA has been correlated with the ultraviolet treatment which permanently "cures" *Euglena* of its chloroplasts (Edelman, Epstein & Schiff, 1964; Edelman, Schiff & Epstein, 1965).

Other characteristics implied by an endosymbiotic origin have been reported for chloroplasts as well: chloroplast-specific RNA complimentary to chloroplast DNA (Eisenstadt & Braverman, 1963); ribosomes (Lyttleton, 1962); and chloroplast ribosomal RNA; evidence for DNA-dependent RNA synthesis; and the uptake of radioactive amino acids into isolated chloroplasts (Gibor & Granick, 1964).

In summing up both the classical observations on replication of chloroplasts and the recent work on the biochemistry of the organelle, Gibor and Granick provide convincing evidence that: the plastids (like the mitochondria) contain DNA and RNA; they are self-duplicating bodies which do not arise *de novo*; the DNA represents a multigenic hereditary system which is not derived from the nucleus and is, in part, responsible for the biochemical properties of the organelle; and that the differentiation of the mature chloroplast from the proplastid is an adaptive system responsive to visible light (criterion (1), section 3.3).

Some mechanisms which assure distribution of at least one copy of the plastid to each daughter of the host throughout the life cycle in plants are summarized in Table 3 (criterion (2), section 3.3).

Examples of organisms which lack plastids but are clearly counterpart to plastid-containing cells are very well known (e.g. *Astasia*, *Polytoma*, etc.). Blue-green algae themselves may be considered free-living prokaryote counterparts of plastids (criterion (6), section 3.3). Despite the search for them, neither fossil nor extant examples of intermediate organisms between the plastid-lacking blue-greens and the plastid-containing eukaryotes have ever been found (criterion (3), section 3.3).

"Cytoplasmic heredity" was first discovered in cases of uniparental chloroplast inheritance. The literature has been well reviewed (Jinks, 1964; Granick, 1962) (criterion (5), section 3.3). Characterization of the phenomena known collectively as "cytoplasmic heredity" is based on the non-chromosomal (and therefore non-mendelian) inheritance of genes. The logical explanation for cytoplasmic non-mendelian genetic systems is based on their legacy as genes of once free-living cells which have become organelles, i.e. hereditary endosymbiosis (Lederberg, 1952).

Apparently another major piece of circumstantial evidence for the endosymbiotic origin of these organelles is their size. We now know that replication takes place strictly on the molecular level; the flagellum itself is the size (and indeed the shape) of a respectable prokaryote, as are the mitochondria and plastids. The self-replication of such organelles of course involves synthesis of all the non-nucleic acid components for perpetuation. The probability that bodies of such enormous size (from a molecular point of view) would form multigenic systems in the cytoplasm is extremely low. The explanation that ". . . certain mutations could occur independently in each organellar DNA unit, and these mutations could be carried along so that, when drastic environmental changes occurred, there could be selection for the most suitable organelles. Thus, a multiple number of mutated organelles per cell could provide for more rapid evolutionary change" (Gibor & Granick, 1964) is patently Lamarckian. These were, however, the reasons given by Gibor & Granick for the large number of remarkably consistent experimental results implying genetic continuity of mitochondria and plastids.

4. Some Predictions

It is likely that the classifications presented in the phylogenetic tree (Fig. 1) err in that the author lacks first-hand knowledge of most of the organisms. It is also true that flagella and chloroplasts can be secondarily lost. However, if the theory presented in this paper is correct, all eukaryotes should ultimately be classified completely, correctly, and consistently according to their position in the origin of mitosis. Consistent with this view is the empirical fact that the flagellar attachment has proven to be a reliable taxonomic criterion.

Analogous to the quantitative relationship between DNA and ploidy, satellite DNA correlated specifically with the various organelles should be found in cells in direct proportion to the number of organelles in these cells. No eukaryotic cell having flagellar basal bodies, cilia, centrioles, centromeres, or any other of the homologues can lack $(9+2)$ homologue-specific DNA. It is likely that this DNA has evaded detection because it has very little metabolic responsibility and needs to be only a few cistrons long to code for its few specific proteins. Identification of $(9+2)$ homologue-specific DNA and a complete characterization of its RNA and of its limited biochemical functions should eventually be possible. This is true of mitochondrial and plastid DNA as well.

If these organelles did indeed originate as free-living microbes, our advancing technology should eventually allow us to supply all growth factors requisite for *in vivo* replication of all three of them—the *coup de grace* to genetic autonomy.

It is likely that biochemical pathways entirely unique to eukaryotes (e.g. steroid synthesis) should have component parts coded for by more than one organellar genome. [For example, it is possible that the biochemical pathway to squalene in steroid synthesis is under nuclear control, and that the subsequent members of the biosynthetic chain—those admittedly tightly bound to particulate elements in the cell—are under the control of the mitochondrial genome (Block, 1965).] The component parts of metabolic pathways in plant metabolism—analogous to starch synthesis in *Peliaina* (Lederberg, 1952) (section 3.2) for example—may eventually be understood in terms of "complementation", e.g. syntheses made possible by the presence in the cell of at least two genomes—host and plastid.

In keeping with the hypothesis, the following organisms should have evolved: a free-living complex flagellar counterpart; a free-living mitochondrion counterpart; and a heterotrophic prokaryote capable of ingesting cells. Free-living cells co-descendant with eukaryotic organelles might still contain cistrons homologous to those in $(9+2)$ homologues, mitochondria, and plastids. For example, we may one day find different types of blue-green algae that are co-descendant with typical chrysophsean and rhodophysean plastids, which contain DNA with cistrons homologous to those in the plastids.

If the theory is correct all eukaryotic cells must be seen as multi-genomed systems. This implies that a goal of cellular chemistry is understanding the way in which all biochemical reactions are coded off the nucleic acid of the nucleus and the subcellular organelles. All eukaryotes must contain at least three specific types of DNA: nuclear, mitochondrial, and $(9+2)$ homologue. An additional DNA that is associated with the chloroplasts must be found in all eukaryotic plants. The plastids of organisms sharing similar photosynthetic metabolic characteristics (e.g. dinoflagellates, brown algae, and diatoms; red and blue-green algae) should have homologous plastid-specific nucleic acid. At the same time, those organisms which demonstrate "host homologies" (e.g. non-pigmented and pigmented dinoflagellates, *Schizogoniacea* and *Porphyra*) should share homologous nuclear, and not necessarily plastid, DNA's. This is quite analogous to the presence of two distinct DNA bands of nearly equal size found in CsCl density gradient runs on DNA isolated from *Paramecium bursaria* (Sagan, L., 1964, unpublished data). In this experiment, the larger DNA band corresponded to a base ratio of $G+C = 29\%$, presumably, quite characteristic of the ciliates (Schildkraut, Mandel, Levisohn, Smith-Sonneborn & Marmur, 1962). The smaller band corresponded to 60% $G+C$, characteristic of chlorellae, and is presumably due to the presence of the photosynthetic zoochlorellar endosymbiont (Sueoka, 1961). The relative abundance of the chlorella band with respect to the nuclear, perhaps indicates the symbiosis is quite recent. Consistent with other data, it is possible that when our technology is more advanced, the quantitative relationship between the amounts of "main band" and "satellite band" DNA's on a CsCl gradient will directly indicate the relative amounts of metabolic responsibility relegated to the nuclear (main band) genome and, hence, to the age of the symbiosis in general. When we know the precise metabolic pathways and their genetic basis, we will be able to calculate directly the number of generations elapsed since each organism evolved from a common ancestor—from the number of mutational sites at which they differ.

Some searches will continue to be futile: for example, attempts to find eumitosis in all eukaryotes (if found, it will clearly be analogous, rather than homologous, to eumitosis in higher eukaryotes, cf. reports of sexuality in *Noctiluca*, a dinoflagellate); "missing links" in the origin of the ancestral phytoflagellate, such as organisms containing chloroplasts but no mitochondria; eumitotic organisms with bacterial flagella, or eumitotic fossils dating from anaerobic times.

The cytological basis of most "cytoplasmic inheritance" will probably be related either to mitochondria, (9+2) homologues, plastids, or other much less generally distributed cytoplasmic organelles (e.g. *Kappa* in *Paramecium aurelia*). As an example, the inheritance of the sr-500 gene in *Chlamydomonas* with the "plus" mating type (mt+) suggests that a correlated uniparental inheritance of chloroplast-specific DNA should be sought (Sager & Tsubo, 1961).

As in the past, future attempts to relate the various classes of algae directly to each other will be futile. For example, the ancestor of the relatively recent and isolated algae the *Bacillaracea* (diatoms), might be better sought among primitive metazoans than the chrysophytes or dinoflagellates. A simple flagellated organism with characteristic rhodophyte pigments probably never existed. A flagellated ancestor to the true molds and premitotic amoebae probably did not either.

The range of DNA base ratios in all plants and animals is quite limited compared to that in prokaryotes (Sueoka, 1961).

". . . The base of the Cambrian period (0.6×10^9 years ago) is marked in marine sediments around the world by the appearance of abundant animal life. . . . This sudden appearance of diverse animal stocks has been the most vexing riddle in paleontology" (Fischer, 1965).

The evolution of mitosis in ancestral heterotrophic amoeboflagellates, making possible mendelian genetic patterns and the tissue and organ level of biological organization, is presumably at the basis of both observations: the small variation in DNA base ratio in all plants and animals and the plethora of new forms of life marking the dawn of the fossil record.

The author acknowledges with pleasure the aid, encouragement and criticism of Drs J. William Schopf, M. Ptashne and Professors J. D. Bernal, E. Barghoorn, C. Sagan, Y. Ben-Shaul, A. O. Klein, T. N. Margulis, D. Hawkins, P. Morrison, P. E. Cloud and especially Professor R. A. Lewin. She is immensely grateful to Mrs J. R. Williams and Mr George Cope for editorial and production assistance.

REFERENCES

ABELSON, P. (1963). *J. Wash. Acad. Sci.* **53**, 105.
BARGHOORN, E. S. & SCHOPF, J. WM. (1966). *Science, N.Y.* **150**, 758.
BELAR, K. (1915). *Arch. Protistenk.* **36**, 24.
BERNAL, J. D. (1957). The Problem of Stages in Biopoesis *in* "The Origin of Life on the Earth", Symposium of the International Union of Biochemistry, Moscow. New York: Macmillan.
BLOCK, K. (1965). *Science, N.Y.* **150**, 19.
BRACHET, J. (1957). "Biochemical Cytology". New York: Academic Press.
BUCHNER, P. (1930). "Tier and Pflanze in Symbiose". Berlin: Borntraeger.
BUCHNER, P. (1953). "Endosymbiose der Tier mit Pflanzlichen Microorganisms". Basil & Stuttgart: Birkhauser.

CALKINS, G. N. (1909). "Protozoology", p. 31. New York & Philadelphia: Lea & Febiger.
CLEVELAND, L. B. (1956). *J. Protozool.* **3**, 78.
CLEVELAND, L. B. (1963). Function of Flagellate and Other Centrioles in Cell Reproduction *in* "The Cell in Mitosis" (L. Levine, ed.). New York: Academic Press.
CLOUD, P. E., JR. (1965). *Science, N.Y.* **148**, 27.
COPELAND, H. F. (1956). "The Classification of the Lower Organisms". Palo Alto, California: Pacific Books.
CRONQUIST, A. (1960). *Bot. Rev.* **26**, 425.
DARLINGTON, C. D. (1958). "The Evolution of Genetic Systems". New York: Basic Books, Inc.
DILLON, L. S. (1962). *Evolution*, **16**, 102.
DOFLEIN, F. & REICHENOW, E. (1929). "Lehrbuch der Protozoenkunde: eine Darstellung der Naturgeschicte der Protozoen mit besonderer Beruchsichtigung der parasitischen und pathogenen Formen". Jena: Fischer.
DOUGHERTY, E. & ALLEN, M. B. (1960). Is Pigmentation a Clue to Protistan Phylogeny? *in* "Comparative Biochemistry of Photoreaction Systems". New York: Academic Press.
DUBUY, H., MATTERN, C. F. T. & RILEY, F. (1964). *J. Cell Biol.* **23**, 26A.
ECHLIN, P. & MORRIS, I. (1965). *Biol. Rev.* **40**, 193.
EDELMAN, M., EPSTEIN, H. T. & SCHIFF, J. A. (1964). *Proc. natn. Acad. Sci. U.S.A.* **52**, 1214.
EDELMAN, M., SCHIFF, J. & EPSTEIN, H. T. (1965). *J. molec. Biol.* **11**, 769.
EISENSTADT, J. & BRAVERMAN, G. (1963). *Biochem. biophys. Acta*, **26**, 319.
FEWSON, C. A., AL-HAFIDH, M. & GIBBS, M. (1962). *Pl. Physiol.* **37**, 402.
FISCHER, A. G. (1965). *Proc. natn. Acad. Sci. U.S.A.* **53**, 1205.
FRITSCH, F. E. (1935). "The Structure and Reproduction of the Algae", Vol. 1. London & New York: Cambridge University Press.
GABRIEL, M. (1960). *Am. Nat.* **94**, 257.
GIBBONS, I. R. (1963). *Proc. natn. Acad. Sci. U.S.A.* **50**, 1002.
GIBOR, A. & GRANICK, S. (1964). *Science, N.Y.* **145**, 890.
GOLDSCHMIDT, R. (1907). *Archiv Protistenk.* **8**, 84.
GOLDSCHMIDT, R. & POPOFF, M. (1907). *Archiv Protistenk.* **8**, 321.
GRANICK, S. (1962). The Chloroplasts: Inheritance, Structure and Function *in* "The Cell", Vol. II (J. Brachet & A. Mirsky, eds.). New York: Academic Press.
HALDANE, J. B. S. (1954). The Origins of Life *in* "New Biology, No. 16" (M. L. Johnson, M. Abercrombie & G. E. Fogg, eds.). London: Penguin Books.
HARTMANN, M. (1921). *Archiv Protistenk.* **43**, 223.
HOFFMAN, E. J. (1965). *J. Cell Biol.* **25**, 217.
JINKS, J. L. (1964). "Extrachromosomal Inheritance." Englewood Cliffs, N.J.: Prentice-Hall.
KARAKASHIAN, S. J. (1963). *Physiol. Zoöl.* **36**, 52.
LASCELLES, J. (1964). "Tetrapyrrole Biosynthesis and its Regulations", p. 17. New York: W. A. Benjamin.
LEDERBERG, J. (1952). *Physiol. Rev.* **32**, 403.
LEFF, J., MANDEL, M., EPSTEIN, H. T. & SCHIFF, J. A. (1963). *Biochem. biophys. Res. Comm.* **13**, 126.
LYMAN, H., EPSTEIN, H. T. & SCHIFF, J. A. (1961). *Biochim. biophys. Acta*, **50**, 301.
LYTTLETON, J. W. (1962). *Expl. Cell Res.* **26**, 312.
MARSH, H. V., GALMICHE, J. M. & GIBBS, M. (1964). *Rec. chem. Prog.* **25**, 260.
MERECHOWSKY, M. (1910) & MINCHIN, E. A. (1915) *in* "The Cell in Development and Heredity", by E. B. Wilson (1925). New York: Macmillan.
MINCHIN, E. A. (1912). "An Introduction to the Study of the Protozoa", p. 113. London: Edward Arnold.
PICKEN, L. (1962). "The Organization of Cells and Other Organisms", p. 259. Oxford, England: Clarendon Press.
PIRIE, N. M. (1959). Chemical Diversity and the Origin of Life *in* "The Origin of Life on the Earth", International Union of Biochem. New York & London: Pergamon Press.

PRESCOTT, D. M. (1964). Cellular Sites of RNA Synthesis *in* "Progress in Nucleic Acid Research and Molecular Biology" (J. N. Davidson & W. E. Cohn, eds.). New York: Academic Press.
RENAUD, F. H. & SWIFT, H. (1964). *J. Cell Biol.* **23,** 339.
RIS, H. & PLAUT, W. (1962). *J. Cell Biol.* **12,** 383.
RUTTEN, M. G. (1962). "The Geological Aspects of the Origin of Life on Earth." Amsterdam & New York: Elsevier Publishing Co.
SAGAN, C. (1961). *Radiat. Res.* **15,** 174.
SAGAN, C. (1965a). Ultraviolet Synthesis of Nucleoside Phosphates *in* "Origin of Prebiological Systems" (S. Fox, ed.). New York: Academic Press.
SAGAN, C. (1965b). Origins of the Atmospheres of the Earth and Planets *in* Section I of "International Dictionary of Geophysics" (S. K. Runcorn, ed.-in-chief). London: Pergamon Press.
SAGAN, L., BEN-SHAUL, Y., SCHIFF, J. A. & EPSTEIN, H. T. (1965). *J. Pl. Physiol.* **40,** 1257
SAGER, R. & TSUBO, Y. (1961). *Zeitschrift fur Vererbungslehre,* **92,** 439.
SCHILDKRAUT, C., MANDEL, M., LEVISOHN, S., SMITH-SONNEBORN, J. & MARMUR, J. (1962). *Nature, Lond.* **196,** 795.
SEAMAN, G. R. (1960). *Exp. Cell Res.* **21,** 292.
SIEGEL, R. W. (1960). *Expl. Cell Res.* **19,** 239.
SIEGEL, R. W. & KARAKASHIAN, S. J. (1959). *Anatomical Record,* **134,** 639.
SLEIGH, M. (1962). "Biology of Cilia and Flagella." Oxford: Pergamon Press.
SMILLIE, R. (1963). *Can. J. Bot.* **41,** 123.
STANIER, R., DOUDEROFF, M. & ADELBERG, E. (1963). "The Microbial World", 2nd Ed. Englewood Cliffs, N.J.: Prentice-Hall Inc.
SUEOKA, N. (1961). *J. molec. Biol.* **3,** 31.
TRAGER, W. (1964). The Cytoplasm of Protozoa *in* "The Cell", Vol. VI (J. Brachet & A. Mirsky, eds.). New York: Academic Press.
UREY, H. C. (1959). "Primitive Planetary Atmospheres and the Origin of Life" *in* "The Origin of Life on the Earth", International Union of Biochemistry. New York & London. Pergamon Press.
WALLIN, J. E. (1922). *Am. J. Anat.* IV, in Wilson (1925).
WALLIN, J. E. (1927). "Symbionticism and the Origin of Species". Baltimore: Williams & Wilkins.
WEINRICH, D. H. (1954). Sex in Protozoa: A Comparative Review *in* "Sex in Microorganisms" (D. H. Weinrich, chairman and ed.). Publication of the American Association for the Advancement of Science, Washington.
WENYON, C. M. (1926). "Protozoology: A Manual for Medical Men, Veterinarians and Zoologists." London: Bailliere, Tindall & Cox.
WILSON, E. B. (1925). "The Cell in Development and Heredity." New York: Macmillan.
YOW, F. W. (1961). *J. Protozool. Suppl.* **8,** 20.

ERRATUM

The first line of the note at the bottom of article page 267 should read "Recent elegant studies of *Myxotricha paradoxa* indicate symbiotic . . . ."

Origins of Prokaryotes, Eukaryotes, Mitochondria, and Chloroplasts

A perspective is derived from protein and nucleic acid sequence data.

Robert M. Schwartz and Margaret O. Dayhoff

Many proteins and nucleic acids are "living fossils" in the sense that their structures have been dynamically conserved by the evolutionary process over billions of years (1). Their amino acid and nucleotide sequences occur today as recognizably related forms in eukaryotes and prokaryotes, having evolved from common ancestral sequences by a great number of small changes (2). These sequences may still carry sufficient information for us to unravel the early evolution of extant biological species and their biochemical processes. There are two principal computer methods that can be used to treat sequence data in order to elucidate evolutionary history. These were first described more than 10 years ago (3, 4) and have been used to construct a vertebrate phylogeny from each of a number of proteins (5–7). This phylogeny is generally consistent with both the fossil record and morphological data. Only recently has enough sequence information become available from diverse types of bacteria and blue-green algae, and from the cytoplasm and organelles of eukaryotes, for us to attempt the construction of a biologically comprehensive evolutionary tree. These sequences include ferredoxins, 5S ribosomal RNA's, and c-type cytochromes.

We describe here an evolutionary tree derived from sequence data that extends back close to the time of the earliest divergences of the present-day bacterial groups.

Knowledge of the evolutionary relationships between all species would have great predictive advantage in many areas of biology, because most systems within the organisms would show a high degree of correlation with the phylogeny. Knowing the relative order of the divergences of prokaryote types and their protein constituents is important to understanding the evolution of metabolic pathways. With such information the long-standing question of how eukaryote organelles originated might be resolved.

Before we consider the phylogenies based on sequence data, we will briefly review the fossil record and describe the time scale during which the various prokaryote groups diverged (8, 9). The early fossil record is sparse and subject to some uncertainty of interpretation. The oldest known bacterium-like structures that could possibly be biogenic are preserved in the Swaziland sediments and are more than 3.1 billion years old (8,

---

Dr. Schwartz is a senior research scientist at the National Biomedical Research Foundation. Dr. Dayhoff is an associate professor in the Department of Physiology and Biophysics at Georgetown University Medical School and Associate Director of Research at the National Biomedical Research Foundation at Georgetown University Medical Center, Washington, D.C. 20007.

10). Stromatolitic structures have been dated at nearly 3.0 billion years old (*11*). These structures are widely assumed to be evidence for the antiquity of blue-green algae but might equally well be interpreted as products of communities of photosynthetic bacteria, gliding flexibacteria, or non-oxygen-releasing ancestors of the blue-green algae (*8*). Both coccoid and filamentous microfossils from 2.3 billion years ago have been identified. The fossil record of microorganisms from about 2 billion years ago clearly shows a great abundance and diversity of morphological types, many resembling present-day bacteria and blue-green algae. Geological evidence suggests that free oxygen began to accumulate in the atmosphere about 2.0 billion years ago (*12*). Eukaryote cells may have originated as early as 1.4 billion years ago, because there is an abrupt increase in cell size and diversity in microfossils of this age (*13*). The multicellular eukaryote kingdoms, plants, animals, and fungi, are thought to have diverged between 1 billion and 700 million years ago. Fossils of metazoans nearly 700 million years old have been found.

There is good evidence that, in many bacterial sequences of basic metabolic importance, the rate of accumulation of point mutations accepted in the wild-type population is even slower than it is in higher plants and animals (*14*). Thus, even though the time span for the evolution of prokaryotes is more than two times as long, we hope to infer correct evolutionary trees for them as well as we do for eukaryotes. Because the morphological evidence of biologists has proved to be inadequate to the task of organizing the major groups of bacteria and because the fossil record is difficult to interpret, sequence data may prove to be essential.

In this article, we assume that the major types of bacteria have conserved the integrity of the groups of sequences performing basic metabolic functions; we also assume that the substitution of a new sequence for one already functioning in a group through genetic transfer is sufficiently rare to be discounted. Frequent transfer between closely related species should not impair our ability to deduce the course of evolution of the major bacterial types. Only sequences that were transferred will lead to conflicting evolutionary histories for the species involved; sequences from any of the close species would be equally useful in deducing the evolutionary position of the bacterial type.

Reconstructing Evolution on the Basis of Sequence Data

Evolutionary history can be conveniently represented by a tree on which each point corresponds to a time, a macromolecular sequence, and a species within which the sequence occurred. Although we may not yet be able to infer its exact location, there is one point that corresponds to the earliest time and the original ancestral sequence and species. Time advances on all branches of the tree emanating from this point. During evolution, sequences in different species have gradually and independently accumulated changes, yielding the sequences found today in the extant species represented at the ends of the branches. The absolute chronology, of course, cannot be inferred from sequences; the topology of the branches gives the relative order of events. On the evolutionary trees in this article the lengths of the branches are proportional to the inferred amount of change in the sequences. It is not usually possible to infer the position of the point of earliest time on a tree from the sequence data. An exception occurs when a gene doubling produces a reiterated sequence that is ancestral to all the sequences on the tree and when this duplication has been well preserved. For some biological groups, such as the vertebrates, the fossil record can be used to fix the location of the point of earliest time as well as its approximate chronological date.

In constructing an evolutionary tree on the basis of sequence data, we treat each amino acid or nucleotide residue in a sequence as an inherited biological trait. This assumption implies our ability

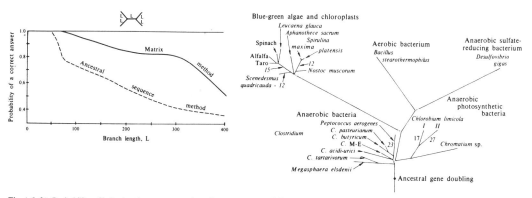

Fig. 1 (left). Probability of inferring the correct topology from sequences of simulated evolutionary connection. A history of five equal intervals of mutational distance was used, as shown at the top of the figure. Ten sets of sequences of 100 residues were generated for each mutational distance (L = 25, 50, 75, 100, 125, 150, 175, 200, 225, 250, 300, 400). Random events were assigned according to the average mutability and mutation pattern of each amino acid (*40*). In totaling the number of correct topologies inferred, a single correct answer counted 1, a wrong answer 0, a two-way tie 0.5, and a three-way tie 0.33. Smoothed curves through the data points are shown. The matrix method is clearly superior. This figure is adapted from (*2*). Fig. 2 (right). Evolutionary tree derived from ferredoxin sequences. Two subtrees were constructed separately on the basis of matrices of percentage differences between sequences from bacteria and from plants and the blue-green algae. A matrix based on an alignment omitting multiple-residue insertions and deletions was used to estimate the evolutionary distance and the connection between the subtrees. Mutiple-residue insertions in the sequences were omitted in constructing the bacterial subtree. The numbers on the tree are evolutionary distances in units of accepted point mutations per 100 residues. The order of divergence of the *Bacillus* and *Desulfovibrio* lines is unclear, as indicated by the dashed connection of the *Desulfovibrio* branch. An insertion of three residues in only these two sequences is consistent with the topology shown, but the topology with the *Desulfovibrio* branch coming directly off the ancestral line to the anaerobic bacteria does have nearly as short an overall branch length. For this alternative topology, the calculated branch lengths in the neighborhood of the *Desulfovibrio* connection would be slightly different.

to align sequences so that changes reflect substitutions of one residue for another during the course of evolution. Insertions and deletions of genetic material affect our ability to align sequences. Superimposed and parallel point mutations limit the accuracy with which we can infer the amount of evolutionary change. (Superimposed mutations are multiple changes at the same site in a sequence; parallel mutations are independent changes in two or more sequences resulting in the same amino acid or nucleotide at corresponding positions.) Experimental and methodological errors in sequencing present further difficulties. These problems affect the overall perspective less as more data become available.

One of the computer methods used for constructing phylogenetic trees proceeds by generating ancestral sequences (15); the other produces a least-squares fit to a matrix of evolutionary distances between the sequences (2, 4). The ancestral sequence method is a problem in double-minimization of inferred changes. For each possible configuration of the evolutionary tree, a set of ancestral sequences is determined that minimizes the number of inferred changes. Of these configurations, the one that minimizes the total number of changes between ancestral and known sequences is selected as the best representation of the evolutionary tree. This method also yields a set of ancestral sequences corresponding to the branch points of the tree.

In the matrix method, used to construct the trees described here, we begin by calculating a matrix of percentage differences between sequences in an overall alignment. Large unmatched regions, either internal or at the ends of sequences, do not correspond to point mutations and therefore are omitted from our calculations. The matrix elements are corrected for inferred parallel and superimposed mutations according to a scale based on the average way amino acids change during evolution (16); this gives evolutionary distances in accepted point mutations per 100 residues. For a given matrix, the determination of the best tree is a problem in double-minimization. For each possible configuration, a set of branch lengths is determined that provides a weighted least-squares fit between the distances given by the reconstructed matrix and those of the original matrix. The configuration that has the minimal total branch length is selected as the best solution.

To test the accuracy of both of these methods, we produced a number of families of sequences that were related through simulated evolutionary change that included amino acid replacements but no insertions or deletions (2). The results (Fig. 1) show that for sparse trees of distantly related sequences the matrix method is clearly superior to the ancestral sequence method. Because this is precisely the type of tree with which we are concerned, all of the individual trees shown here were constructed by the matrix method.

The matrix method is much more accurate for distantly related sequences because the information utilized is degraded more slowly by superimposed and parallel mutations than that utilized in the ancestral sequence method. If genetic material has been deleted or inserted, there is an additional factor favoring the use of the matrix method: the results obtained are less affected by variations in gap placement. The number of matching residues is not critically dependent on the exact alignment of two sequences. Typically, several alignments varying in the placement of gaps give the same number, and there are usually many more alignments corresponding to slightly smaller numbers of matches.

Our alignments are based on a computer program that determines the best alignment of two sequences (17, 18). The residue-by-residue comparisons match amino acids according to a model of the point mutation process that takes into account amino acid mutabilities and replacement probabilities (16). These pairwise sequence comparisons are adjusted to produce a comprehensive alignment. In the alignments used here, the relative magnitudes of terms in the matrices of percentage differences closely reflect the order of similarity of the pairs of sequences. Other criteria can be used in determining gap placement. For example, overall conformation appears to be well conserved by evolution. Dickerson et al. (19) propose an alignment of c-type cytochromes that matches residues according to their positions in the three-dimensional structure of the molecules. Alignments based on mutations are more appropriate here because the programs for reconstructing phylogeny seek a minimum number of genetic events.

The topological configuration obtained is not very sensitive to the correction method used for the matrix elements. In the simulated problems (Fig. 1), the curve of correctly inferred topologies obtained directly with the matrix of percentage differences is almost identical with that obtained with the values corrected for presumed superimposed and parallel mutations. The reconstructed branch lengths approached an asymptote when the matrix of percentage differences was used, whereas they were correct on the average (but only very approximately in any particular case) when the corrected matrix was used.

Four protein superfamilies include sequences from several prokaryotes and eukaryotes: ferredoxin, 5S ribosomal RNA, c-type cytochromes, and azurin-plastocyanin. Two other superfamilies, flavodoxin and rubredoxin, have sequences that are known from at least four types of bacteria in common with the first superfamilies. Each of these groups of sequences can be used to construct an evolutionary tree; generally, the information that an individual tree provides corresponds closely to the evolution of the biochemical system within which the molecule functions. Plastocyanin, for example, functions in oxygen-releasing photosynthesis, and these sequences provide information about the evolution of photosynthesis in the blue-green algae and in the chloroplasts of higher plants. Together with azurin sequences they also depict the divergence of the blue-green algae from the bacteria.

None of these individual trees, however, gives an overall picture of the course of evolution and the development of new biochemical adaptations from the appearance of the earliest living forms to the divergence of the eukaryote kingdoms. For example, no one tree contains sequences from both cytoplasm and chloroplasts of higher plants; thus, individual trees leave unresolved the question of whether or not the eukaryote organelles had symbiotic origins. Understanding the development of new biochemical pathways requires information that cannot come from a tree derived from data on a single type of molecule. Fortunately, the groups of organisms and eukaryote organelles from which sequences are available overlap in such a way that the trees can be correlated and a composite tree can be constructed. This composite tree depicts more fully the relationships between the major developments in early biological evolution.

Ferredoxins

The ferredoxins are small, iron-containing proteins that are found in a broad spectrum of organisms and that participate in such fundamental biochemical processes as photosynthesis, oxidation-reduction respiratory reactions, nitrogen fixation, and sulfate reduction. The amino acid sequences of these proteins have been elucidated by a number of workers, including particularly K. T. Yasunobu, H. Matsubara, and their co-workers [see (20, 21)]. The tree (Fig. 2) derived from these sequences provides a framework for the events outlined in the other evolutionary trees presented here; moreover, a gene-doubling shared by all the ferredoxin sequences makes it possible to deduce the point of earliest time in these trees. The clostridial-type ferredoxins, in particular, are still very similar in sequence to the extremely ancient protein that duplicated. Most of these ferredoxins are composed of fewer than the 20 coded amino acids, and they lack those amino acids that are thermodynamically least stable, such as tryptophan and histidine (22).

The clostridial-type ferredoxins show the strongest evidence of gene-doubling. Using our computer program RELATE (17), we compared all pairs of different segments that were 15 residues long within each of these proteins and calculated a probability of less than 10^{-9} that the repetitive character of the two halves occurred by chance. From an alignment of the first and second halves of these sequences, we inferred an ancestral half-chain sequence. This sequence, doubled, was included in the computations of the evolutionary tree. Because all of the ferredoxin sequences show some evidence of gene-doubling, this event must have occurred prior to the species divergences shown here. The doubled sequence is therefore located at the base of the tree. All organisms near the base of the tree (*Clostridium*, *Megasphaera*, and *Peptococcus*) are anaerobic, heterotrophic bacteria (23). Most species of these groups lack heme-containing proteins, such as the cytochromes and catalase. It has long been thought that, of the extant bacteria, these species most closely reflect the metabolic capacities of the earliest species (24). In Fig. 2, *Chlorobium* and *Chromatium* are pictured as having diverged very early from the ancestral heterotrophic bacteria, although the exact point of this divergence is not clearly resolved. *Chromatium* and *Chlorobium* are anaerobic bacteria capable of photosynthesis using H_2S as an exogenous electron donor. Of the anaerobic bacteria shown, only *Chlorobium limicola* cannot live fermentatively, and it is reasonable to suppose that this ability is primitive and was lost by this bacterium. The two ferredoxins in *Chlorobium* are the result of a gene duplication within this line.

In Fig. 2, the *Bacillus* and *Desulfovibrio* lines diverge next from the line leading to the blue-green algae. Members of the genus *Bacillus* are either strictly aerobic or facultatively aerobic, capable of respiring aerobically but living fermentatively under anaerobic conditions. *Desulfovibrio* is a sulfate-reducing bacterium; it respires anaerobically using sulfate as the terminal electron acceptor. The use of sulfate by *Desulfovibrio* contrasts sharply with the use of oxygen as the terminal electron acceptor of respiration by *Bacillus*. This difference suggests that the divergence of these bacteria occurred after some components in the respiratory chain had developed. The topology pictured here indicates that the final components in the chain evolved separately in the *Bacillus*, *Desulfovibrio*, and blue-green algal lines.

The plant-type ferredoxins are all very closely related. Those from the green alga *Scenedesmus* and the higher plants are found in the chloroplasts of these organisms. *Spirulina* and *Nostoc* belong to one major division of the blue-green algae, the filamentous type; *Aphanothece* represents the other major division, the coccoid type (25).

The tree in Fig. 2 indicates that the structure of ferredoxin has changed much more in some lines than in others; this, at least in part, reflects changes in ferredoxin function. The clostridial sequences are little changed over the entire time represented, possibly more than 3 billion years. The adjustment to bacterial photosynthesis required somewhat more change, the adjustment to an aerobic metabolism required still more, and the most change occurred in the adaptation to oxygen-releasing photosynthesis. Unlike the situation in eukaryotes, the rate of acceptance of point mutations in prokaryote sequences is very uneven and cannot provide a useful evolutionary clock.

Origins of Eukaryote Organelles

There are two schools of thought concerning the origin of the eukaryote mitochondria and chloroplasts: one is that they arose by the compartmentalization of the DNA within the cytoplasm of an evolving protoeukaryote (26); the other is that they arose from free-living forms that established symbiotic relationships with host cells (27). According to the first theory, all genes arose within a single ancestral line, and homologs found in both the nucleus and the organelles arose by gene duplication. Thus, in the evolutionary tree for ferredoxin, the animals and fungi would appear together with the higher plants in the upper portion of the tree, after the divergence of the blue-green algae.

According to the symbiotic theory, the chloroplasts descended from free-living blue-green algae; other symbionts would include the mitochondrion, which was originally a free-living aerobic bacterium, and the flagellum and mitotic apparatus, which may have descended from spirochetes. It is proposed that these prokaryotes separately invaded protoeukaryote host cells with which they became symbiotic and continued to evolve to their current status as organelles. If the symbiotic theory is accurate, mitochondrial and chloroplast genes should show evidence of recent common ancestry with the separate types of contemporary free-living prokaryote forms. The host and organelles should occur on different branches of the tree that also contain free-living forms. The tree of plant-type ferredoxins would thus depict the radiation of blue-green algae followed by the development of symbiosis between one of these organisms and an ancestor of *Scenedesmus* and the higher plants. Although the appearance of the ferredoxin tree, particularly its branch lengths, is more consistent with this theory than with the first explanation, the tree by itself does not enable us to distinguish between the two theories because no ferredoxin sequences are available from the eukaryote cytoplasm or mitochondria.

5S Ribosomal RNA

The 5S ribosomal RNA molecule has a low molecular weight and is about 120 nucleotides in length. It is associated with the larger ribosomal subunit and is thought to function in the nonspecific binding of transfer RNA to the ribosome during protein synthesis (28). Because this function is independent of the kind of amino acid, this type of molecule could be extremely ancient, predating the contemporary form of the genetic code. Sequences of 5S ribosomal RNA have been determined by a number of workers, including B. J. Forget, S. M. Weissman, and C. R. Woese [see (29, 30)]; they have been taken from a wide variety of sources, including aerobic and anaerobic bacteria, blue-green algae, and the cytoplasm of several eukaryotes. The cytoplasmic sequences, in particular, present the possibility that an evolutionary tree based on this molecule will provide further insight into the origin of the eukaryotes.

Aligning nucleotide sequences in a way that reflects their evolution is more difficult than aligning amino acid sequences because there are only four kinds of bases. However, knowledge of the secondary structure alleviates the alignment problem somewhat because we can assume that positions involved in the base-paired regions of the molecule were highly conserved during evolution. We have aligned the known sequences to reflect their natural division into groups from prokaryotes and eukaryotes and to match a model of their secondary structure (29) adapted from Nishikawa and Takemura (31).

We derived an evolutionary tree (Fig. 3) on the basis of this alignment and placed its base on the branch to the anaerobic bacterium *Clostridium* in conformance with the ferredoxin tree. All the eukaryote 5S ribosomal RNA's were isolated from cytoplasmic ribosomes, one of the three ribosomal systems found in eukaryotes.

The branch leading to these cytoplasmic sequences diverges from the prokaryotes at a point that is close to the origin of the *Bacillus* branch. Like some members of the genus *Bacillus*, *Escherichia* is also facultatively aerobic and has both fermentative and respiratory metabolisms. Unless aerobic respiration arose separately in the *Bacillus* and *Escherichia* lines, their most recent common ancestor had this bimodal metabolic capacity also.

The two lines leading to organisms that are capable of oxygen-releasing photosynthesis appear on opposite sides of this tree. One leads to rye and *Chlorella*, a eukaryote green alga, the other to *Anacystis*, a blue-green alga. *Anacystis* is grouped in the same family with *Aphanothece* (25), which appears on the ferredoxin tree. These coccoid blue-green algae are certainly more closely related than the blue-green algal orders represented by *Aphanothece* and *Spirulina*. Thus, we predict that *Aphanothece* would be found to diverge near the end of the *Anacystis* branch, preceded slightly by the divergence of the chloroplast branches. The very separate history of the cytoplasmic sequences points to a symbiotic origin of the chloroplasts, with the cytoplasmic sequences representing the evolution of the organism that was invaded by a blue-green alga.

C-Type Cytochrome

The evolutionary tree based on c-type cytochromes (Fig. 4) is important to an understanding of the origin and evolution of the mitochondrion. R. Ambler, E. Margoliash, D. Boulter, M. Kamen, E. Smith, and G. Pettigrew, in particular, are responsible for sequencing many of these proteins [see (32)]. Cytochrome c is coded in the nucleus but functions in the mitochondrion. This is usually explained in the symbiotic theory by transfer of genetic information, including the gene for cytochrome c, from the invading aerobic bacterium to the protoeukaryote host during the development of their current relationship. In the nonsymbiotic theory, genetic rearrangement is also an essential feature.

The eukaryote mitochondrion is placed in the portion of the tree (Fig. 4) that includes the aerobic bacteria after their divergence from the blue-green algae and chloroplasts. The cytochrome c sequences are on a branch that most recently diverged from cytochrome c_2 of the nonsulfur, purple, photosynthetic bacteria; together these diverged from the branch leading to cytochrome c_{551} from strict aerobes such as *Pseudomonas*. This contrasts with the evolution of the eukaryote cytoplasmic constituents depicted by the tree derived from 5S ribosomal RNA (Fig. 3). There the eukaryotes diverged with the facultatively aerobic bacteria, such as *Bacillus*, from the line leading toward the blue-green algae. *Pseudomonas* diverged from this line somewhat later. This contrasting

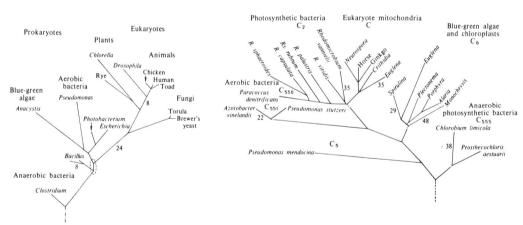

Fig. 3 (left). Evolutionary tree derived from 5S ribosomal RNA. This tree was derived by the matrix method. Branches are drawn proportional to the amount of evolutionary change they represent; selected branch lengths are indicated in numbers of accepted point mutations per 100 residues. The order of divergence for the branches leading to the eukaryotes and to *Bacillus* is not clearly resolved; the tree whose topology reverses the order of these branches has nearly as short an overall length. The prokaryote species shown are *Clostridium pasteurianum*, *Bacillus megaterium*, *B. licheniformis*, *Escherichia coli*, *Pseudomonas fluorescens*, and *Anacystis nidulans*. The *Chlorella* species is *C. pyrenoidosa*. Fig. 4 (right). Evolutionary tree derived from c-type cytochromes. The subtrees pictured here, such as the cytochrome c_6 tree, were derived separately from matrices of percentage differences between the complete sequences. Branches connecting the subtrees were estimated from a matrix calculated from an alignment that omitted multiple-residue insertions and deletions. Branch lengths are given in accepted point mutations per 100 residues. The points of earliest divergence in the individual subtrees cannot be precisely determined and, therefore, the bases of the subtrees are represented by dashed lines. Cytochrome c_{550} and all of the c_2's differ from the other c-type cytochromes in having a deletion close to the heme-binding cysteine in their sequences and on this basis were placed on a branch separate from the cytochrome c sequences. *Euglena* and *Crithidia* have been placed on a single branch together, a configuration slightly less than optimal, because they share a unique mutation of the active cysteine. The connections of the c_{551} and c_{555} sequences have been centered. The genera *Rhodospirillum* and *Rhodopseudomonas* are abbreviated Rs. and R., respectively. Cytochrome c_6 sequences were taken from the following species: *Spirulina maxima*, *Monochrysis lutheri*, *Porphyra tenera*, *Euglena gracilis*, *Alaria esculenta*, and *Plectonema boryanum*; cytochrome c sequences were taken from protists *Euglena gracilis* and *Crithidia oncopelti*. This tree is adapted from (2).

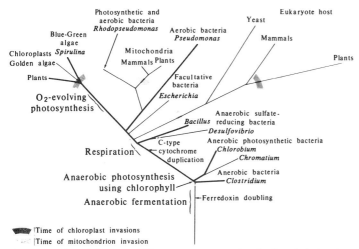

Fig. 5. Composite evolutionary tree. This tree presents an overview of early evolution based on ferredoxin, c-type cytochromes, and 5S ribosomal RNA sequences. The heavy lines represent a tree calculated from a matrix of evolutionary distances combining two or more of the individual trees. The lighter lines represent branches scaled from a single tree and added to the combined tree. The point in evolution at which the mitochondrial symbiosis occurred is stippled; the chloroplast symbiosis is shaded.

picture reinforces arguments favoring a symbiotic origin for the mitochondrion. The cytoplasmic 5S ribosomal RNA sequences appear to describe the evolution of the protoeukaryote host. The subtree for animals that was derived from cytochrome c (5) is consistent with that derived from 5S ribosomal RNA sequences. According to the symbiotic theory, this would be because these divergences were subsequent to the mitochondrial invasion of host cells.

Cytochrome c_2 is found in bacteria such as *Rhodomicrobium* and *Rhodopseudomonas*. These bacteria photosynthesize anaerobically but respire aerobically. *Paracoccus denitrificans* possesses cytochrome c_{550}, which is very similar to cytochrome c_2 along its entire length. The position of *Paracoccus* in the tree suggests that it arose from a nonsulfur photosynthetic bacterium by the loss of its photosynthetic ability (19).

Two different c-type cytochromes, c_5 and c_{551}, are found in *Pseudomonas*, a nonphotosynthetic bacterium. These may be the result of a gene duplication early in the tree with subsequent loss of the c_5 gene in some lines. For species in which the gene was not lost, we would expect that the topology of the main tree subsequent to the duplication would be reiterated on the branch that included the c_5 sequence. A transfer of genetic material from a bacterium that contained the c_5 gene to *Pseudomonas* could also explain this branch.

The presumed duplication of the c-type cytochrome gene in *Pseudomonas* brings up a problem in interpreting evolutionary trees that is especially acute here. If we are treating two products of an unsuspected gene duplication, we will assume that the two lines of protein evolution correspond to a single pattern of species evolution. This can lead us to believe that two closely related organisms are quite distant from one another. For example, if cytochrome c_{551} had gone undetected in species of *Pseudomonas*, the tree in Fig. 4 would suggest that *Azotobacter* and *Pseudomonas* were very distantly related. At present, metabolic function guides our selection of homologous proteins and helps avoid this difficulty. In the tree of c-type cytochromes, protein function has changed, and this presents an added difficulty in interpretation. As a more complete picture of the protein complement of each species becomes known, this problem will become unimportant.

Cytochrome c_6 is found in the photosynthetic lamellae of blue-green algae and in the chloroplasts of eukaryotes where it functions in the electron transport chain between photosystems I and II. As in the ferredoxin tree, there is a close similarity between the sequences from the blue-green alga *Spirulina* and the various eukaryote algal chloroplasts; as in the 5S ribosomal RNA tree, the blue-green algae are most closely related to strictly aerobic bacteria, such as *Pseudomonas*. It is not possible to locate precisely the point at which the main tree connects to the subtree of cytochrome c_6. However, the topology and branch lengths reflect a symbiotic origin for photosynthesis in eukaryotes. There are separate branches leading to the two filamentous blue-green algae, *Spirulina* and *Plectonema*, intermixed with the eukaryote algae branches. The most direct explanation for this is that the cytochrome c_6 subtree reflects, at least in part, the evolutionary relationships among the blue-green algae that became symbionts rather than the speciation of eukaryotes. Some of the eukaryote algal chloroplasts appear to be derived from different symbiotic associations, as Raven has suggested (33).

The point of earliest time on the tree for c-type cytochromes was placed near the divergence of *Chlorobium* and *Prosthecochloris*; both of these are anaerobic, obligate, photosynthetic bacteria. This is consistent with the position of the sequence from *Chlorobium* in the ferredoxin tree. As in the tree of 5S ribosomal RNA's, prokaryote branch lengths are strikingly unequal, probably reflecting changes in protein function, and cannot be used to estimate time reliably.

Composite Evolutionary Tree

Each of the individual trees we have presented contains information about the early course of biological evolution. We have used the topologies and evolutionary distances derived for these trees to construct a composite tree (Fig. 5) which, although it is based on sparse data from a few species, begins to provide a coherent evolutionary framework that can be expanded as new sequence data become available.

We constructed the composite evolutionary tree from a composite matrix according to the same methods that we used to construct the individual trees. The matrix included the six species that appear on at least two of the three individual trees (their branches are represented by the heavy lines in Fig. 5). First, the trees were scaled so that distances were comparable. To do this, we compared the overall length of the distances each tree had in common with each other tree. These ratios were adjusted slightly to give a consistent set of scale factors. A combined matrix of distances between the six species was calculated by averaging the scaled contributions from each of the individual trees.

As previously, for each possible configuration of the combined trees, a set of branch lengths was determined that provided a weighted least-squares fit between the matrix of distances between species and a matrix reconstructed from the tree. The configuration with the shortest overall branch length was chosen. This configuration is the one that is also consistent with all of the individual trees. Finally, branches found in only one tree were scaled in length and added to the composite tree, maintaining the relative internodal distances (these are represented by light lines in Fig. 5).

The genetic doubling of the clostridial-type ferredoxins allows us to locate the base on this tree. Moreover, because the species whose sequences have changed least since this doubling event were all anaerobic, heterotrophic bacteria, it is likely that the ability to live fermentatively is primitive.

The composite tree describes the evolution of photosynthesis using chlorophyll, starting with the development of the ability to synthesize this class of compounds. Three families of photosynthetic bacteria are represented: Chromatiaceae, Chlorobiaceae, and Rhodospirillaceae. The divergence of the Chromatiaceae and the Chlorobiaceae from the other anaerobic bacteria was quite early, and it is clear that this type of photosynthesis arose at a very early stage in evolution and has not changed much. The Rhodospirillaceae provide an example of the confusion that morphological criteria can cause. As Stanier et al. (23) point out, these bacteria are indistinguishable in structure and pigments from members of the Chromatiaceae under anaerobic conditions; however, when grown under strictly aerobic conditions, they appear to be the same as nonphotosynthetic bacteria of similar form, such as the pseudomonads. The c-type cytochrome sequences clearly place them in a portion of the tree surrounded by strictly aerobic forms on a branch leading to *Pseudomonas*.

If we assume that it is very hard to achieve a photosynthetic metabolism with its many coordinated macromolecules, but relatively easy to lose one through any one of many genetic changes, then it is reasonable to suppose that the main trunk of this tree represents a continuum of photosynthetic forms. We would not be surprised to find photosynthetic forms branching off at any point. Except for the early anaerobes, all nonphotosynthetic bacteria would be descended from a few ancestral forms that have independently lost their photosynthetic ability. This is very different from biological classifications where all photosynthetic forms are grouped together, separate from the nonphotosynthetic forms.

The next major event shown on the main trunk of the tree is the development of aerobic respiration. As we noted in discussing the ferredoxin evolutionary tree, the divergence of *Bacillus* and *Desulfovibrio* probably marks the appearance of some components of this adaptation. The final elements in this adaptation were evolved separately because these groups differ in their terminal electron acceptor in respiration. Additionally, the divergence of cytochrome c_5 occurred just prior to this time. In the tree of the c-type cytochrome, it is unclear whether this divergence represents a gene duplication or a recent genetic transfer; in the context of the evolution of a respiratory metabolism, a duplication of the ancestral gene could have provided the genetic material and relaxed evolutionary constraints necessary for the development of aerobic respiration.

The clearest and most direct interpretation of the sequence data is provided by the symbiotic theory for the origin of the eukaryotes. The branch we identify as the eukaryote host diverged at about the same time as *Bacillus* and *Escherichia*. Both of these bacteria are facultative aerobes. This suggests that the ancestral eukaryote host was also facultatively aerobic at the time of its divergence.

The bacterium that became the mitochondrion was most closely related to the third family of photosynthetic bacteria, the Rhodospirillaceae. The topology of the tree invites the speculation that this ancestral bacterium was photosynthetic until shortly before or just after it invaded the host. Because a single cytochrome c is found in most eukaryotes, it seems reasonable to suppose that the aerobic respiratory metabolism of this invading protomitochondrion was more effective than that of the host and that the host lost any primitive system that it might have had.

The final biochemical adaptation depicted here is the development of photosystem II. The blue-green algae and chloroplasts of the eukaryotes are capable of oxygen-releasing photosynthesis, and the available sequence data point to this capacity having evolved only once. It appears to have combined the new biochemical adaptation, photosystem II, with proteins modified from two earlier adaptations, bacterial photosynthesis and respiration. The chloroplasts, like the mitochondria, are grouped together with free-living prokaryotes, a result consistent with the symbiotic theory.

It is frequently suggested that aerobic respiration developed after oxygen-releasing photosynthesis as a protective mechanism in response to atmospheric oxygen produced by blue-green algae (24, 34). The composite tree clearly suggests that many components of the respiratory chain predate oxygen-releasing photosynthesis; all of the organisms on the tree above *Desulfovibrio* are aerobic, whereas the earlier branches all lead to anaerobic forms. Although possible, it seems unlikely that the use of oxygen as the terminal electron acceptor in respiration evolved separately in the lines leading to the blue-green algae, the facultative aerobic bacteria, and the strictly aerobic bacteria after the development of photosystem II in the blue-green algal line. As Schopf (8) has pointed out, it is difficult to imagine the development of oxygen-releasing photosynthesis prior to the development of a rudimentary mechanism for coping with oxygen. Oxygen is produced intracellularly in photosynthesis, and it is intracellular components that must be protected from oxidation. A rudimentary form of aerobic respiration most probably arose at a time near the divergence of the *Bacillus* line from that leading to the blue-green algae.

The composite tree makes it particularly clear that the three branches that contribute to the eukaryote host and organelles are distinctly separate; each is closely related to free-living prokaryotes. The chloroplasts share a recent ancestry with the blue-green algae; the mitochondrion shares a recent ancestry with certain respiring and photosynthetic bacteria, the Rhodospirillaceae; whereas the eukaryote host diverged from the other groups at a considerably earlier time along with *Bacillus* and *Escherichia*.

Corroborative Sequence Data

A limited amount of sequence data from four or more species is available from other proteins. Azurin and plastocyanin are blue, copper-containing proteins whose sequences show statistical evidence of their common evolutionary origin (20). Moreover, they appear to be parts of homologous electron transport systems because each of them exchanges electrons with c-type cytochromes. Azurin is thought to exchange an electron with cytochrome c_{551} in bacterial respiration (35); plastocyanin exchanges an electron with cytochrome c_6 in the electron transport chain between photosystems I and II.

Plastocyanin sequences from both eukaryote chloroplasts and a blue-green alga have been determined by a number of workers (20, 36) including D. Boulter and A. Aitken, in particular; azurin sequences from aerobic bacteria have been determined by R. Ambler (20). We constructed an evolutionary tree (Fig. 6) based on these sequences. The relationships depicted are consistent with those presented in the other trees. The aerobic bacteria, on the right side of the tree, are very closely related, and the extents of their divergences are comparable to those of the higher plants, shown on the left side of the tree. *Anabaena*, a filamentous blue-green alga, is closely related to chloroplasts of the eukaryote green alga *Chlorella* and of the higher plants. The topology of this tree again is consistent with the symbiotic origins for chloroplasts. The divergences of the higher plant chloroplasts are close enough to classical phylogenies to be explained by a single symbiotic association that predated plant divergences. Lewin (37) has recently proposed a new division, the Prochlorophyta, for a group of bright green, generally spherical, prokaryote algae. These had previously been classified with the blue-green algae. However, like the higher plant chloroplasts and unlike the blue-green algae they contain both chlorophylls a and b and no detectable bilin pigments. Sequences from these organisms might be especially helpful in extending our understanding of the evolution of higher plant photosynthesis.

Flavodoxins are low-molecular weight proteins that have been isolated so far from only bacteria and algae. They have one flavin mononucleotide prosthetic group per molecule and substitute for ferredoxins in a variety of reactions, including the phosphoroclastic splitting of pyruvate and the photosynthetic reduction of nicotinamide-adenine dinucleotide phosphate. Flavodoxin sequences are known from three anaerobic bacteria, *Clostridium, Megasphaera,* and *Desulfovibrio,* and an aerobic bacterium, *Azotobacter*; most of these sequences were determined by K. T. Yasunobu, J. L. Fox, and their co-workers [see (38)]. The matrix of the percentage differences between these flavodoxin sequences (Table 1) reveals the close relationship of the highly conserved *Megasphaera* and *Clostridium* sequences and of the more highly evolved *Desulfovibrio* and *Azotobacter* sequences, supporting the phylogeny we have drawn.

Rubredoxin, another protein that participates in electron transport, has also been used as a basis for constructing an evolutionary tree (20, 39). This tree includes sequences from *Megasphaera elsdenii, Peptococcus aerogenes, Clostridium pasteurianum, Pseudomonas oleovorans, Desulfovibrio gigas,* and *D. vulgaris,* and is also consistent with the ones we have constructed except that the closely related *Desulfovibrio* sequences diverge close to the earliest point on the tree and are very highly conserved. That the placement of the *Desulfovibrio* branch is inconsistent with that depicted in the trees for ferredoxin and flavodoxin suggests that there might be an unsuspected gene duplication in the rubredoxins, a rare occurrence of an accepted gene transfer, or a misleading concatenation of evolutionary events in these short sequences.

Summary

If current estimates of the antiquity of life on earth are correct, bacteria very much like *Clostridium* lived more than 3.1 billion years ago. Bacterial photosynthesis evolved nearly that long ago, and it seems reasonable, in view of our composite tree, to attribute the most ancient stromatolites, formed nearly 3.0 billion years ago, to early photosynthetic bacteria. Blue-green algae appear to have evolved later. The tree shows that by the time oxygen-releasing photosynthesis originated in the blue-green algal line, there must have been a great diversity of morphological types, including bacteria that are ancestral to most of the major groups pictured on the composite tree. This time probably corresponded to the great increase in complexity of the fossil record about 2 billion years ago. Our composite tree suggests that aerobic respiration preceded oxygen-releasing photosynthesis. This may mean that the formation of oxygen from water in the upper atmosphere was important to evolving prokaryotes prior to 2 billion years ago. Oxygen-releasing photosynthesis arose later and was, in large measure, responsible for the final transition to the present-day oxygen level in the atmosphere. Judging from the relative branch lengths on the tree, the mitochon-

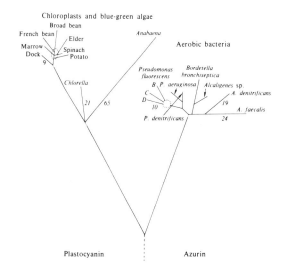

Table 1. Flavodoxins

	Number of differences			
Megasphaera elsdenii		76	107	155
Clostridium MP	55		104	152
Desulfovibrio vulgaris	71	69		151
Azotobacter vinelandii	85	83	82	
	Percentage difference			

Fig. 6. Evolutionary tree derived from azurin and plastocyanin. Subtrees for the two proteins were derived separately from matrices calculated from complete sequences. In order to estimate the evolutionary distance between the two families, we used a matrix calculated from an alignment that omitted multiple-residue insertions and deletions between the families. Branch lengths are drawn proportional to amounts of evolutionary change in the units of accepted point mutations per 100 residues. The positions of the connecting branch within the subtrees cannot be precisely determined from the data, and therefore the connections are represented by dashed lines. The order of divergence of the three biotypes of *Pseudomonas fluorescens* could not be resolved. The order of divergence of the chloroplasts of higher plants is also unclear; the topology with the minimal overall length is shown. The algal species shown are *Chlorella fusca* and *Anabaena variabilis*.

drial invasion occurred during this transition. Finally, perhaps 1.1 billion years ago, several independent symbioses between protoeukaryotes and various blue-green algae gave rise to photosynthetic eukaryotes; some of these developed into modern eukaryote algae, whereas a single line, possibly from an ancestral green alga, appears to have evolved into the higher plants.

By combining the information from evolutionary trees based on several types of sequences, we have developed a broad outline of early events in the emergence of life that can be refined as new sequence information becomes available. The schema presents a working hypothesis for relating the many observations from metabolic and morphological studies of bacteria and from paleogeology. Eventually all of the biochemical components of intermediary metabolism may be correlated with the development of the prokaryote types and their metabolic capacities.

References and Notes

1. V. Bryson and H. J. Vogel, Eds., *Evolving Genes and Proteins* (Academic Press, New York, 1965); M. O. Dayhoff and R. V. Eck, Eds., *Atlas of Protein Sequence and Structure* (National Biomedical Research Foundation, Washington, D.C., 1968), vol. 3.
2. M. O. Dayhoff, *Fed. Proc. Fed. Am. Soc. Exp. Biol.* **35**, 2132 (1976).
3. R. V. Eck and M. O. Dayhoff, Eds., *Atlas of Protein Sequence and Structure* (National Biomedical Research Foundation, Washington, D.C., 1966), vol. 2.
4. W. M. Fitch and E. Margoliash, *Science* **155**, 279 (1967).
5. P. J. McLaughlin and M. O. Dayhoff, *J. Mol. Evol.* **2**, 99 (1973).
6. W. M. Fitch, *ibid.* **8**, 13 (1976); M. Goodman, G. W. Moore, J. Barnabas, G. Matsuda, *ibid.* **3**, 1 (1974).
7. M. O. Dayhoff, L. T. Hunt, P. J. McLaughlin, D. D. Jones, in *Atlas of Protein Sequence and Structure*, M. O. Dayhoff, Ed. (National Biomedical Research Foundation, Washington, D.C., 1972), vol. 5, p. 17.
8. J. W. Schopf, *Annu. Rev. Earth Planet. Sci.* **3**, 213 (1975).
9. P. Cloud, *Paleobiology* **2**, 351 (1976).
10. J. W. Schopf, *Origins Life* **7**, 19 (1976).
11. T. R. Mason and V. Von Brunn, *Nature (London)* **266**, 47 (1977).
12. P. E. Cloud, Jr., *Science* **160**, 729 (1968).
13. J. W. Schopf and D. Z. Oehler, *ibid.* **193**, 47 (1976).
14. M. O. Dayhoff, W. C. Barker, P. J. McLaughlin, *Origins Life* **5**, 311 (1974).
15. M. O. Dayhoff, C. M. Park, P. J. McLaughlin, in *Atlas of Protein Sequence and Structure*, M. O. Dayhoff, Ed. (National Biomedical Research Foundation, Washington, D.C., 1972), vol. 5, p. 7; W. M. Fitch and J. S. Farris, *J. Mol. Evol.* **3**, 263 (1974).
16. M. O. Dayhoff, Ed., *Atlas of Protein Sequence and Structure* (National Biomedical Research Foundation, Washington, D.C., 1976), vol. 5, suppl. 2, p. 311.
17. M. O. Dayhoff, in *ibid.*, p. 1.
18. S. B. Needleman and C. D. Wunsch, *J. Mol. Biol.* **48**, 443 (1970).
19. R. E. Dickerson, R. Timkovich, R. J. Almassy, *ibid.* **100**, 473 (1975).
20. W. C. Barker, R. M. Schwartz, M. O. Dayhoff, in *Atlas of Protein Sequence and Structure*, M. O. Dayhoff, Ed. (National Biomedical Research Foundation, Washington, D.C., 1972), vol. 5, p. 51. Respiratory protein sequences and our alignments of them are collected here as are tabulations of computer alignment scores for the ferredoxins.
21. T. Hase, K. Wada, H. Matsubara, *J. Biochem. (Tokyo)* **79**, 329 (1976); M. Tanaka, M. Haniu, K. T. Yasunobu, K. K. Rao, D. O. Hall, *Biochemistry* **13**, 5284 (1974); _____; *ibid.* **14**, 5535 (1975); K. Wada, T. Hase, H. Tokunaga, H. Matsubara, *FEBS Lett.* **55**, 102 (1975); M. Tanaka, M. Haniu, K. T. Yasunobu, M. C. W. Evans, K. K. Rao, *Biochemistry* **14**, 1938 (1975); T. Hase, N. Ohmiya, H. Matsubara, R. N. Mullinger, K. K. Rao, D. O. Hall, *Biochem. J.* **159**, 55 (1976); T. Hase, K. Wada, M. Ohmiya, H. Matsubara, *J. Biochem. (Tokyo)* **80**, 993 (1976).
22. R. V. Eck and M. O. Dayhoff, *Science* **152**, 363 (1966).
23. R. E. Buchanan and N. E. Gibbons, Eds., *Bergey's Manual of Determinative Bacteriology* (Williams & Wilkins, Baltimore, Md., ed. 8, 1974); R. Y. Stanier, M. Doudoroff, E. A. Adelberg, *The Microbial World* (Prentice-Hall, Englewood Cliffs, N.J., ed. 3, 1970). Nutritional and ecological descriptions of the bacterial genera used in the text were taken from these references.
24. E. Broda, *The Evolution of the Bioenergetic Process* (Pergamon, Oxford, 1975).
25. T. V. Desikachary, in *The Biology of Blue-Green Algae*, N. G. Carr and B. A. Whitton, Eds. (Univ. of California Press, Berkeley, 1973), p. 473.
26. R. A. Raff and H. R. Mahler, *Science* **177**, 575 (1972); T. Uzzell and C. Spolsky, *Am. Sci.* **62**, 334 (1974).
27. L. Margulis, *Origin of Eukaryotic Cells* (Yale Univ. Press, New Haven, Conn., 1970); _____, in *Handbook of Genetics*, R. C. King, Ed. (Plenum, New York, in press), vol. 1.
28. R. Monier, in *Ribosomes*, M. Normura, A. Tissieres, P. Lengyel, Eds. (Cold Spring Harbor Laboratory, Cold Spring Harbor, N.Y., 1974), p. 141.
29. R. M. Schwartz and M. O. Dayhoff, in (*16*), p. 293. Sequences of 5S ribosomal RNA's, a model of their secondary structure, and our alignment of them are collected here, as are tabulations of computer alignment scores.
30. C. R. Woese, C. D. Pribula, G. E. Fox, L. B. Zablen, *J. Mol. Evol.* **5**, 35 (1975); C. D. Pribula, G. E. Fox, C. R. Woese, *FEBS Lett.* **64**, 350 (1976); H. A. Raue, T. J. Stoff, R. J. Panta, *Eur. J. Biochem.* **59**, 35 (1975); J. Benhamou and B. R. Jordan, *FEBS Lett.* **62**, 146 (1976); P. I. Payne and T. A. Dyer, *Eur. J. Biochem.* **71**, 33 (1976).
31. K. Nishikawa and S. Takemura, *J. Biochem. (Tokyo)* **76**, 935 (1974).
32. M. O. Dayhoff and W. C. Barker, in (*16*), p. 25. Sequences of c-type cytochromes and our alignment of them are collected here, as are tabulations of computer alignment scores; R. P. Ambler, T. E. Meyer, M. D. Kamen, *Proc. Natl. Acad. Sci. U.S.A.* **73**, 472 (1976); A. Aitken, *Nature (London)* **263**, 793 (1976).
33. P. H. Raven, *Science* **169**, 641 (1970).
34. L. Margulis, J. C. G. Walker, M. Rambler, *Nature (London)* **264**, 620 (1976).
35. M. Brunori, C. Greenwood, M. T. Wilson, *Biochem. J.* **137**, 113 (1974).
36. B. Haslett, C. R. Bailey, J. A. M. Ramshaw, D. Scawen, D. Boulter, *Biochem. Soc. Trans.* **2**, 1329 (1974); J. A. M. Ramshaw, D. Scawen, D. Boulter, *Biochem. J.* **141**, 835 (1974); M. D. Scawen and D. Boulter, *ibid.* **143**, 237 (1974); M. D. Scawen, J. A. M. Ramshaw, D. Boulter, *ibid.* **147**, 343 (1975); A. Aitken, *ibid.* **149**, 675 (1975).
37. R. A. Lewin, *Nature (London)* **261**, 697 (1976).
38. W. C. Barker and M. O. Dayhoff, in (*16*), p. 67. Flavodoxin sequences are collected here; M. Tanaka, M. Haniu, K. T. Yasunobu, D. C. Yoch, *Biochem. Biophys. Res. Commun.* **66**, 639 (1975).
39. H. Vogel, M. Bruschi, J. Le Gall, *J. Mol. Evol.* **9**, 111 (1977).
40. M. O. Dayhoff, R. V. Eck, C. M. Park, in *Atlas of Protein Sequence and Structure*, M. O. Dayhoff, Ed. (National Biomedical Research Foundation, Washington, D.C., 1972), vol. 5, p. 89.
41. Supported by NASA contract NASW 3019 and NIH grants GM-08710 and RR-05681. We thank W. C. Barker and B. C. Orcutt for their helpful discussions and criticism and M. J. Gantt for technical assistance.

A Multiple Origin for Plastids and Mitochondria

Many independent symbiotic events may have been involved in the origin of these cellular organelles.

Peter H. Raven

In view of much accumulating evidence, it now seems almost certain that the plastids and mitochondria in eucaryotic cells originated as free-living procaryotes which found shelter within primitive eucaryotic cells and eventually were stabilized as permanent symbiotic elements within them (1). This evidence, however, has been largely ignored in the construction of phylogenetic classifications of living organisms, such as the excellent one recently presented by Whittaker (2). What implications for the phylogenetic relationships between groups of organisms does a symbiotic origin of plastids and mitochondria have?

Evidence for Symbiotic Origin

The evidence for a symbiotic origin of mitochondria and plastids will be presented in two parts. First, we shall review the properties of these cellular organelles themselves, and then we shall consider the nature and occur-

The author is affiliated with the Department of Biological Sciences, Stanford University, Stanford, California 94305.

rence of analogous symbioses that occur at the present day.

Both mitochondria and plastids have the capacity for semiautonomous growth and division which is only partially controlled by the nuclear DNA. They arise only from preexisting mitochondria (3) and plastids, respectively. Both contain unique base compositions and configurations of DNA and RNA (4, 5), both of which can be synthesized and replicated within the organelle. The DNA is known in the mitochondria of a number of multicellular animals to be present in the form of double-stranded circles, each with a molecular weight of about 9 to 10×10^6 (5). In other organisms, such as the higher plants and *Neurospora*, the mitochondrial DNA has a much higher molecular weight and has not yet been shown to exist in a circular form (5).

In yeast, circular DNA appears to be only a minor component of mitochondrial DNA (6). The amount of DNA in a mitochondrion may amount to about 0.01 of that present in a cell of *Escherichia coli*, or sometimes considerably more (5). In chloroplasts, there is nearly a hundred times as much, almost as much as in *Escherichia coli* (7). It is not known whether chloroplast DNA is divided into molecules or exists as one continuous piece (8). In bacteria, mitochondria, and chloroplasts, the DNA is histone-free and bound to membranes; in eucaryotic chromosomes, it is associated with histones and not bound to membranes (5, 9).

The extent to which plastids and mitochondria of DNA act as templates for transcription of specific messenger RNA's is currently the subject of investigation in a number of laboratories. Both chloroplasts and mitochondria are able to incorporate amino acids into proteins in vitro (5, 10). In the cell, chloroplast DNA plays a role in the synthesis of at least some of the characteristic proteins of chloroplasts, but nuclear DNA likewise participates to a large extent in these syntheses (11). One of the most significant findings has been that mutant *Euglena* strains which lack the ability to synthesize chloroplasts also lack DNA of characteristic low guanosine-cytosine content (12). In mitochondria, a few, but certainly not all (5), of the characteristic proteins are produced within the mitochondria from a template of mitochondrial DNA (13). For example, Woodward and Munkres (14) have found differences in at least one amino acid residue in mitochondrial structural protein in two "cytoplasmic" mutants of *Neurospora*. In mitochondria, the replication of DNA seems to occur independently of that in the nucleus (15). Many species of transfer RNA are found only in the mitochondria of rat livers and not elsewhere in the same cells (16). The DNA polymerase which has been found in mitochondria (17) may therefore also be produced within these organelles. Furthermore, *n*-formylmethionyl transfer RNA is known to be present only in the mitochondria of eucaryotes and in bacteria (18).

Chloroplast ribosomes from higher plants resemble bacterial ribosomes in their sedimentation behavior and the sizes of their RNA components (19), and the fact that their ability to incorporate amino acids into proteins is inhibited by chloramphenicol (20), as it is in mitochondria (21). In contrast, cytoplasmic ribosomes are larger than chloroplast ribosomes (22) and are insensitive to chloramphenicol both in vitro (23) and in vivo (24). Hybridization studies have shown that ribosomal RNA probably originates from organelle DNA, not from nuclear DNA as in eucaryotic systems (25).

Therefore, mitochondria and plastids clearly resemble entire procaryotic cells more closely than other components of the eucaryotic cells in which they occur (26). They have a higher degree of autonomy than any other cellular component. They likewise resemble procaryotic cells in size (typically 1 to 5 micrometers).

Mitochondria and plastids differ from all other cellular organelles in being bound by a double membrane. The inner membrane is convoluted into a series of folds, greatly increasing the "working area" of the cell on which enzymes are located. Similarly, in bacterial cells, the plasma membrane is often extensively folded into the interior of the cell. The outer cell wall of bacteria appears to be a specialized structure, whereas the outer layer of mitochondria, at least, may be in effect an extension of the endoplasmic reticulum (26). Photosynthetic bacteria and the blue-green algae have membranous vesicles and lamellae upon which many of the photosynthetic pigments are located, and these are presumably homologous with the inner membranes of mitochondria and chloroplasts.

A Common Origin for Mitochondria and Plastids?

DuPraw (27) has pointed out that there are remarkable similarities between mitochondria and chloroplasts, many of which have been stressed in the preceding discussion. The structural proteins of mitochondria and chloroplasts are similar—both may contain an actomyosin-like contractile protein, and both show contraction dependent on adenosine triphosphate. Both mitochondria and chloroplasts carry out the phosphorylation of adenosine diphosphate coupled to electron transport phenomena, with similar electron carriers. These considerations have led to the hypothesis that mitochondria and chloroplasts may have had a common origin.

In any case, it is virtually certain that procaryotes with the properties of chloroplasts evolved before those with the properties of mitochondria, even though mitochondria are found in all eucaryotic cells, chloroplasts only in some. Free oxygen is utilized by mitochondria as an acceptor for electrons, but chloroplasts can carry out their functions anaerobically. It now seems likely that photosynthesis had already evolved by the time of deposition of the first known fossils, at least 3.2×10^9 years ago (28). Blue-green algae appear to be at least 2.7×10^9 years old as a group. On the other hand, eucaryotic cells probably did not evolve much more than 1.1×10^9 years ago (29). Other evidence strongly suggests that all the oxygen in the atmosphere has been derived from the process of photosynthesis. The achievement of current concentrations of oxygen in the atmosphere is often coupled with the invasion of the land by plants, animals, and fungi, an event that seems to have taken place about 4.5×10^8 years ago (30). At any rate, it is likely that there was insufficient oxygen in the atmosphere to allow the evolution of cells with the properties of mitochondria until perhaps 1.5×10^9 years ago, and perhaps much more recently than that.

If plastids and mitochondria did have a common origin, then, it appears almost certain that mitochondria are derived from plastids. In that case, an autotrophic procaryote presumably became a symbiote in the cells of a larger anaerobe maintaining itself with energy from glycolysis. This symbiote was then functionally equivalent to a chloroplast in the cells of the anaerobic host. Later, modification of its properties in the course of evolution led to its acquiring the oxidative capabilities associated with mitochondria.

It appears more likely, however, that, although mitochondria and plastids did have a common origin in the sense that both are derived from procaryotic symbiotes that became stabile elements in eucaryotic cells, they originated from separate symbiotic events. First, there are impressive differences between them. Chloroplasts are larger than mitochondria, have a different characteristic shape, contain nearly a hundred times as much DNA, and differ greatly biochemically. Second, mitochondria occur in all eucaryotic cells, including protists, plants, animals, and fungi. Plastids, on the other hand, are found only in plants. It appears simplest to visualize the acquisition of mitochondria by the common ancestor of all eucaryotes and the subsequent acquisition of plastids by one or more lines of eucaryotes as separate events. The simplicity with which intracellular symbioses apparently become established and are maintained argues in favor of a separate symbiotic origin of mitochondria and chloroplasts.

A Multiple Origin for Mitochondria?

Did mitochondria themselves arise from a series of separate symbiotic events? Symbioses that are observed at the present time invariably involve autotrophs and are thus analogous to the events that presumably led to the evolution of chloroplasts. All eucaryotic cells already have mitochondria, and a symbiosis involving additional mitochondrion-like particles would appear to have no selective value. In the past, this may not have been the case. If eucaryotic cells acquired characteristic mitochondria after their pattern of nuclear organization was already established, and they were in a sense already eucaryotic, one could visualize a series of symbiotic events involving mitochondrion-like organisms analogous to those involving autotrophs we see at the present day. It seems likely, however, that only one sort of procaryote was involved, because of the impressive similarities between mitochondria in all eucaryotic cells that have been investigated. In this connection, it is interesting to note that the ciliate *Paramecium aurelia*, for example, harbors a number of endosymbiotic gram-negative bacteria of uncertain adaptive value in nature, among them kappa, mu, lambda, and sigma particles (*31*). In any event, it is difficult to evaluate the question of a common or multiple origin for mitochondria by studying contemporary organisms.

Contemporary Symbioses and the Origin of Plastids

Photosynthetic algae are, at the present time, symbiotic in a very wide variety of organisms. Symbiotic relationships involving the procaryotic blue-green algae will be reviewed below, but first a brief survey of the symbiotic relationships of eucaryotic algae will be presented here to illustrate the diversity of these interactions.

In addition to the symbioses involving eucaryotic algae with higher plants, other algae, and vertebrates, the relationships between the algal and fungal components of lichens have been much discussed and studied experimentally. In addition to their participation in all of these kinds of systems, autotrophic algae are known to occur as symbiotes in more than 150 genera of invertebrates, representing eight phyla (*32, 33*). In the vast majority, if not all, of these instances, the symbiotic relationships arose independently. The symbiotes include three orders of green algae as well as the dinoflagellates and diatoms.

All stages in reduction of the cell wall in the symbiotic algae are represented among these relationships. For instance, in the dinoflagellate *Platymonas convolutae*, symbiotic in the marine acoelous turbellarian *Convoluta roscoffensis*, the symbionts lack the cell wall, capsule, flagella, and stigma of the free-living forms of the same species (*31*). They likewise have finger-like extensions 1 to 2 micrometers long that greatly increase the area of contact with the host cytoplasm.

A number of opisthobranch gastropods feed on siphonaceous green algae and have symbiotic chloroplasts, apparently derived from these algae, which are confined to the cells that form the hepatic tubules (*32*). These functional and clearly autonomous chloroplasts play an important role in the nutrition of their hosts.

The diversity of contemporary symbiotic relations attests both to the high selective value of such relations for heterotrophs and to the ease with which they are established. Thousands of such symbioses of varying age exist at the present day, and have been established even in the face of competition from established autotrophs, both unicellular and multicellular.

Symbiotic Origin of Mitochondria and Plastids

In view of the similarities between existing mitochondria and plastids, on the one hand, and procaryotic cells, on the other, it would be very difficult to conclude that they had originated separately. When the array of demonstrable symbiotic relationships between eucaryotic and, as we shall see shortly, procaryotic autotrophs and various kinds of heterotrophic organisms is taken into account, the case becomes overwhelming. Some symbiotic blue-green algae can at the present day be distinguished only with the greatest difficulty from chloroplasts—and, indeed, the distinction seems to be a false one. The case for the symbiotic origin of plastids appears overwhelming, and some of the implications of this for a phylogenetic understanding of the relationship between organisms will now be considered.

Mitochondria, like 9-plus-2 flagella and a differentiated nucleus (but see *34*), appear to be characteristic of all eucaryotic cells, but they may well have had a multiple origin, as suggested above. They have apparently lost most of their DNA content subsequently. Whether the spindle apparatus of eucaryotic cells likewise had a symbiotic origin, as suggested by Sagan (*35*), is a separate question that will not be considered in this article.

Symbiosis and the Origin of Plastids

The numerous symbiotic associations that can be observed in living organisms, together with their diversity, strongly suggest that such relations arise with relative ease. They would have had an even higher selective value when no eucaryotic cells were autotrophic, a fact that lends credence to the view that plastids arose not once, but many times. The implications of this view for the phylogenetic relationships of the major groups of autotrophic procaryotes will now be considered.

Blue-Green Algae as Symbionts

One group of living procaryotes is biochemically and structurally (36) similar to the chloroplasts of certain eucaryotes. These are the blue-green algae, which, in addition to chlorophyll *a*, contain phycobilins, two unusual porphyrins related to the bile pigments (37) in their cells (38). Outside of the cells of the blue-green algae, phycobilins occur in the red algae (37), in several genera of flagellated Cryptophyta (39), and in the anomalous hot-spring alga *Cyanidium caldarium* (40). They function as accessory pigments, and, because of their unusual structure, it is difficult to imagine that they evolved independently in these four groups, which are markedly dissimilar when judged on other criteria—so much so that it is extremely difficult to imagine a direct phylogenetic connection between them.

Symbiotic relationships involving blue-green algae are very common. These procaryotes are frequently found as symbiotic components in the cells of amoebae; flagellated protozoa; green algae such as *Gloeochaete, Glaucocystis,* and *Cyanoptyche* that lack chloroplasts; and diatoms (41). Some are even associated with fungi such as the phycomycete *Geosiphon pyriforme* (42). In all of these organisms, the blue-green algae play the role of chloroplasts, bringing the characteristic biochemistry of their procaryotic free-living relatives with them. In the symbiont of *Glaucocystis nostochinearum, Skujapelta nuda*, the cell walls are nearly or entirely lacking, a condition that Hall and Claus (43) consider an apparent adaptation to the intracellular habitat and symbiotic association.

These relationships suggest that the simplest hypothesis to account for the biochemical similarities between the groups of algae mentioned above would be that the chloroplasts of red algae, flagellated Cryptophyceae, and *Cyanidium* are in fact blue-green algae that entered into symbiotic relationships with the ancestors of these organisms. For each group, there is evidence for and against the proposed hypothesis.

In the blue-green algae, chlorophyll *a* occurs with several carotenoids and phycobilins. No other chlorophyll is present. The presence of a cell wall in the blue-green algae and its absence in chloroplasts poses no problem for, in at least two instances (44), known symbiotic blue-green algae lack a cell wall.

In red algae, chlorophyll *d* has been reported in many, but not all forms examined. Chlorophyll *d*, when present, is usually only a trace constituent; it has never been detected in the absorption spectra of living red algal cells or thalli; and it may merely be an oxidation product derived from chlorophyll *a* in vitro (45). This hypothesis has recently been strengthened by the detection of a pigment with the spectral qualities of chlorophyll *d* in extracts derived from the green alga *Chlorella pyrenoidosa* (46). The phycobilins of red algae are slightly different from those in blue-green algae, but, all in all, there seems to be no compelling biochemical reason not to regard the chloroplasts of red algae as ancient symbiotic blue-green algae. Moreover, the structure of the chloroplasts in red algae is extremely simple. These chloroplasts appear to contain a single thylakoid (47) and thus to be virtually identical to entire cells of the blue-green algae. The blue-green algae have been in existence for some 3 billion years (48), the red algae probably for less than 650 million. If the hypothesis presented here is accepted, the relationship to be considered is not that between the procaryotic blue-green algae and the eucaryotic red algae, but between the former and the chloroplasts of the latter. Aside from their chloroplasts, the cells of red algae have nothing in common with blue-green algae in organization or biochemistry (49).

Blue-green algae, lacking cell walls, are in fact the functional chloroplasts in one member of the Cryptophyta, *Cyanophora paradoxa* (43). The flagellated Cryptophyta—cryptomonads—are essentially Protozoa with a proteinaceous pellicle and a contractile vacuole. Some of them are heterotrophic and ingest food particles. They are not plantlike in any of their characteristics except for the biochemistry of their chloroplasts. The chloroplasts are relatively simple and similar in structure to those of the red algae (47) and to entire cells of the blue-green algae. Could these chloroplasts be ancient symbiotic blue-green algae?

The chief objection to this hypothesis appears to be the presence of chlorophyll *c*, together with carotenoids and phycobilins, in the chloroplasts of at least some cryptomonads (50). In structure, chlorophyll *c* differs widely from other chlorophylls (51). It occurs as an accessory pigment in the diatoms and some other Chrysophyta, in the dinoflagellates (Pyrrhophyta), and in the brown algae (Phaeophyta). These groups seem totally unrelated, and perhaps chlorophyll *c* evolved separately in the "chloroplasts" of cryptomonads. It may well be that the Cryptophyta are polyphyletic, as a number of phycologists have suggested. In a review of the algae (52), some are said to have "a cellulose membrane" and there are said to be two types of chloroplasts—"small, blue-green bodies" (symbiotic blue-green algae?) and "one or two parietal plates." As the rather numerous genera of cryptomonads are investigated further, it should be possible to determine which have chloroplasts of blue-green algal origin.

There appears to be no valid reason not to regard the large lobed chloroplast of *Cyanidium caldarium* as a symbiotic blue-green algal cell that has become stabilized in this role. These chloroplasts contain chlorophyll *a*, carotenoids, and phycobilins, thus closely approximating the biochemistry of living blue-green algae, to which they are likewise similar in structure. If this is the case, then biochemical evidence from the chloroplasts should not be taken into account in deciding where *Cyanidium* should be placed among the groups of organisms (53).

The Origin of Plastids in Other Algae

If similar reasoning is applied to the biochemistry of the chloroplasts in other groups of algae, certain relationships become evident (54). For example, the green algae (Chlorophyta) and euglenoids (Euglenophyta) have chloroplasts that are essentially identical biochemically. These chloroplasts resemble those of the land plants in containing chlorophyll *b* as an accessory pigment. In other respects, the green algae—which are "typical" plants with a cellulose cell wall—could not be more different from the euglenoids—which are "typical" flagellate protozoa with a proteinaceous pellicle and a contractile vacuole, and which at times ingest solid food particles (55). The simplest assumption to account for the similarity between the two groups is that both harbor the remnants of an ancient line of procaryotes, now presumably extinct, which had both chlorophyll a and b. These hypothetical procaryotes, then, would have been the group in which the photosynthetic apparatus characteristic of the land plants evolved (56).

The remaining algae—Phaeophyta, Chrystophyta, Xanthophyta, and Pyrrhophyta are the major groups—have chloroplasts in which chlorophylls a and c are associated with various carotenoids. If we can allow for the evolution and diversification of carotenoids after the establishment of symbiotic relationships in each case, then the biochemical similarities between the chloroplasts characteristic of these taxa might be accounted for by postulating a third group of procaryotes with chlorophyll a, in this case accompanied by chlorophyll c, which became symbiotic in cells ancestral to these groups (34, 57).

To summarize, the simplest way to account for the similarities and differences between the biochemistry of the procaryotic blue-green algae on the one hand and the chloroplasts of the various algal divisions on the other is as follows. First, chlorophyll a and a system of photosynthesis that led to the evolution of oxygen evolved in one line of procaryotes. In these, chlorophyll a and probably certain carotenoids that served as accessory pigments were probably arranged on photosynthetic lamellae within their cells. This line gave rise to at least three biochemically distinct derivatives: (i) the living blue-green algae, in which evolved phycobilins; (ii) the "green procaryotes," in which evolved chlorophyll b; and (iii) the "yellow procaryotes," in which evolved chlorophyll c. All of these groups entered into symbiotic associations with primitive eucaryotic cells, probably more than once, and groups (ii) and (iii) no longer exist as free-living organisms. From symbiotes of group (i) were derived the chloroplasts of the Rhodophyta, Cryptophyceae, and the anomalous genus *Cyanidium*. From symbiotes of group (ii) were derived the chloroplasts of the Chlorophyta (and through them, the bryophytes and vascular plants) and Euglenophyta. From symbiotes of group (iii) were derived the chloroplasts of Phaeophyta, Chrysophyta, Xanthophyta, and Pyrrhophyta.

Prospects for the Future

The relationships discussed here suggest several promising lines of investigation. First, the diversity of base-pair ratios in the blue-green algae (58) might provide a clue as to which might have been most likely to have given rise to chloroplasts in particular groups during the course of evolution, although the base ratios may of course have been altered subsequently by the operation of distinctive DNA polymerases. If we may take them more or less at face value, however, they may be useful; for example, the base ratios in the endosymbiote *Cyanocyte korschikoffiana*, which occurs in the cryptomonad *Cyanophora paradoxa*, strongly support the placement of *Cyanocyta* (10) in the order Chroococcales of the blue-green algae (58).

Second, in searching for homologies between contemporary blue-green algae and chloroplasts, it may not be appropriate to consider the vascular plants, in which the chloroplasts are very likely not homologous with blue-green algae. When DNA-hybridization experiments are more widely applied to the problem of the origin of mitochondria and plastids and their homologies, the hypotheses presented here invite several specific avenues of attack. For example, it would seem appropriate to compare certain contemporary blue-green algae with the chloroplasts of the red algae, preferably by DNA hybridization.

Even though it appears almost certain that mitochondria and plastids had independent, multiple, symbiotic origins, their functions are shared to varying degrees by nuclear DNA in contemporary eucaryotes. The hypothesis presented here suggests that the degree to which the nuclei have taken over what were presumably once independent functions may vary widely in different groups of eucaryotes. If DNA has actually been lost in the course of evolution from the symbiotes, perhaps becoming incorporated in the nucleus in some, this might have also altered radically the base-pair ratios in their mitochondria and plastids and might lead to unexpected results in DNA-hybridization experiments. It might eventually be possible in some systems to show that the nuclear cistrons responsible for mitochondrial or plastid functions (for example, cytochromes) might have base compositions similar to that of the organelle which they affect.

Summary

The impressive homologies between mitochondria and plastids, on the one hand, and procaryotic organisms, on the other, make it almost certain that these important cellular organelles had their origin as independent organisms. The vast number of symbiotic relationships of all degrees of evolutionary antiquity which have been found in contemporary organisms point to the ease with which such relationships can be established.

In view of this, the similarities between such totally different groups as blue-green algae and red algae, dinoflagellates and brown algae, and green algae and euglenoids can best be explained by postulating an independent, symbiotic origin of the plastids in each instance. A minimum of three groups of photosynthetic procaryotes appears to be necessary to explain the relationships among contemporary Protista and green plants: (i) the blue-green algae, which possess chlorophyll a, carotenoids, and phycobilins; (ii) the "green procaryotes," a hypothetical group characterized by chlorophylls a and b and a distinctive assemblage of carotenoid accessory pigments, but not phycobilins; and (iii) the "yellow procaryotes," a second hypothetical group whose members had chlorophylls a and c and various carotenoids but not phycobilins. There is, however, no reason to think that only three kinds of organisms were involved; numerous symbiotic events presumably occurred in each of these lines.

The "green procaryotes" and "yellow procaryotes" survive today only as chloroplasts from which the characteristics of the original, free-living forms can be deduced only in part. Hybridization between selected plastid DNA's may be helpful in unraveling this story, and is likewise suggested as the key to understanding the relationship between the blue-green algae and the chloroplasts of the red algae and cryptomonads.

It is postulated that the symbiotic organisms have lost various functions to the nucleus in the course of evolutionary time. If mitochondria and plastids have had a multiple origin, as suggested here, it will be necessary to examine the division of function between the two subsets of DNA for a wide variety of organisms before valid conclusions can be obtained.

References and Notes

1. C. Mereschkowsky, *Biol. Zentralbl.* **25**, 593 (1905); in *The Cell in Development and Heredity*, E. B. Wilson, Ed. (Macmillan, London, 1925); S. Granick, *Encycl. Plant. Physiol.* **1**, 507 (1955); A. Famintzin, *Biol. Zentralbl.* **27**, 353 (1907); H. Ris and W. Plaut, *J. Cell Biol.* **13**, 383 (1962); A. Gibor and S. Granick, *Science* **145**, 890 (1964); A. L. Lehninger, *The Mitochondrion* (Benjamin, New York, 1964); P. Echlin and I. Morris, *Biol. Rev.* **40**, 193 (1965); L. Sagan (now L. Margulis), *J. Theor. Biol.* **14**, 225 (1967); M. Edelman, D. Swinton, J. A. Schiff, H. T. Epstein, B. Zeldin, *Bacteriol. Rev.* **31**, 315 (1967); J. T. O. Kirk and R. A. E. Tilney-Bassett, *The Plastids* (Freeman, San Francisco, 1967); P. Borst, A. M. Kroon, C. J. C. M. Ruttenberg, in *Genetic Elements, Properties and Function*, D. Shugar, Ed. (Academic Press, New York, 1967), p. 81; E. J. DuPraw, *Cell and Molecular Biology* (Academic Press, New York, 1968); L. Margulis, *Science* **161**, 1020 (1968); D. B. Roodyn and D. Wilkie, *The Biogenesis of Mitochondria* (Methuen, London, 1968). An increasingly less popular view—that these organelles have evolved independently in eucaryotes—has been defended recently by A. Allsopp [*New Phytol.* **68**, 591 (1969)]. The arguments necessary to defend such a view in the face of contemporary evidence are far more complex than those supporting what is now clearly the majority opinion.
2. R. H. Whittaker, *Science* **163**, 150 (1969). However, there are exceptions; see, for example, L. Margulis, *J. Geol.* **77**, 606 (1969).
3. G. Shatz, *Biochemistry* **8**, 322 (1969); D. J. L. Luck, *J. Cell Biol.* **16**, 483 (1963); *ibid.* **24**, 445 (1965); *ibid.*, p. 461.
4. For *Neurospora* mitochondria see D. J. L. Luck and E. Reich [*Proc. Nat. Acad. Sci. U.S.* **52**, 931 (1964)]; for mitochondria of the myxomycete *Physarum* see J. E. Cummins, H. P. Rusch, T. E. Evans [*J. Mol. Biol.* **23**, 281 (1967)]; for *Euglena* chloroplasts see J. A. Schiff and H. T. Epstein [in *Reproduction: Molecular, Subcellular, and Cellular*, M. Locke, Ed. (McGraw-Hill, New York, 1965), p. 131].
5. References summarized by M. M. K. Nass [*Science* **165**, 25 (1969)].
6. L. Shapiro, L. I. Grossman, J. Marmur, *J. Mol. Biol.* **33**, 907 (1968).
7. E. J. DuPraw, *Cell and Molecular Biology* (Academic Press, New York, 1968), p. 97.
8. Chloroplast DNA has been shown by G. Brawerman [in *Biochemistry of Chloroplasts*, T. W. Goodwin, Ed. (Academic Press, New York, 1966), vol. 1, p. 301] to differ from nuclear DNA in its much lower content of guanosine and cytosine (about 24–27 percent in green algae) or higher content of guanosine and cytosine (in most vascular plants).
9. References summarized by S. Nass [*Int. Rev. Cytol.* **25**, 55 (1969)].
10. Evidence summarized by E. J. DuPraw [*Cell and Molecular Biology* (Academic Press, New York, 1968), pp. 97–98, 152–153].
11. J. T. O. Kirk [in *Biochemistry of Chloroplasts*, T. W. Goodwin, Ed. (Academic Press, New York, 1966), vol. 1, p. 301] summarizes much evidence that the biosynthesis of many components of the chloroplast depends upon nuclear genes.
12. D. S. Ray and P. C. Hanawalt, *J. Mol. Biol.* **11**, 760 (1965).
13. R. P. Wagner, *Science* **163**, 1026 (1969); see also more indirect evidence summarized in (5).
14. D. O. Woodward and K. D. Munkres, *Proc. Nat. Acad. Sci. U.S.* **55**, 872 (1966).
15. E. W. Guttes, P. C. Hanawalt, S. Guttes, *Biochim. Biophys. Acta* **142**, 181 (1967).
16. C. A. Buck and M. M. K. Nass, *Proc. Nat. Acad. Sci. U.S.* **60**, 1045 (1968); *J. Mol. Biol.* **41**, 67 (1969).
17. E. Wintersberger, *Biochem. Biophys. Res. Commun.* **25**, 1 (1966). Activity of DNA-dependent RNA polymerase has been demonstrated for both mitochondria [by D. J. L. Luck and E. Reich, *Proc. Nat. Acad. Sci. U.S.* **52**, 931 (1964)] and chloroplasts (11). R. R. Meyer and M. V. Simpson [*Proc. Nat. Acad. Sci. U.S.* **61**, 130 (1968)] showed that mitochondrial DNA polymerase was distinct from that located in the nucleus.
18. A. E. Smith and K. A. Marcker, *J. Mol. Biol.* **38**, 241 (1968); J. E. Darnell, *Biochem. Biophys. Res. Commun.* **34**, 205 (1969).
19. J. W. Lyttleton, *Exp. Cell Res.* **26**, 312 (1962); N. K. Boardman, R. I. B. Franki, S. G. Wildman, *J. Mol. Biol.* **17**, 470 (1966); E. Stutz and H. Noll, *Proc. Nat. Acad. Sci. U.S.* **57**, 774 (1967); L. S. Dure, J. L. Epler, W. E. Barnett, *ibid.* **58**, 1883 (1967).
20. E. Stutz and H. Noll, *Proc. Nat. Acad. Sci. U.S.* **57**, 774 (1967); U. E. Loening and J. Ingle, *Nature* **215**, 363 (1967); additional references are summarized by S. Nass (9).
21. G. F. Kalf, *Arch. Biochem. Biophys.* **101**, 350 (1963); E. Wintersberger, *Biochem. Z.* **341**, 409 (1965); G. D. Clark-Walker and A. W. Linnane, *Biochem. Biophys. Res. Commun.* **25**, 8 (1966); *J. Cell Biol.* **34**, 1 (1967).
22. Cytoplasmic and chloroplast ribosomes isolated from the same cells have been shown to differ in this respect in *Euglena* by J. M. Eisenstadt and G. Brawerman [*J. Mol. Biol.* **10**, 392 (1964)] and in *Nicotiana* by R. J. Ellis [*Science* **163**, 477 (1969)].
23. A. Marcus and J. Feeley, *J. Biol. Chem.* **240**, 1675 (1965); B. Parisi and O. Ciferti, *Biochemistry* **5**, 1638 (1966); R. J. Ellis and I. R. MacDonald, *Plant Physiol.* **42**, 1297 (1967).
24. R. J. Ellis, *Phytochemistry* **3**, 221 (1964).
25. Y. Suyama, *Biochemistry* **6**, 2829 (1967); H. Fukuhara, *Proc. Nat. Acad. Sci. U.S.* **58**, 1065 (1967); N. S. Scott and R. M. Smillie, *Biochem. Biophys. Res. Commun.* **28**, 598 (1967).
26. S. Nass, *Int. Rev. Cytol.* **25**, 55 (1969).
27. E. J. DuPraw, *Cell and Molecular Biology* (Academic Press, New York, 1968), p. 154.
28. A. E. J. Engel, B. Nagy, L. A. Nagy, *Science* **161**, 1005 (1968).
29. J. W. Schopf, in *McGraw-Hill Yearb. Sci. Technol.* (1967), p. 47.
30. The reasoning is that the development of a layer of ozone sufficient to protect land organisms from the destructive effects of ultraviolet radiation was necessary before the land could be fully occupied, and once this had occurred, the occupation may have happened rather rapidly.
31. G. H. Beale, A. Jurand, J. R. Preer, *J. Cell Sci.* **5**, 65 (1969).
32. D. Smith, L. Muscatine, D. Lewis, *Biol. Rev.* **44**, 17 (1969).
33. M. Droop, *Symp. Soc. Gen. Microbiol.* **13**, 171 (1963).
34. This argument has recently been strengthened by the demonstration of a form of "mitosis" in the Pyrrhophyta totally distinct from that in any other organism; see D. F. Kubai and H. Ris [*J. Cell Biol.* **40**, 508 (1969)]. Any direct phylogenetic connection between the dinoflagellates and, for example, the brown algae appears extremely implausible.
35. L. Sagan, *J. Theor. Biol.* **14**, 225 (1967).
36. P. Echlin and I. Morris, *Biol. Rev.* **40**, 193 (1965).
37. H. W. Siegelman, D. J. Chapman, W. J. Cole, in *Porphyrins and Related Compounds*, T. W. Goodwin, Ed. (Academic Press, New York, 1968).
38. Review by C. Ó hEocha, in *Chemistry and chemistry of Plant Pigments*, T. W. Goodwin, Ed. (Academic Press, New York, 1968).
39. C. Ó hEocha and M. Raftery, *Nature* **184**, 1049 (1959).
40. M. B. Allen, *Arch. Mikrobiol.* **32**, 270 (1959).
41. M. Droop, *Symp. Soc. Gen. Microbiol.* **13**, 171 (1963); L. Geitler, *Syncyanosen*, in Ruhland's *Handb. Pflanzenphysiol.* **11**, 530 (1959).
42. E. Schnepf, *Arch. Mikrobiol.* **49**, 112 (1964).
43. W. T. Hall and G. Claus, *J. Cell Biol.* **19**, 551 (1963).
44. Summarized in N. Lang, *Annu. Rev. Microbiol.* **22**, 20 (1968).
45. M. B. Allen, *The Chlorophylls*, L. P. Vernon and G. R. Seely, Eds. (Academic Press, New York, 1966), pp. 511–519.
46. M. R. Michel-Wolwertz, C. Sironval, J. C. Goedheer, *Biochim. Biophys. Acta* **94**, 584 (1965).
47. J. T. O. Kirk, in *The Plastids*, J. T. O. Kirk and R. A. E. Tilney-Bassett, Eds. (Freeman, San Francisco, 1967), pp. 30–47.
48. J. W. Schopf, "Antiquity and evolution of Precambrian life," in *McGraw-Hill Yearb. Sci. Technol.* (1967), pp. 47–55.
49. Arguments such as that presented by A. Allsopp [*New Phytol.* **68**, 591 (1969)], who views the red algae as an intermediate group between procaryotic and eucaryotic cells, are therefore considered invalid; they are based entirely on the properties of the chloroplasts of red algae.
50. F. T. Haxo and D. C. Fork, *Nature* **184**, 1051 (1959).
51. L. P. Vernon and G. R. Seely, Eds., *The Chlorophylls* (Academic Press, New York, 1966).
52. G. W. Prescott, *The Algae: A Review* (Houghton Mifflin, Boston, 1968).
53. P. C. Silva, in *Physiology and Biochemistry of Algae*, R. A. Lewin, Ed. (Academic Press, New York, 1962), pp. 827–837; the carotenoid relationships are discussed by T. W. Goodwin [in *Chemistry and Biochemistry of Plant Pigments*, T. W. Goodwin, Ed. (Academic Press, New York, 1965), pp. 130–133].
54. As suggested by L. Sagan, *J. Theor. Biol.* **14**, 252 (1967).
55. J. J. Wolken, *Euglena* (Appleton-Century-Crofts, New York, ed. 2, 1967); G. F. Leedale, *Euglenoid Flagellates* (Prentice-Hall, Englewood Cliffs, N.J., 1967).
56. The carotenoid pigments of most green algae and vascular plants are likewise similar; see T. W. Goodwin, in *Chemistry and Biochemistry of Plant Pigments*, T. W. Goodwin, Ed. (Academic Press, New York, 1965), p. 130.
57. The presence of fucoxanthin in Phaeophyta and Chrysophyta and its absence in Xanthophyta will likewise have to be taken into account in evaluating these relationships; T. W. Goodwin, in *Chemistry and Biochemistry of Plant Pigments*, T. W. Goodwin, Ed. (Academic Press, New York, 1965), p. 133.
58. M. Edelman, *Bacteriol. Rev.* **31**, 315 (1967).
59. I thank R. B. Flavell (Cambridge Plant Breeding Institute) and A. Staehelin (Harvard University) as well as my colleagues P. C. Hanawalt, R. W. Holm, D. O. Woodward, and C. Yanofsky for their helpful comments on this manuscript. Supported in part by NSF grant GB 7949X.

6

Copyright © 1980 by Elsevier/North-Holland Biomedical Press B.V.
Reprinted from *Trends Biochem. Sci.,* June, pp. 146–149 (1980)

Revolutionary concepts in evolutionary cell biology

W. Ford Doolittle

Developments in micropaleontology, RNA and protein sequencing, and the analysis of genome organization suggest a view of early cellular evolution radically different from that accepted as recently as ten years ago.

It is difficult to remember exactly how one thought about cellular evolution ten or fifteen years ago and certainly risky to assume that everyone thought about it in that same way. However, it is probably fair to assume that most biochemists and molecular biologists then accepted the following three notions. (i) The appearance of the first living cell was the cumulative result of very many random events of extremely low probability, and thus the period of biological (cellular) evolution was preceded by a very much longer period of prebiotic evolution involving chance associations of non-biologically synthesized macromolecules accumulating in the Oparin [1] ocean. (ii) The first living cells were what we would now call prokaryotes, and these and their modern prokaryotic descendants, for all their biochemical diversity, represent a single monophyletic assemblage of entities which are essentially similar at the most fundamental levels of cellular organization, genetic organization and expression. (iii) Even though the 'line of demarcation between eukaryotic and prokaryotic cellular organisms is the largest and most profound single evolutionary discontinuity in the contemporary biological world' [2], eukaryotes arose rather recently from among the prokaryotes (Fig. 1). The remarkable differences in genetic organization, mechanisms of gene expression and evolutionary versatility which tempt us to consider eukaryotes more advanced than prokaryotes were to be seen as in part the cause and in part the consequence of the transition between prokaryotic and eukaryotic levels of cellular organization.

Each of these three earlier views can now be seriously challenged. Although none of the points I make below is uniquely mine, none can be taken as proven, and some remain highly controversial, they deserve to be considered together because together they represent a radical revision of the way in which we think about cellular evolution. The points are these. (i) Living cells arose very early in the history of the earth. (ii) Living cells diverged very early into three (organizationally prokaryotic) lineages; the 'archaebacterial' lineage, the 'eubacterial' lineage (as defined by Woese and Fox [3]), and that lineage which gave rise to the nucleus of eukaryotic cells (termed the 'urkaryotic' lineage by Woese and Fox, and here called the nuclear-cytoplasmic lineage). (iii) Eukaryotic cells themselves resulted from the repeated fusion of representatives of two of the three lineages.

Living cells arose very early in the history of the earth

The fossil record for multicellular eukaryotes is some 700 million years old and, until the late 1950s or early 1960s, was the only fossil record we had [4]. Since then, microfossils which are cellular by several criteria have been discovered in deposits of increasing and almost unbelievable antiquity. Organic microstructures from the Swaziland system of S. Africa which, at 3.5 billion years, is more than three-fourths as old as the earth itself, have now been shown by Knoll and Barghoorn [5] to be almost unquestionably biogenic, on five grounds: (i) chemical composition,

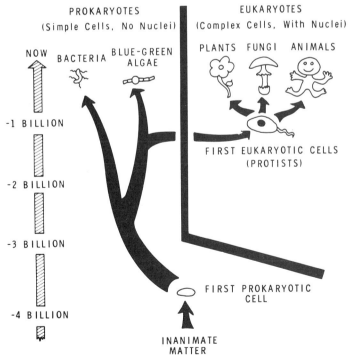

Fig. 1. The view of cellular evolution prevalent in the 1960s and early 1970s.

W. Ford Doolittle is at the Department of Biochemistry, Dalhousie University, Halifax, Nova Scotia, Canada B3H 4H7.

Revolutionary Concepts in Evolutionary Cell Biology 147

(ii) unimodal size distribution, (iii) morphology, (iv) sedimentary context, and (v) the preservation of cells in the process of division. The oldest (3.8 billion years) sedimentary rocks are those of the Isua formation of Greenland. These too *may* show evidence of biological activity [6]. The earth itself is only some 700 million years older than these rocks [7]. Even if we assume that conditions conducive to the accumulation of non-biologically synthesized precursors of biological systems existed from the very beginning, we must conclude that life arose precipitously; a long period of biological (cellular) evolution was preceded by only a *very* much shorter period of prebiotic evolution.

Living cells diverged very early into three lineages

As I remember it, the view of cellular evolution prevalent in the 1960s was that shown in Fig. 1. Life was monophyletic, all living organisms being descended from an ancient anaerobic heterotrophic prokaryote not unlike a modern *Clostridium*. The first eukaryotic cell was probably a unicellular alga, derived from a prokaryotic blue–green alga. From this eukaryotic 'Uralga' arose higher plants and, by loss of photosynthesis, non-photosynthetic protists, fungi and animals. The overwhelming homology of photosynthetic processes in blue–green algae and plants and the basic similarities of cellular and genetic organization in all eukaryotes, whether or not they are capable of photosynthesis, seemed to demand such a specific scenario.

This scenario no longer appears tenable, not only because evidence for the composite nature (symbiotic origin) of eukaryotic cells is now compelling (see below), but because there appear to be in the contemporary biological world two evolutionary discontinuities more profound than that separating prokaryotic and eukaryotic cellular organisms (Fig. 2). In particular, it seems that living things comprise three distinct evolutionary lineages which diverged at a (presumably very early) stage when genetic organization and the mechanisms and machinery of gene expression were still undergoing rapid Darwinian evolution, in the direction of increased efficiency and accuracy. The profound differences between these lineages are best interpreted as independently achieved solutions to the problems of how genes should best be organized and expressed.

This idea, and the evidence supporting it, stems directly from the work of Woese, Fox and collaborators [3,8–11] on the

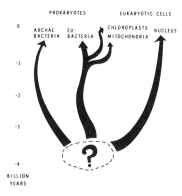

Fig. 2. A view of cellular evolution made plausible by the recent discoveries discussed here.

sequences of 16 S and 18 S ribosomal RNAs (rRNAs). Comparisons, by numerical taxonomic techniques, of the catalogs of sequences of oligonucleotides derived from these molecules by T1-ribonuclease digestion, allow their division into three distinct classes; 'archaebacterial' 16 S, 'eubacterial' 16 S and 'nuclearcytoplasmic' 18 S. Although there are internal divisions reflecting phylogenetic divergence within each class, the classes do not overlap, and each can be further defined by a distinct set of highly-conservative oligonucleotides (frequently showing distinctive post-transcriptional modifications) which presumably occupy functionally constrained regions of the molecule. The simplest interpretation is that given above; archaebacterial, eubacterial and nuclear-cytoplasmic lineages diverged at a time when problems of efficiency, speed and accuracy of translation were incompletely solved, and each lineage has achieved its own independent final solution. We can assume this time to have been at least 3.0 billion, and probably more, years ago, because recognizable blue–green algal fossils first appear at about that time [4], and there is ample evidence from both RNA and protein sequence data that there are divergencies within the eubacteria themselves which pre-date the appearance of the blue–green algae [3,12,13].

There are objections to the radical phylogenetic revision proposed by Woese and Fox [3]. They are in part semantic. Woese and Fox call the three lineages 'primary kingdoms', and suggestions for taxonomic revision at this level are never easily accepted. The term 'archaebacteria' unfortunately suggests that the members of this group are living fossils; the term 'eubacteria' has an earlier and more limited taxonomic meaning [2], and the term 'urkaryote' is at best dyseuphonious. One hopes for better terms.

More serious is the objection that the organisms so far classified as archaebacteria are biochemically and ecologically very dissimilar. They comprise the obligately anaerobic methane-forming bacteria, the obligately aerobic and obligately halophilic halobacteria, which may show certain 'advanced' traits [14], and two thermoacidophiles, *Thermoplasma* and *Sulfolobus* which are, aside from their thermoacidophily, dissimilar. The range of archaebacterial metabolic modes and morphologies is as great as that shown by the 'eubacteria' which, in the terminology of Woese and Fox, comprise most if not all other organisms described in Bergey's manual, as well as the blue–green algae (cyanobacteria), which are still considered by some to be primitive algae. The objections to such a division of the prokaryotic world were strong when its only basis was rRNA sequence data. Objections should weaken as more and more traits which distinguish the archaebacteria from the eubacteria are discovered. To date these include: distinctive tRNA modification patterns, distinctive RNA polymerases, distinctive antibiotic sensitivities, the absence in archaebacteria of peptidoglycan and their possession of unusual lipids, in which ether linkages replace ester linkages and isoprenoid side chains replace fatty acids [10,15,16].

The second radical conclusion derived from the data of Woese and Fox is that the nuclei of eukaryotic cells derive from a lineage, the nuclear-cytoplasmic lineage, which diverged from the eubacterial lineage during the earliest stages of cellular evolution. This will be most objectionable to those who adhere to autogenous hypotheses for the origin of eukaryotic cells, since such hypotheses virtually require that the entire eukaryotic cell be of relatively recent blue–green algal (eubacterial) origin [17,18]. Endosymbiotic hypotheses, which most data now support, are less specific about the origin of the nuclear-cytoplasmic component, although most consider it to be eubacterial [19].

There is no compelling evidence to support such an origin. Differences in genetic organization and expression between the nucleus and prokaryotes (specifically eubacteria, since little is known about the molecular biology of archaebacteria [14]) have been recognized for a long time. It was until recently customary to explain them as an adaptation to the genomic com-

84

partmentalization which defines the eukaryotic condition, and to view eukaryotic genomes as advanced descendants of prokaryotic genomes. The recent discovery [20,21] that genes in prokaryotes and genes in eukaryotes can differ in a very fundamental way – those in prokaryotes being intact and colinear with protein; those in eukaryotes being interrupted by intervening sequences which complicate the process of gene expression in a way which cannot yet be seen as selectively advantageous – prompted a re-examination of this notion. I argued [22] that it was easier to imagine intervening sequences as relics of a very primitive stage in cellular evolution, which have been eliminated in prokaryotic lineages but retained in nuclear-cytoplasmic lineages, than it is to imagine their introduction during the prokaryote–eukaryote transition. Darnell [23] pointed out that many other differences between nuclear-cytoplasmic and eubacterial molecular biologies (such as the paucity of operons, the possession of multiple RNA polymerases and the extensive post-transcriptional processing and modification of messenger RNAs) are 'so profound as to suggest that sequential prokaryote to eukaryote cell evolution seems unlikely'. These features, together with the evidence that eukaryotic cytoplasmic ribosomes contain not only distinct classes of rRNA but function differently in protein synthesis [24,25] can be most easily accommodated by the hypothesis that the nuclear-cytoplasmic components of eukaryotic cells diverged from prokaryotes (or at least from eubacteria) before final solutions to the problems of effective genetic organization and expression had been found.

Eukaryotic cells themselves result from the repeated fusions of representatives of two of the three major cellular lineages

There are two general ways to explain the functional compartmentalization of nuclear, mitochondrial and (in eukaryotic algae and plants) plastid DNA in eukaryotes (Figs. 3 and 4). Neither is new; what is new is the fact that molecular sequence data now at last allow us to choose between them. In 'autogenous origin' hypotheses (Fig. 3), mitochondrial and plastid genomes derive in some fashion from nuclear (or 'protonuclear') genomes [17,18]. In 'endosymbiotic origin' hypotheses, mitochondrial genomes descend from the genomes of once free-living aerobic bacteria trapped in permanent endosymbioses within the cytoplasms of protoeukaryotic cells, while plastid genomes similarly descend from the genomes of once free-living oxygen-evolving photosynthetic prokaryotes similar to modern blue–green algae. The second hypotheses has gained recent renewed prominence through the efforts of Lynn Margulis and others [19]. Not everyone regards it as proven, but molecular sequence data leave little room for autogenous alternatives. T1 oligonucleotide catalogs show strong affinities between plastid and blue–green algal 16 S rRNAs and no affinity between either of these and the cytoplasmic 18 S rRNA encoded by nuclear genes which are present in the same cell [13,26]. Mitochondrial rRNAs can be shown in the same way to be 'generally' prokaryotic (and unrelated to cytoplasmic rRNAs), while cytochrome *c* sequence analyses point to a specific rhodospirillacean (purple non-sulfur bacterial) ancestry for the mitochondrion [27,28]. To accommodate the data on plastids, adherents of autogenous origin hypotheses must assume a blue–green algal origin for the entire eukaryotic cell, with rapid subsequent evolutionary divergence of the nuclear genome and the rRNA which it encodes.

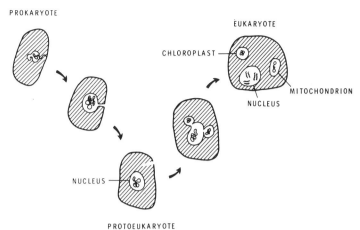

Fig. 3. A simplified version of the autogenous origin hypotheses, in which organellar genomes derive from the genome(s) present in the protoeukaryotic cell.

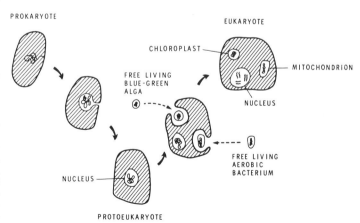

Fig. 4. A simplified version of the endosymbiotic origin hypotheses, in which organellar genomes derive from those of once free-living prokaryotes trapped in permanent endosymbioses within the cytoplasms of protoeukaryotic cells.

To accommodate the data on mitochondria, they must postulate a rhodospirillacean ancestor for the whole eukaryotic cell with, again, rapid nuclear divergence. One cannot have it both ways; RNA and protein sequence data show that blue–green algae and purple non-sulfur bacteria diverged long before plastids and blue–green algae on the one hand, or mitochondria and purple non-sulfurs on the other [12,13,28]. Thus the ancestor of eukaryotic cells cannot have been some prokaryote intermediate between blue–green algae and purple non-sulfur bacteria, as suggested by Uzzell and Spolsky [29]. The rapid divergence of the nuclear genome, which supporters of autogenous hypotheses are obliged to assume, is unmotivated and, if we accept the arguments given above for the antiquity of the nuclear-cytoplasmic lineage, unnecessary.

Implications

Evolutionary issues are always hotly debated. I cannot claim that the views presented here are universally accepted, but will predict that they will become more generally accepted. They raise further interesting questions. (i) How is it possible that cells arose in so short a time, when the chance assortment of non-biologically synthesized materials to form self-replicating cells seems intuitively so much more difficult to accomplish than all the subsequent steps of biological evolution? (ii) Have archaebacteria achieved independent solutions to problems of genetic organization and expression, as they seem to have done for problems of ribosome structure? Do they have intervening sequences or structures even more bizarre? (iii) If prokaryotic and nuclear lineages diverged at an early stage in cellular evolution, is it not likely that mechanisms controlling gene expression in the two lines are independently evolved, and that no prokaryotic regulatory system is a model for any eukaryotic regulatory system in any sense other than a heuristic one? (iv) How deep are divergencies within the nuclear-cytoplasmic lineage itself? Is it possible that eukaryotic cells are polyphyletic in the sense that enclosure of the nuclear genome within a nuclear membrane occurred independently in different sublineages, as well as in the sense that plastids and mitochondria arose from independent endosymbionts? Are there survivors of the nuclear-cytoplasmic lineage which are still organizationally prokaryotic?

References

1. Oparin, A. I. (1938) *The Origin of Life*, Macmillan, New York
2. Stanier, R. Y., Adelberg, E. A. and Ingraham, J. L. (1976) *The Microbial World*, Prentice-Hall, Englewood Cliffs, New Jersey
3. Woese, C. R. and Fox, G. E. (1977) *Proc. Natl. Acad. Sci. U.S.A.* 74, 5088–5090
4. Schopf, J. W. (1975) *Endeavour* 34, 51–58
5. Knoll, A. H. and Barghoorn, E. S. (1977) *Science* 198, 396–398
6. Nisbet, E. G. (1980) *Nature (London)* 284, 395–396
7. Schidlowski, M., Appel, P. W. U., Eichmann, R. and Junge, C. E. (1979) *Geochim. Cosmochim. Acta* 43, 189–199
8. Fox, G. E., Magrum, L. J., Balch, W. E., Wolfe, R. S. and Woese, C. R. (1977) *Proc. Natl. Acad. Sci. U.S.A.* 74, 4537–4541
9. Magrum, L. J., Luehrsen, K. R. and Woese, C. R. (1978) *J. Mol. Evol.* 11, 1–8
10. Woese, C. R., Magrum, L. J. and Fox. G. E. (1978) *J. Mol. Evol.* 11, 245–252
11. Woese, C. R. and Fox, G. E. (1977) *J. Mol. Evol.* 10, 1–6
12. Schwartz, R. M. and Dayhoff, M. O. (1978) *Science* 199, 395–403
13. Bonen, L. and Doolittle, W. F. (1976) *Nature (London)* 261, 669–673
14. Bayley, S. T. (1979) *Trends Biochem. Sci.* 4, 223–225
15. Tornabene, T. G., Langworthy, T. A., Holzer, G. and Oró, J. (1979) *J. Mol. Evol.* 13, 73–83
16. Zillig, W., Stetter, K. O. and Janekovic, D. (1979) *Eur. J. Biochem.* 96, 597–604
17. Cavalier-Smith, T. (1975) *Nature (London)* 256, 463–468
18. Taylor, F. J. R. (1976) *Taxon* 25, 377–390
19. Margulis, L. (1970) *Origin of Eukaryotic Cells*, Yale University Press, New Haven
20. Gilbert, W. (1978) *Nature (London)* 271, 501
21. Crick, F. (1979) *Science* 204, 264–270
22. Doolittle, W. F. (1978) *Nature (London)* 272, 581–582
23. Darnell, J. E. (1979) *Science* 202, 1257–1260
24. Kozak, M. (1979) *Nature (London)* 280, 82–85
25. Hagenbüchle, O., Santer, M., Steitz, J. A. and Mans, R. J. (1978) *Cell* 13, 551–563
26. Bonen, L., Doolittle, W. F. and Fox, G. E. (1979) *Can. J. Biochem.* 57, 879–888
27. Bonen, L., Cunningham, R. S., Gray, M. W. and Doolittle, W. F. (1977) *Nucleic Acids Res.* 4, 663–671
28. Almassy, R. J. and Dickerson, R. E. (1978) *Proc. Natl. Acad. Sci. U.S.A.* 75, 2674–2678
29. Uzzell, T. and Spolsky, C. (1974) *Am. Sci.* 62, 334–343

Part II

MECHANISMS NECESSARY FOR SERIAL SYMBIOSIS

Editors' Comments
on Papers 7 through 12

7   **CUTTING and SCHULMAN**
The Biogenesis of Leghemoglobin. The Determinant in the Rhizobium-Legume Symbiosis for Leghemoglobin Specificity

8   **LORCH and JEON**
Rapid Induction of Cellular Strain Specificity by Newly Acquired Cytoplasmic Components in Amoebas

9   **SMITH**
From Extracellular to Intracellular: The Establishment of a Symbiosis

10   **SONEA**
Bacterial Plasmids Instrumental in the Origin of Eukaryotes?

11   **MARTIN and FRIDOVICH**
Excerpts from *Evidence for a Natural Gene Transfer from the Ponyfish to Its Bioluminescent Bacterial Symbiont* Photobacter leiognathi. *The Close Relationship Between Bacteriocuprein and the Copper-Zinc Superoxide Dismutase of Teleost Fishes*

12   **FAIRFIELD, MESHNICK, and EATON**
Malaria Parasites Adopt Host Cell Superoxide Dismutase

Obligate symbiotic associations between two cells must be characterized by levels of efficiency in nutrient and energy use and in reproductive capacity that somehow surpass those of the individual cells growing separately in the same environment. That is, the whole symbiotic unit must be greater than the sum of its parts, under a given set of conditions. An efficient symbiotic association can be maintained by the following:

1. mechanisms by which nutrients or other compounds may be efficiently transferred from one partner to another (Paper 9)
2. mechanisms by which functions that are redundant or unnecessary for the association are lost, ultimately by the loss of vestigial genes encoding those functions

3. mechanisms by which the genetic material controlling the association is centralized for efficiency in replication, transcription, and translation. This centralization is accomplished by enzymatic control by one partner over another partner's gene functions, or by transfer of genes from one partner to another. Regulatory mechanisms of this type may be used to coordinate reproduction of the partners, so that they neither outgrow each other nor become lost during divisions (Papers 7–12, 14, 17, 18, and 23).

7

Copyright © 1971 by Elsevier/North-Holland Biomedical Press B.V.
Reprinted from Biochim. Biophys. Acta **229**:58-62 (1971)

THE BIOGENESIS OF LEGHEMOGLOBIN. THE DETERMINANT IN THE RHIZOBIUM-LEGUME SYMBIOSIS FOR LEGHEMOGLOBIN SPECIFICITY

J. A. Cutting and H. M. Schulman

SUMMARY

By disc electrophoresis of leghemoglobins produced in the nodules of various legumes inoculated with specific strains of the bacterial genus Rhizobium, we have shown that:

1. A single plant type treated with rhizobial strains of diverse origins always produces leghemoglobins having the same electrophoretic mobilities.

2. Pairs of plant types produce effective nodules when inoculated with single rhizobial strains. The leghemoglobins are always plant-type specific.

3. Representative species of the pea cross-inoculation group, effectively nodulated with a single rhizobial strain, each produced a different electrophoretic pattern of leghemoglobin.

These results are interpreted as evidence that the type of leghemoglobin produced in a given rhizobium–legume symbiosis, is plant specific. It is speculated that this effect is dependent on the genetic information, defining the amino acid sequence for apoleghemoglobins, being resident in the plant genome.

INTRODUCTION

Leghemoglobin is the heme protein found in root nodules of legumes having an effective symbiosis with a strain of the bacterial genus Rhizobium. Electropherograms of leghemoglobins, extracted from the nodules of field-grown legumes, manifest a low number of distinct bands. The electrophoretic properties of the leghemoglobins appear to be plant species specific[1]. However, the rhizosphere of such field-grown legumes can be shown to harbor a large number of rhizobial types[2]. The apparent plant specificity obtained in the electropherograms may thus reflect a stringent selection by the plant for that rhizobial symbiont which carries genetic information for the "optimal" leghemoglobins for that particular plant species. In this regard we have recently demonstrated that in the soybean–rhizobium symbiosis, the bacterial partner synthesizes heme which can be inserted into apoleghemoglobin[3]. This paper reports our attempts to define the origin of the globin moiety of leghemoglobin. After the work presented here was completed, DILWORTH[4] reported a similar analysis, from which he concluded that the genetic specification for hemoglobins in legume root nodules is a property of the plant. Our results extend his findings.

MATERIALS AND METHODS

Bacterial culture and isolation

Some rhizobial strains were isolated from nodules produced on plants grown in local soils. The nodules were surface sterilized for 20 sec with a 1.5% sodium hypochlorite solution, and then copiously washed in sterile 0.1 M sodium phosphate buffer (pH 7.0). Using standard sterile procedure, one end of the nodule was cut off, and material from the medulla streaked out onto plates prepared from the medium previously described[3], solidified with 1.75% agar. After growth, single colonies were isolated and used for an inoculum in the liquid medium. Harvesting and seed inoculation were as previously described[3].

The following lists the rhizobial strains employed, their sources and designate hosts:

Strain	Source	Host
3I1b125	U.S. Department of Agriculture, Beltsville, Md.	*Glycine max*
3I1b138	Same as above	*Glycine max*
61A76	Nitragin Co, Milwaukee, Wisc.	*Glycine max*
505	Bacteriology Department, University of Wisconsin, Madison, Wisc.	*Glycine max*
Joo1	San Diego, Calif.	*Glycine max*
10/2/3	Same as above	*Phaseolus vulgaris*
Lo1	Same as above	*Lathyrus odoratus*
Legume Aid	Atlee Burpee Co., Riverside, Calif.	Commercial mixture for field and garden legumes

Plants

Plant culture was as before[3], except as herein described. Growth containers were sterilized by immersion in 1.5% sodium hypochlorite solution for at least 1 h, followed by copious washing in demineralized water. Growth supportants and culture media were sterilized by autoclaving. Each plant container was surrounded by a 9-inch-high wall of heavy gauge polyethylene sheeting.

Before germination, seeds were surface sterilized for 20 sec in a 1.5% sodium hypochlorite solution and then copiously washed with sterile water. Plants were germinated and grown in a controlled environment chamber, having a 24-h cycle with a 15-h light period at an intensity of illumination of 2000 ftcandles, and with a temperature cycle of 90°F day and 75°F in the dark period. Control uninoculated containers with plants were included in each experiment. Nodulation did not occur in these legumes for the duration of the experiment.

The following seeds were obtained from Atlee Burpee Seed Company, Riverside: *Glycine max* var. bansei, *Phaseolus vulgaris* var. pencil pod, red kidney and Kentucky wonder, *Lathyrus latifolius* var. mixed colors, *Lathyrus odoratus* var. Bijou sweet peas, all colors, *Vicia faba* var. long pod fava, *Pisum sativum* var. Alaska. *Lens esculenta* and *Cicer ariatinum* were purchased from a local supermarket.

Nodules were harvested at the appearance of the first flowers and stored frozen at $-80°$ until needed.

Nodule extraction

This was essentially as previously described[3]. Frozen nodules were crushed with pestle and mortar, under H_2 at a pressure of 15 cm Hg and in the presence of 3.0 M $(NH_4)_2SO_4$. After removing the debris by centrifugation, the pink supernatant was made 3.8 M with crystalline $(NH_4)_2SO_4$. The pink precipitate was dissolved in a minimum volume of water and then dialyzed against a large excess of distilled water for 4 h. Any precipitate that formed was removed by centrifugation.

Disc electrophoresis

This was performed on a vertical gel apparatus as described by AKROYD[5]. The gel container was 9 cm deep and 0.5 cm wide. The gel and buffer systems were those of DAVIS[6], and were employed as previously described[3]. Vertical depths of 6-cm running gel and 1-cm focus gel were used. A maximum of 200 μg of each nodule extract was applied to the gel on a 0.5 cm × 0.5 cm surface area, in a volume not greater than 100 μl. Electrophoresis was for 2 h at about 25 V/cm at a maximum power of 1 W/cm² of the surface to which the sample was applied.

Staining and recording of electropherograms

The pink bands in the electropherograms were stained for heme using the *o*-dianisidine reagent of OWEN *et al.*[7]. The gels were first washed at 0° in 0.15 M sodium acetate buffer (pH 4.7) for 30 min; then treated with the dye (100 mg *o*-dianisidine in 0.15 M sodium acetate buffer (pH 4.7) containing 70% ethanol and freshly diluted with 2% 100 vol. H_2O_2) at room temperature for 30 min; and finally washed with more of the acetate buffer. The line drawings in this paper are from tracings made against photographs of the stained gels.

EXPERIMENTAL

Fig. 1 depicts the result of a search for heterogeneity in the electrophoretic pattern of leghemoglobin from the soybean plant *Glycine max*. The experiment was performed by using rhizobial strains obtained from widely diverse origins. In all, six different strains (from the sources indicated in MATERIALS AND METHODS) induced effective nodules, and as the figure indicates, all of these produced electrophoretically identical leghemoglobin types.

Fig. 2–4 are the results of experiments in which two or more plant species were separately infected with the same rhizobial strain. Fig. 2 and 3 show that the strain used in each experiment resulted in the formation of a different electrophoretic pattern of the leghemoglobins extracted from the nodules of each plant species employed.

BIOSYNTHESIS OF LEGHEMOGLOBIN

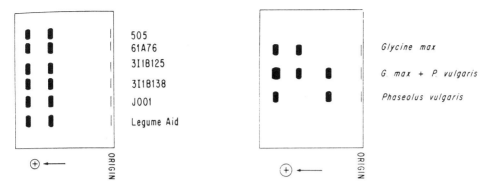

Fig. 1. Electropherogram of the leghemoglobins produced in soybean plants (*Glycine max*) separately inoculated with the rhizobial strains indicated. (Sources of the strains are given in MATERIALS AND METHODS.)

Fig. 2. Electropherogram of the leghemoglobins produced in the plant species indicated upon infection with the rhizobial strain 505.

TABLE I

RESULTS OF INOCULATION OF PEA CROSS-INOCULATION GROUP MEMBERS WITH THE RHIZOBIAL STRAIN Lo1

Plant	Result
Lathyrus latifolius	Small effective nodules
Lathyrus odoratus	Many effective nodules
Vicia faba	Many ineffective nodules
Cicer ariatinum	No nodulation
Lens esculenta	Many effective nodules
Pisum sativum	Many effective nodules

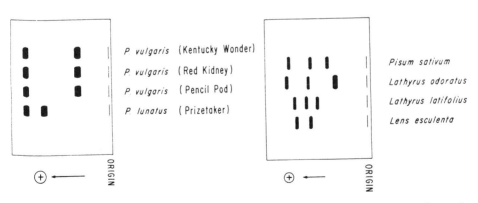

Fig. 3. Electropherogram of the leghemoglobins produced in the plant species indicated upon infection with the rhizobial strain 10/2/3.

Fig. 4. Electropherogram of the leghemoglobins produced in the indicated species of the pea cross-inoculation group, upon infection with the rhizobial strain Lo1.

In Table I are recorded the results of inoculation of six members of the pea cross-inoculation group[2]. The strain employed, L01, was isolated from *Lathyrus odoratus*, grown in local soil. As indicated in Fig. 4, the four plant species that were effectively nodulated by L01 (Table I) gave separate and distinct leghemoglobins in the electropherograms.

DISCUSSION

Our results are indicative of the plant being the determinant for the leghemoglobin type(s) appearing in a given effective legume–rhizobium symbiosis. The same electrophoretic pattern is always produced by a given plant species, even when the infecting rhizobial strains have originated in widely separated areas (Fig. 1). When a single rhizobial strain is able to form effective nodules on more than one plant species, the electropherograms are plant-type specific (Fig. 2–4).

The most straightforward explanation for these results is that the plant symbiont harbors the genetic information determining the primary structure for apoleghemoglobin. An alternative explanation that the plant symbiont is selecting "optimal" leghemoglobins—leghemoglobin being actually coded for by the bacterial genome—seems unlikely in view of the results in Fig. 4. With this hypothesis, Strain L01 would have to code for at least 10 leghemoglobins (this being the number of different leghemoglobins bands on the whole electropherogram of Fig. 4).

Our results confirm those of DILWORTH[4], and in the light of our recent findings[3], we are lead to postulate that leghemoglobin is a true product of the symbiosis; the heme moiety is synthesized by the rhizobia, while the primary structure of the globin is determined by the plant genome.

ACKNOWLEDGMENTS

We wish to thank the following persons for the gifts of rhizobial strains: Dr. Ura Mae Means, U.S. Soil Laboratory, U.S. Department of Agriculture, Beltsville, Md. for 3I1b125 and 3I1b138; Dr. J. C. Burton, Director of Research, The Nitragin Company, Milwaukee, Wisc. for 61A76; Professor O. N. Allen, Department of Bacteriology, University of Wisconsin, Madison, Wisc. for 505.

This work was supported in part by a National Institute of Health Grant AM08250.

REFERENCES

1 E. MOUSTAFA, *Nature*, 199 (1963) 1189.
2 E. B. FRED, I. L. BALDWIN AND E. MCCOY, *Root Nodule Bacteria and Leguminous Plants*, University of Wisconsin Press, Madison, Wisc., 1932.
3 J. A. CUTTING AND H. M. SCHULMAN, *Biochim. Biophys. Acta*, 192 (1969) 486.
4 M. J. DILWORTH, *Biochim. Biophys. Acta*, 184 (1969) 432.
5 P. AKROYD, *Anal. Biochem.*, 19 (1967) 399.
6 B. J. DAVIS, *Ann. N.Y. Acad. Sci.*, 121 (1964) 404.
7 J. A. OWEN, H. J. SILBERMAN AND C. GOTT, *Nature*, 182 (1958) 1373.

Rapid Induction of Cellular Strain Specificity by Newly Acquired Cytoplasmic Components in Amoebas

I. J. Lorch and K. W. Jeon

We previously reported bacterial endosymbionts of *Amoeba proteus* that changed from parasites (*1*) to required cytoplasmic components within a few years (*2*). These symbiotic bacteria are capable of infecting other amoebas and of causing the latter to become dependent on the symbionts in 200 cell generations (18 months) (*3*). We now report that within 4 weeks (10 to 15 generations) of infection with the symbiotic bacteria, host amoebas may become genetically distinct from the original strain.

Amoebas do not reproduce sexually, and the only way to test genetic compatibility between two strains is through nuclear transplantation. Exchange of nuclei between amoebas of the same strain (homotransplants) results in the formation of viable cells, whereas heterotransplants (that is, between amoebas of different strains) are nonviable because of nuclear-cytoplasmic incompatibility (*4*). Heterologous nuclei also exert a strong lethal effect when implanted into amoebas of different strains (*5*) or genera (*6*). Thus, when a heterokaryon is produced from different strains, both donor and host nuclei become nonviable within minutes, in contrast to homokaryons which produce viable clones. The lethal effect of heterologous nuclei has been attributed to a strain-specific lethal factor (*5*), which consists of high-molecular-weight proteins (*7*). Thus an internuclear lethal effect is one criterion for determining cell variation and strain specificity.

Taking advantage of the specificity of this phenomenon, we studied the role of newly acquired cell components, namely, endosymbionts of the xD strain of *A. proteus* and their ability to effect permanent cell changes in their hosts. The xD strain arose in 1966 after spontaneous infection of the D strain with rod-shaped X bacteria (*1*). The xD strain now requires these endosymbionts and its nuclei are not viable in their absence (*2*). Our studies were designed to determine (i) if the D and xD strains are sufficiently different to display nuclear incompatibility, (ii) how many cell generations are required before such incompatibility is expressed, and (iii) whether the lethal effect is dependent on the continued presence of the endosymbiotic bacteria. Changes in compatibility between D and xD amoebas are easily studied, since symbiosis between D amoebas and X bacteria can be established at will (*8*). We report here that xD amoeba nuclei are unilaterally lethal to D amoebas (that is, D nuclei are nonlethal to xD amoebas). Furthermore, the introduction of endosymbionts into a D amoeba induces irreversible changes in the host progeny in 10 to 15 generations (about 4 weeks), and these changes render the nuclei of new xD amoebas lethal to D amoebas when they are transplanted into these cells.

First, to test mutual lethal effects, we inserted nuclei of one strain into amoebas of the other by means of a de Fonbrune micromanipulator on agar-coated slides (*9*). After 5 minutes we removed the nuclei micrurgically. In previous studies, this period was sufficient for heterologous nuclei to exert lethal effects (*5*). Homokaryons were studied concurrently. The amoebas were cultured singly in watch glasses (U.S. Bureau of Plant Industry; Thomas, Philadelphia) containing modified Chalkley's medium (*2*), and we observed the amoebas until they either formed viable clones (five or more divisions) or died. The amoebas were fed three times a week with *Tetrahymena* (*10*) and kept at 20°C.

As shown in Table 1, 84.4 percent of the D hosts that had contained xD nuclei for 5 minutes died, whereas only 14.5 percent of the corresponding homokaryons died. The lethal effect exerted by xD nuclei after only a 5-minute stay in D amoebas was somewhat weaker than that exhibited by those that remained in D cells for 24 hours or longer (not shown in the table). However, we decided to remove heterologous nuclei after 5 minutes because it became impossible to distinguish them from the host nuclei after this

Table 1. Strain specificity of D and xD amoebas, as determined by the lethal effects of their nuclei. The subscripts n and c represent the nucleus and cytoplasm, respectively.

Strain combination*	Number of cells studied	Number of cells dividing	Number of clones obtained	Lethal effect of nuclei (%)
$(xD_n)D_{nc}$	128	50	20	84.4
$(D_n)D_{nc}$	131	121	112	14.5
$(xD_n)xD_{nc}$	56	51	48	14.3
$(D_n)xD_{nc}$	71	69	66	7.1

*The parentheses denote that the donor nuclei were removed micrurgically from the host cells after 5 minutes. The host amoebas contained their own nuclei.

Table 2. Lethal effect of newly obtained xD amoeba nuclei on D amoebas. The prefix N indicates a newly established amoeba strain.

Strain combination*	Time in symbiosis	Number of cells studied	Number of cells dividing	Number of clones obtained	Lethal effect of nuclei (%)
$(NxD_n)D_{nc}$	4 weeks	107	33	4	96.3
$(NxD_n)D_{nc}$	9 weeks	38	13	6	84.2
$(xD_n)D_{nc}$	14 years	128	50	20	84.4

*The parentheses denote that the donor nuclei were removed micrurgically from the host cells after 5 minutes. The host amoebas contained their own nuclei.

Table 3. Compatibility between nuclei of newly obtained xD amoebas and D amoeba cytoplasm (back-transfer).

Strain combination*	Time in symbiosis	Number of cells studied	Number of cells dividing	Cells forming clones	
				Number	Percentage
D_nD_c		27	26	26	96.3
NxD_nD_c	4 weeks	40	15	2	5.0
NxD_nD_c	15 weeks	22	1	1	4.5
xD_nD_c	14 years	84	6	1	1.2

*The donor nuclei were transplanted into enucleated cytoplasm, and left in the latter.

time. Amoebas that failed to form viable clones showed symptoms of ensuing death [hybrid syndrome (4)]. In addition, their nuclei appeared abnormal because of clumping of nucleoli. The interval between the first appearance of these symptoms and cell death varied between 3 and 20 days. Over 85 percent of the progenies of $(D_n)D_{nc}$ homokaryons and $(D_n)xD_{nc}$ heterokaryons formed viable clones without showing these symptoms (Table 1), indicating that D nuclei had no adverse effect on either D or xD amoebas.

Next, we used newly infected amoebas (designated NxD) as the nuclear donors to determine how long it took for the nuclei of newly infected D amoebas to acquire the lethal effect. The majority of D nuclei became lethal to other D amoebas as early as 4 weeks after becoming infected with X bacteria (Table 2). Nuclei of xD amoebas in symbiosis for only 4 to 9 weeks exerted as strong a lethal effect as those from the original xD strain that had been in culture for 14 years.

We then asked whether nuclei from symbiont-depleted xD amoebas retained the lethal effect on D amoebas after losing their symbionts. Symbiotic bacteria disappeared if xD amoebas were kept for 7 days at 26.5°C (3), with amoeba nuclei assuming an abnormal appearance with clumped nucleoli (11). The symbiont-depleted amoebas were kept at 26.5°C for another 7 days before their nuclei were tested for lethal effects, to assure that the nuclei were free from the influence of symbiotic bacteria. Of the 44 D amoebas that had hosted heat-treated xD nuclei for 5 minutes, 20 divided once or more, but only five formed permanent clones (that is, 88.6 percent of the amoebas that had been exposed to xD nuclei died). Thus, xD nuclei that had been in the symbiont-free cytoplasm for 7 days retained their lethal effect on D amoebas, indicating that the nuclear change was permanent.

Since the nuclei of newly infected D amoebas became lethal to D amoebas in 4 weeks, it was of interest to see if these changed nuclei were still compatible with the original D cytoplasm. Thus, we back-transferred the nuclei of NxD amoebas that had been in symbiosis for 4 to 15 weeks into the cytoplasm of D amoebas (Table 3). Most of the amoebas with nuclear transplants (that is, NxD_nD_c) failed to form viable clones, indicating that the D nuclei became incompatible with D cytoplasm after reproducing in xD cytoplasm for 4 to 15 weeks. However, when symbionts were removed from these newly established xD amoebas by culturing them at 26.5°C, most of them continued to grow and form viable cultures, confirming earlier results (3). Thus, it appeared that the nuclei of newly established NxD_n amoebas did not become fully dependent on symbionts and were able to remain viable if the symbionts were removed gradually.

Our study of the interaction between D and xD strains of amoebas shows that (i) nuclei of xD amoebas exert a lethal effect when inserted into D cells, (ii) a newly established xD amoeba strain may acquire a lethal activity within 4 weeks, (iii) the xD nuclei retain a lethal effect even after their symbionts have been removed, and (iv) most of the nuclei of newly established xD amoebas are incompatible with the original D cytoplasm after a 4-week stay in cytoplasm containing X bacteria.

Establishment of the original xD strain of amoebas took several hundred cell generations (a few years) (2). Biologists who are used to regarding cellular changes on a long geologic time scale have considered this as an exceptional case of rapid symbiont integration (12). Our present data demonstrate that an even smaller number of generations is required for symbiont integration and cell divergence.

The mechanism whereby bacterial symbionts cause rapid changes in cellular strain specificity is not known, although the observed lethal effect of xD nuclei on D amoebas is similar in etiology to that previously described for other strains of amoebas (5). By inference, it is likely that xD amoebas produce a new protein—the lethal factor (7)—as a consequence of becoming symbiotic. Involvement of such a factor in the killing phenomenon is further supported by our study in which we were able to neutralize the lethal effect of xD nuclei by "washing" them for 5 minutes in the cytoplasm of D amoebas before testing their ability to kill intact D amoebas. Eleven out of 13 D amoebas that hosted washed xD nuclei for 5 minutes remained viable, whereas none of the D amoebas in which xD nuclei had been washed formed clones.

The new strain-specific lethal factor could be synthesized by xD amoebas either as a result of acquiring new DNA templates or by altered expression of existing chromosomal genes. Further work is needed to distinguish between these possibilities, but the latter mechanism would be simpler in that products of symbiotic bacteria could effect an alteration of gene expression of amoebas without involving any transfer of their own DNA's. Such progressive changes in nuclear synthetic activities caused by cytoplasm occurs regularly during embryonic development of metazoans.

References and Notes

1. K. W. Jeon and I. J. Lorch, *Exp. Cell Res.* **48**, 236 (1967).
2. K. W. Jeon and M. S. Jeon, *J. Cell. Physiol.* **89**, 337 (1976); K. W. Jeon, *Science* **176**, 1122 (1972).
3. K. W. Jeon and T. I. Ahn, *Science* **202**, 635 (1978).
4. K. W. Jeon and I. J. Lorch, in *The Biology of Amoeba*, K. W. Jeon, Ed. (Academic Press, New York, 1973), p. 549; I. J. Lorch and J. F. Danielli, *Q. J. Microsc. Sci.* **94**, 461 (1953).
5. K. W. Jeon and I. J. Lorch, *Exp. Cell Res.* **57**, 223 (1969); E. E. Makhlin, *Tsitologiya* **13**, 1020 (1971); *ibid.* **16**, 1406 (1974).
6. E. E. Makhlin and A. L. Yudin, *Tsitologiya* **11**, 744 (1969).
7. K. W. Jeon and I. J. Lorch, *J. Cell. Physiol.* **75**, 193 (1970).
8. T. I. Ahn and K. W. Jeon, *ibid.* **98**, 49 (1979).
9. K. W. Jeon and I. J. Lorch, *Nature (London)* **217**, 463 (1968); K. W. Jeon, *Methods Cell Physiol.* **4**, 179 (1970).
10. L. Goldstein and C. Ko, *Methods Cell Biol.* **13**, 239 (1976).
11. I. J. Lorch and K. W. Jeon, *J. Protozool.* **27**, 423 (1980).
12. L. Margulis, *Exp. Parasitol.* **39**, 277 (1976); D. C. Smith, *Proc. R. Soc. London Ser. B* **204**, 115 (1979); F. J. R. Taylor, *ibid.*, p. 267; J. M. Whatley, P. John, F. R. Whatley, *ibid.*, p. 165.
13. We thank A. M. Jungreis and W. S. Riggsby for their critical reading of the manuscript. Experimental work was carried out during I.J.L.'s leave of absence at the University of Tennessee. This work was supported by grant PCM7684382 from the National Science Foundation to K.W.J. and by a Canisius College faculty fellowship awarded to I.J.L.

25 July 1980; revised 5 November 1980

From extracellular to intracellular: the establishment of a symbiosis

By D. C. Smith, F.R.S.

Department of Botany, University of Bristol, Woodland Road, Bristol BS8 1UG, U.K.

The colonization of host cells by modern symbionts is surveyed. The morphological distinction between extracellular and intracellular symbionts is not sharp, and the various kinds of association can be arranged in a graded series of increasing morphological integration of the symbiont into the host cell. Apart from some aggressive parasitic infections, the great majority of symbionts are enclosed by a host membrane in a vacuole. Those not enclosed in a host vacuole usually cannot be cultivated outside the cell. It is therefore surmised that encirclement by a vacuolar membrane would only disappear, if at all, in the later stages of the evolution of intracellular symbiosis.

Recognition mechanisms between host and symbiont occur, but have been little studied. In some associations, recognition at surface contact occurs, and there is evidence for the involvement of lectins in certain cases. In other associations, recognition may occur wholly or in part after the entry of symbiont into host cells. After entry, special mechanisms for the biotrophic transfer of nutrients from symbiont to host develop. Both the symbiont population size and its rate of increase are strictly regulated by the host cell; symbiont metabolism may be controlled likewise. Rates of evolution of intracellular symbionts are probably very rapid, owing in part to responses of the host cell to its symbiont.

Introduction

The concept of symbiosis most appropriate both to this paper and to this symposium is that based upon an ecological approach. This may be explained as follows. There are three main habitats for living organisms in the world: land, water and other living organisms. Any organism whose normal habitat for part or all of its life is another living organism can be described as a symbiont, using the term in the original broad sense of De Bary. This paper, and this symposium, are concerned with the particular case where it is individual cells which become colonized by symbionts.

If such colonization is to the selective disadvantage of the host, it is usually called parasitism. If it is of advantage, it is often called mutualism; the benefit to the host is generally clear and obvious, but the benefit to the colonizing microbe is sometimes no more than that of an extension or change of habitat. However, whether initially mutualistic or parasitic, some of the processes and principles involved in the intracellular colonization and existence of symbionts are similar.

Furthermore, parasitic as well as mutualistic associations can evolve into a state where there is a stable and permanent relation between symbiont and host. Indeed, evolutionary processes can lead to such a degree of morphological modification and integration of the symbiont into the cellular habitat that it becomes no longer easily recognizable as a foreign intrusion. Such phenomena raise the question of the extent to which existing cell structures are relics of ancient colonizations and, indeed, the extent to which originally foreign DNA has become an integral part of the host cell genetic machinery.

This paper considers the beginnings rather than the end products of these evolutionary processes. All organisms or relics of organisms currently within living cells have had, at some previous evolutionary stage, an existence outside the cell. The purpose of this paper is to explore the problems involved in the initial period of colonization, and the sequence of events that occurs when an organism whose existence was once entirely extracellular evolves to a stage where all or part of its existence is intracellular. It will be concerned with associations that are stable rather than transient and temporary.

This evolutionary progression cannot be followed through for any one symbiosis. As with so many kinds of phylogenetic study, a variety of existing symbiotic associations showing differing degrees of colonization have to be considered, and then from such disparate and fragmentary evidence, determine whether any coherent picture or set of general principles can be discerned.

MORPHOLOGICAL CONSIDERATIONS
What is intracellular?

The morphological relations between a symbiont and its host are various. Modern symbiotic associations can be arranged into a graded series of increasing morphological integration with the host (figure 1). There need be no sharp transition from the external environment to intracellular existence, and various intermediate steps may occur such as existence within a host organ surrounded by, but not within, cells. Even when the relation between an individual symbiont and host cell is considered, the distinctions between extracellular and intracellular may be blurred.

To illustrate this, the specific example of relations of symbiotic algae to their various types of host cells (figure 2) will be considered in some detail. At one extreme, lichens present an example of a genuine extracellular relation. There is normally no penetration of the alga by the fungus, except in a very few species. Collins & Farrar (1978) calculate that, in the typical and common lichen, *Xanthoria parietina*, just over 20 % of the algal cell surface is in contact with fungal hyphae. They give detailed calculations to show that this degree of contact is sufficient for the observed mass transfer of nutrients to occur between the symbionts by diffusion.

The stage where the alga becomes completely surrounded by host cells and in

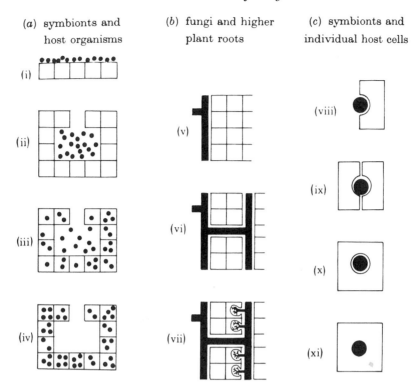

FIGURE 1. Types of morphological integration between symbionts and hosts. (a) Symbionts (black circles) and host organisms (squares): (i) symbionts entirely on surface (e.g. epiphytic or epizootic microbes); (ii) symbionts within organism (e.g. digestive tract microorganisms such as rumen microbes); (iii) symbionts within organism, partly extracellular, partly intracellular, as occurs with blue-green algae in tropical marine sponges (Wilkinson 1978); (iv) symbionts entirely intracellular. (b) Fungal hyphae and higher plant roots: (v) rhizosphere fungi which do not penetrate roots; (vi) ectotrophic mycorrhizal fungi of forest trees, which penetrate between but not into root cells; (vii) fungi which penetrate the root and into cells by haustoria as in vesicular/arbuscular mycorrhizas; the haustoria are enclosed by a host membrane. (c) Symboints and individual host cells (see figure 2 for a more detailed example): (viii) symbionts in close physical contact with surface of host cells, but also in contact with external environment (e.g. lichens, *Prochloron* on ascidians); (ix) symbionts enclosed by host cells, but not within them; (x) symbiont within host cells, but enclosed by a host vacuole (most intracellular symbionts); (xi) symbionts lying free in cytoplasm (e.g. endonuclear bacteria).

such total contact with them as to be virtually another cell of the tissue is represented by some acoelous turbellarians. The simplest situation occurs in *Amphiscolops langerhansii* which contains the symbiont *Amphidinium* (Taylor 1971); the alga remains recognizable as such, and even retains its flagella. In another flatworm, *Convoluta roscoffensis*, relations are more difficult to discern. The algal symbiont, *Platymonas*, which can exist as a free-living flagellate unicell in the phytoplankton, loses its flagella and cell wall when it enters symbiosis, and becomes a naked, membrane-bound cell. Some ultrastructural observations suggest that it

118   D. C. Smith (Discussion Meeting)

is intracellular (Dorey 1965), while others maintain that it is extracellular but presents a confusing picture because it interdigitates extensively between the cells of the worm (Oschman 1966).

Unequivocal examples of algae enclosed within cells are shown by the *Chlorella* symbionts of green hydra and *Paramecium bursaria*. Here, the algae are each enclosed in a vacuole and so are sharply segregated from the rest of the cellular environment. After division of the algal symbionts, each daughter cell becomes

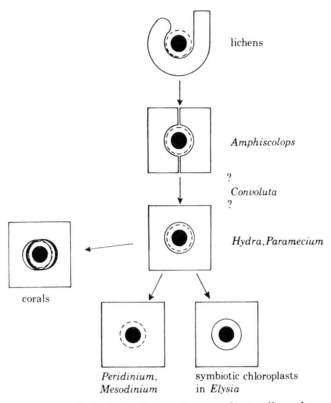

FIGURE 2. Types of morphological relations between host cells and symbiotic algae. The names on the diagram are of representative hosts at each stage. Solid lines, host cell; ○, outer limiting wall or membrane of alga; ●, chloroplast.

sequestered into its own vacuole, and at no stage do the algae ever come into contact with host cell cytoplasm.

The question then arises of whether it is normal for all other algal symbionts to be segregated from the rest of the cellular habitat by a host membrane. There are at least two problems in trying to answer this question from the wealth of ultrastructural literature. The first arises in studying dinoflagellate symbionts, by far the commonest algal symbionts of marine invertebrates, and universally present, e.g. in all reef-building corals. When examined by electron microscopy, the symbionts appear bounded by a complex of membranes, and it is difficult to

The establishment of a symbiosis

disentangle whether any or all of the encircling membranes are of host origin. Recently, Duclaux (1977) made developmental observations suggesting that most of the encircling membranes are in fact of host origin, and appear to be progressively formed as if in response to symbiont presence.

The second problem is that, in some groups of lower organisms, satisfactory fixation for electron microscopy is sometimes difficult, so that it becomes impossible to determine if a host vacuole is present. However, whenever adequate techniques have been used, symbiotic algae are almost always contained within a host vacuole. This is found even in some of the more extreme situations where there has been controversy over whether intracellular structures are chloroplasts or endosymbiotic blue-green algae. One such example is *Cyanophora paradoxa*, which is believed by some to be a colourless flagellate alga containing from two to four symbiotic blue-green algae. Others (Stanier & Cohen-Bazire 1977) consider these objects to be chloroplasts, since their genome size, of molecular mass 10^8, is only 5–10 % of the smallest of the free-living unicellular blue-green algae. A recent ultrastructural study by Trench *et al.* (1978) shows quite clearly that these structures, be they 'semi-autonomous symbionts' or 'semi-autonomous organelles', are enclosed in a distinct host vacuole and, indeed, they have a detectable prokaryotic cell wall.

Situations where encircling host membranes are absent are very rare, but do occur. Taylor (1979, this volume) shows clearly that the cryptic eukaryotic symbionts in dinoflagellates such as *Peridinium balticum* are not within host vacuoles and lie free in the cytoplasm.

The extreme situation is where it is just the chloroplast and not the whole cell that becomes symbiotic. For example, a number of elysioid sacoglossan molluscs possess symbiotic chloroplasts within their digestive cells. The chloroplasts are acquired during normal feeding from their host food plants, and may remain actively photosynthetic for up to 8–12 weeks after entry into the cells (Hinde & Smith 1972), their photosynthesis making a quantitatively important contribution to host animal nutrition (Hinde & Smith 1975). Preliminary studies by Trench *et al.* (1973) suggested at first that, while some chloroplasts in the digestive cell were bounded by an animal membrane, it was equally common for others to lie free in the cytoplasm. A more detailed and recent study by C. R. Hawes (unpublished) shows, however, that the great majority of chloroplasts are bounded by a host membrane, and only a small minority are not.

Nearly all algal symbionts are thus segregated from the rest of the host cell by a membrane, and this raises the question of whether this is true of intracellular microbial symbionts in general. In so far as bacteria are concerned, a study of those instances where satisfactory fixation procedures have been used suggests that the following generalization can be made. Symbionts are always enclosed in host membranes if: (*a*) their relation to the host cell has become stabilized (as distinct from, e.g. an agressive parasitic infection), and (*b*) they retain the ability or at least the potential for independent growth. Many symbionts falling outside these categories may also be enclosed by host membranes as outlined by the extensive review on host–parasite interfaces by Bracker & Littlefield (1973), but it is not universal. For example, some of the endosymbionts in *Paramecium aurelia*, such as kappa and other particles (now believed to be the relics of gram-negative bacteria; cf. Preer *et al.* 1974) lie free in the cytoplasm without an enclosing membrane. The only particle which can be cultured extracellularly for long periods of time is lambda; yet lambda (and its close relative, sigma) are the only ones enclosed in host vacuoles. Endonuclear bacteria represent another category not enclosed by host vacuoles.

It is not known why the enclosure of stable endosymbionts by host membranes should be so widespread. The symbiont might get some advantage in being protected from destructive host enzymes. In so far as the host is concerned, sequestration of symbionts into vacuoles may offer some form of control, and would certainly enable restriction of symbiont demands upon nutrient pools.

Viruses have not been mentioned so far, since they are outside the scope of this paper. However, it is interesting to note that viruses in animal cells often become coated with host membrane-like material, with simply the nucleic acid passing free into the cytoplasm. In plant cells, by contrast, they tend to lie free in the cytoplasm.

Modes of entry to host cell

It may be concluded that, in the early stages of the evolution of a permanent stable intracellular symbiosis, the symbiont would almost always be enclosed in a host vacuole, the latter disappearing, if at all, only in the later stages of evolution. Relevant to the origin of these vacuoles is the question of how symbionts originally entered their host cells. It is consequently useful to consider mechanisms of entry by those modern symbionts which can exist both inside and outside the host cell. There is a clear distinction between animal and plant host cells.

Animal cells lack a cell wall, and there are three possible modes of entry. (*a*) Passive entry via the normal phagocytotic processes of digestion so that the symbiont becomes enclosed in a vacuole as it enters. This method is adopted by many mutualistic symbionts and explains their common occurrence in digestive cells. (*b*) Induction of phagocytosis or engulfment by the host cell, so that the symbiont induces the formation of a vacuole around it. An example of this is induction of parasitophorous vacuoles formed during the invasion of red cells of the blood by *Plasmodium*, described by Bannister (1979, this volume). (*c*) Active penetration of the host cell and its membrane by the symbiont which comes to lie free in the cytoplasm. However, the host cell usually dies and stable associations rarely result.

The fact that two of these modes of entry involve the initial engulfment of the symbiont in a vacuole helps to explain how symbionts become enclosed in host membranes.

The plant cell presents different problems for the entry of the symbiont.

Phagocytosis is absent, and contact with cell membrane blocked by the cell wall. This may partly explain why there are fewer stable intracellular symbionts in plants. Entry to the host cell has to involve some method of crossing the barrier formed by the outer wall. In higher plants, the only natural gaps in this barrier are plasmodesmata, but these are small and do not usually permit the passage of organisms larger than viruses. Some symbionts enter through breaches caused by injury or attack by organisms such as insects, while others actively disrupt the wall by chemical or physical means. Having penetrated the wall, the symbiont or its intrusions (such as haustoria) are nevertheless again normally surrounded by a host cell membrane which persists except where aggressive infections lead to death.

In some types of mutualistic association the host cell itself participates in transporting symbionts across the wall. As described by Beringer *et al.* (1979, this volume), *Rhizobium* bacteria enter legume roots along an 'infection thread', a tube formed by self-invagination and inward growth of host cell walls. A similar process appears to occur during the formation of nitrogen-fixing nodules in non-leguminous plants, where the symbionts are presumed to be actinomycetes (Strand & Laetsch 1977). In both cases, the symbionts become confined within vacuoles in host cells.

After entry of symbionts into the host cell, whether animal or plant, such morphological changes as occur tend towards simplification, and include suppression of unnecessary structures such as flagellae, as well as reduction in cell walls.

RECOGNITION

For any organism that has a dual existence inside and outside a cell, there are recognition mechanisms between symbiont and host. In parasitic associations, it is usually a mechanism for symbiont recognizing host, but for mutualistic associations the ability of the host cell to recognize is also involved, sometimes as the major mechanism.

The problem of cell recognition currently attracts much attention, but the bulk of published work relates to recognition of 'like-for-like', mating type for mating type, or gamete for gamete, between cells of the same species. The problem in symbiotic recognition is that of how a cell recognizes something specifically and very *unlike* itself. This leads to the question of whether the mechanisms are nevertheless similar. Although recognition mechanisms of 'like-for-like' are not completely understood, it is clear that molecules such as glycoproteins, lectins, antigens, etc. play a key role, and that physical contact between cell surfaces is also important. For some symbiotic associations, there is indeed good evidence for the occurrence of such mechanisms. Plant lectins and bacterial antigens are probably involved in the recognition between legume roots and *Rhizobium* bacteria (Beringer *et al.* 1979, this volume). Lockhart, Rowell & Stewart (1978) have recently shown that lectins isolated from lichen fungi bind to blue-green algae that are potential lichen symbionts, but not to those which are not. The establishment of almost any symbiosis involved a stage of physical contact, even some positive adhesion of symbiont to host. This is manifest both in the attachment of some protozoan parasites to their hosts (Bannister 1977) and in the attachment of spirochaetes to protozoa (Margulis et al. 1979, this volume). It is thus reasonable to assume that those cell surface molecules capable of binding and of showing fine specific differences might be involved in recognition phenomena.

However, other types of mechanism may be involved in certain associations. Where symbionts enter host cells by the normal processes of digestion and phagocytosis, problems would arise if, at the initial contact phase, there was a mechanism that stopped phagocytosis in general if recognition did not occur. Here, either there is a recognition at first contact which does not affect entry but signals the host cell to accept the symbiont after phagocytosis, or the recognition mechanism operates after phagocytosis and might have to be by diffusible signals rather than physical contact since the symbiont would be inside a vacuole. Muscatine & Pool (1979, this volume) discuss the problem for recognition of *Chlorella* by *Hydra*.

Specificity

Whatever the type of recognition mechanisms, the level of specificity seems higher for intracellular than for extracellular associations. Indeed, for almost any group of microorganisms, the number of species which are successful intracellular symbionts is always much less than those which are extracellular. Thus, various genera of nitrogen-fixing bacteria may occur on the root surface in the rhizosphere, but only *Rhizobium* can occur intracellularly. Many genera of algae can be extracellular symbionts, but only *Chlorella* in fresh water and *Gymnodinium* in marine habitats are at all common intracellularly. The fungi of ectotrophic mycorrhizas are distributed among a range of genera from several families of fungi; in vesicular–arbuscular mycorrhizae, where host cells are penetrated, they are largely confined to four or five genera in the single family *Endogonaceae*. This general phenomenon as it applies to parasites is discussed in detail by Moulder (1979, this volume).

NUTRIENT TRANSFER

A characteristic feature of symbiosis is the movement of substances between host and symbiont in one or both directions (figure 3). The only type of movement to have received appreciable study at the molecular level is movement from symbiont to host. Most evidence relates to movement of fixed carbon from photosynthetic endosymbionts, a process which has been studied in most of the associations shown in figure 2, and where the following generalizations have been found to apply (Smith 1974). (*a*) Movement of photosynthate from symbiont to host is substantial, and typically involves 40–80 % of the photosynthetically fixed carbon. (*b*) Only one or a few types of molecule move and this has the presumed advantage to the symbiont of regulating efflux because the number of channels through which

The establishment of a symbiosis

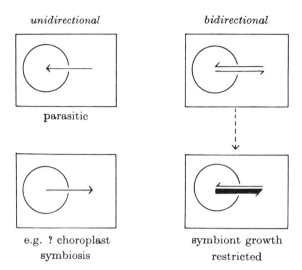

FIGURE 3. Types of nutrient movement in intracellular symbiosis. Symbionts are shown as circles, and host cells as rectangles; arrows indicate direction of nutrient flow. Symbiotic chloroplasts cannot grow and divide in host cells, so are shown as releasing, but not taking up nutrients.

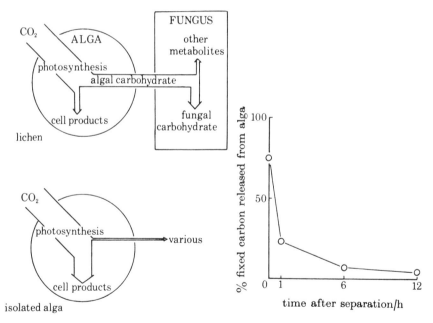

FIGURE 4. Effects of symbiosis on lichen algae. After separation from the fungus, the release of fixed carbon from the alga (normally as a single, simple carbohydrate) declines sharply. The graph, showing decline in release of fixed carbon from algae after isolation from lichen, refers to isolation of *Coccomyxa* from the lichen *Peltigera aphthosa*, and is based on the data of Green (1970).

export can occur is restricted. (c) The mobile molecules are of three general kinds: glycerol or polyhydric alcohols, neutral amino acids, and the sugars, glucose and maltose. (d) Release of mobile molecules ceases soon after isolation of the symbionts from symbiosis; the situation in lichens is summarized in figure 4. For all symbionts, the fact that release ceases soon after isolation illustrates the dangers of assuming that symbiont behaviour in culture is the same as in symbiosis.

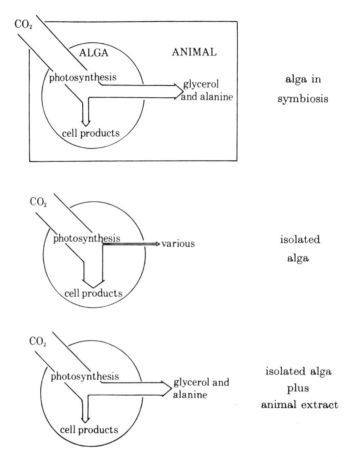

FIGURE 5. Effects of symbiosis on nutrient release by symbiotic algae in marine invertebrates. Within host cells, the algae release substantial amounts of fixed carbon, predominantly glycerol and alanine. This release declines on isolation but can be restored by adding an aqueous extract of host animal material (see Trench 1971; Muscatine et al. 1972).

The question therefore arises of how existence in symbiosis induces the massive efflux of photosynthate from algae. In hosts which are marine invertebrates, release-stimulating 'factors' are produced by host cells in response to the presence of symbionts (Muscatine et al. 1972; Gallop 1974; Smith 1974) (figure 5). These 'factors' are water soluble and thermolabile, but remain otherwise uncharacterized. Such 'factors' have not been detected in lichens, where it is possible that the

The establishment of a symbiosis

stimulus might be physical rather than chemical. It is not known to what extent the massive efflux results from changes in membrane transport or changes in internal metabolism which generates a large accumulation of mobile molecules at the sites of the export.

In animal cells, symbiotic algae may also take up nitrogenous waste products, converting them to amino acids which are then released back to the animal; *Convoluta roscoffensis* (figure 6) is a good example of this. Growth factors such as

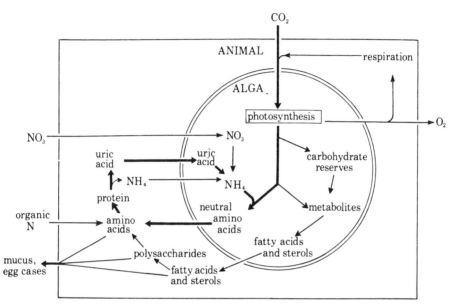

FIGURE 6. Recycling of nitrogen by the algal symbiont of *Convoluta roscoffensis*. The main pathways are indicated by the thicker arrows. The principal nitrogenous waste product of the animal is uric acid. This is taken up and broken down to ammonia by the alga (*Platymonas convoluta*), and is then used to form amino acids. Neutral amino acids (especially glutamine, but also glycine, serine and alanine) are released back to the animal. For further details, see Boyle & Smith (1975) and Holligan & Gooday (1975).

sterols may also be supplied to the animal. Thus, there may be an efflux of a number of types of molecule from the symbiont; it is tempting to assume that the same fundamental mechanism underlies the release of each of them, but there is no proof of this. There is also movement in the opposite direction, into the symbiont, of nutrients essential for its survival and growth. There is no evidence that such movement from the host is unduly massive, and indeed for most organic nutrients there may be competition between symbiont and host.

For all types of intracellular symbiont including algae, there is therefore likely to be a complex though well regulated bidirectional flux of substances between it and the host cell. For permanent intracellular inhabitants, it is easy to visualize the development of special transport mechanisms at the interface with the host.

REGULATION

An essential feature of the evolution of a stable intracellular symbiosis is the development of mechanisms for the control and regulation of the symbionts by the host cell. For parasitic symbionts, host control mechanisms are largely those of digestion, destruction or ejection of invaders. For mutualistic symbionts the problems are more complex. Host cell mechanisms for repelling or destroying unwanted invaders may still exist, and have to be overcome. The symbiont may either repress them (cf. *Chlorella* in *Paramecium bursaria*; Karakashian & Karakashian 1973), resist them (as is probable for symbiotic chloroplasts; Trench *et al.* 1973) or be moved away from them (as in green *Hydra*; see Muscatine *et al.* 1979, this volume).

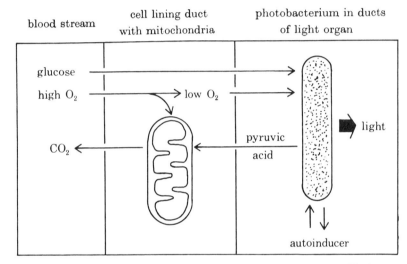

FIGURE 7. Hypothetical model of the light organ of a fish (after Ruby & Nealson 1976). The light organs of fish such as *Monocentris japonica* comprise tubular ducts densely packed with a single species of bacterium, *Photobacterium fischeri*. Luciferase synthesis occurs when an autoinducer secreted by the bacterium reaches a critical concentration in the confined space of the duct. Light emission increases at low oxygen levels, thought to arise by the excretion of pyruvic acid by the bacterium (cultured bacteria release pyruvate when grown on glucose, present in the bloodstream). Pyruvate increases mitochondrial oxygen consumption in cells lining the duct.

As well as overcoming host defence mechanisms, growth and division of the symbiont must be regulated, for at least three reasons: (*a*) to prevent the symbiont outgrowing host cell and perhaps bursting it; (*b*) to reduce the pressure of competitive demands between the symbiont and the rest of the cell for scarce nutrients; and (*c*) to increase the flow of substances from the symbiont (i.e. substances that might otherwise be devoted to symbiont growth). Indeed, in all cases of stable associations where symbionts can also live outside cells, maximum recorded growth rates of symbionts within cells is always less than that outside cells. So far, it is not known for any association how symbiont division is controlled. In green *Hydra*, Muscatine *et al.* (1979, this volume) speculate that limitation of nutrient supply may be involved, and McAuley (1979) has demonstrated a close linkage between symbiont and host cell mitosis.

In so far as regulation of symbiont metabolism is concerned, the induction of nutrient release has already been discussed. There is also the problem of the development of the optimum micro-environment within the cell for the symbiont to function to the best advantage of the host, and nitrogen-fixing symbionts are a good example of this. The process of nitrogen fixation requires much energy, and the enzyme responsible, nitrogenase, is sensitive to oxygen. The creation of optimum conditions requires that the host can simultaneously supply energy and maintain lower oxygen tensions. Beringer *et al.* (1979, this volume) describe how this is achieved by the legume nodule.

Another example is presented by the light-emitting organs of certain marine fish. The emission of light is due to luminous bacteria, present as a more or less pure culture in parallel tubular ducts of the organ. The luminous bacteria also occur living free in seawater, but display scarcely any luminosity. Ruby & Nealson (1978) have studied the biochemistry of the symbiotic interactions between *Photobacterium fischeri* and the host fish *Monocentris japonica*, and produced the hypothetical scheme shown in figure 7. An autoinducer is excreted by the symbiont and, upon reaching a critical concentration induces luciferase synthesis. The confined space and dense culture promotes the effect of the autoinducer. In culture, light production increases at low oxygen levels; in symbiosis, the bacteria are believed to excrete pyruvate which is respired by the mitochondria in the cells lining the duct to give a consequent lowering of oxygen tension.

RATES OF EVOLUTION

Unlike non-living habitats, living habitats such as the cell manifest their own dramatic reactions to colonization by symbiotic microbes. Consequently, the colonizing microbe is subject to greatly increased selection pressure, perhaps analogous to that originally noticed by Darwin for plants and animals under domestication. A recent observation by Jeon & Jeon (1976) of a naturally occurring bacterial infection of a laboratory strain of *Amoeba proteus* illustrates the speed of such symbiotic selection. The infection was originally harmful to the amoebae in that they grew more slowly and showed other adverse effects of infection, but after only 1000 generations the harmful effects had disappeared and the amoebae had become obligately dependent upon the bacteria.

As the sometimes rapid integration of a symbiont into the host cell evolves, it has a number of consequences. The more closely the symbiont adapts to a cellular existence, the less easy it becomes to culture outside the cell. If such unculturability is accompanied by the marked morphological reductions which often occur, then the symbiont becomes no longer identifiable. Indeed, the number of such partly or completely unidentified foreign objects in cells is appreciable (table 1). The universality of the genetic code presumably means that there will be a tendency for unnecessary duplication of DNA to be eliminated, with a consequential transfer of genetic control between symmbiont and host. The fact that segments of DNA can be transferred to or from the main genetic store of the host cell nucleus can make the origin of cytoplasmic DNA even more obscure.

TABLE 1. EXAMPLES OF ENDOSYMBIONTS OF UNKNOWN OR TENTATIVE IDENTITY

Blochman bodies in insect mycetomes
gamma particles in *Blastocladiella*
'DNA-containing bodies' in *Amoeba*
hydrogenosomes of anaerobic trichomonads (?bacteria)
Paramecium particles: kappa, pi, mu, lambda, sigma, gamma, tau, delta, nu, alpha (?gram negative bacteria)
Euplotes particles: omikron, epsilon (?bacteria)
endophyte of non-legume nodules (?actinomycete)
binucleate dinoflagellates (?chrysophyte)

This is where the study of the cell as a habitat becomes much more confusing than the study of non-living habitats. In non-living habitats, an organism either exists or it does not. In the cell habitat, an invading organism can progressively lose pieces of itself, slowly blending into the general background, its former existence betrayed only by some relic. Indeed, one is reminded of Alice in Wonderland's encounter with the Cheshire Cat. As she watched it, 'it vanished quite slowly, beginning with the tail, and ending with the grin, which remained some time after the rest of it had gone'. There are a number of objects in a cell like the grin of the Cheshire Cat. For those who try to trace their origin, the grin is challenging and truly enigmatic.

REFERENCES (Smith)

Bannister, L. H. 1977 The invasion of red cells by *Plasmodium*. In *Symposia of the British Society for Parasitology* no. 15 (ed. A. E. R. Taylor & R. Muller), pp. 27–55. Oxford: Blackwell ScientificPublications.
Bannister, L. H. 1979 The interactions of intracellular Protista and their host cells, with special reference to heterotrophic organisms. *Proc. R. Soc. Lond.* B **204**, 141–163 (this volume).
Beringer, J. E., Brewin, N., Johnston, A. W. B., Schulman, H. M. & Hopwood, D. A. 1979 The *Rhizobium*–legume symbiosis. *Proc. R. Soc. Lond.* B **204**, 219–233 (this volume).
Boyle, J. E. & Smith, D. C. 1975 Biochemical interactions between the symbionts of *Convoluta roscoffensis*. *Proc. R. Soc. Lond.* B **189**, 121–135.
Bracker, C. E. & Littlefield, L. J. 1973 Structural concepts of host–pathogen interfaces. In *Fungal pathogenicity and the plant's response* (ed. R. J. W. Byrde & C. V. Cutting), pp. 159–318. London and New York: Academic Press.
Collins, C. R. & Farrar, J. F. 1978 Structural resistances to mass transfer in the lichen *Xanthoria parietina*. *New Phytol.* **81**, 71–83.
Dorey, A. E. 1965 The organization and replacement of the epidermis in acoelous turbellarians. *Q. J. micr. Sci.*, **106**, 147–172.

Duclaux, G. N. 1977 Recherches sur quelques associations symbiotique d'algues et de metazoaires. Thèse de Doctorat d'État ès Sciences Naturelles, L'Université Pierre et Marie Curie, Paris.

Gallop, A. 1974 Evidence for the presence of a 'factor' in *Elysia viridis* which stimulates photosynthate release from its symbiotic chloroplasts. *New Phytol.* **73**, 1111–1117.

Green, T. G. A. 1970 The biology of lichen symbionts. D.Phil. thesis, University of Oxford.

Hinde, R. & Smith, D. C. 1972 Persistence of functional chloroplasts in *Elysia viridis* (Opisthobranchia, Sacoglossa). *Nature, new Biol.* **239**, 30–31.

Hinde, R. & Smith, D. C. 1975 The role of photosynthesis in the nutrition of the mollusc *Elysia viridis*. *Biol. J. Linn. Soc.* **7**, 161–171.

Holligan, P. M. & Gooday, G. W. 1975 Symbiosis in *Convoluta roscoffensis*. In *Symbiosis* (29th Symp. Soc. exp. Biol.) (ed. D. M. Jennings & D. L. Lee), pp. 205–227. Cambridge University Press.

Jeon, K. W. & Jeon, M. S. 1976 Endosymbiosis in amoebae: recently established endosymbionts have become required cytoplasmic components. *J. Cell Physiol.* **89**, 337–344.

Karakashian, M. W. & Karakashian, S. J. 1973 Intracellular digestion and symbiosis in *Paramecium bursaria*. *Expl Cell Res.* **81**, 111–119.

Lockhart, C. M., Rowell, P. & Stewart, W. D. P. 1978 Phytohaemagglutinins from the nitrogen-fixing lichens *Peltigera canina* and *P. polydactyla*. *FEMS Microbiol. Lett.* **3**, 127–130.

McAuley, P. J. 1979 Regulation in the green hydra symbiosis. Ph.D. thesis, Bristol University.

Margulis, L., Chase, D. & To, L. 1979 Possible evolutionary significance of spirochaetes. *Proc. R. Soc. Lond.* B **204**, 189–198 (this volume).

Moulder, J. W. 1974 Intracellular parasitism: life in an extreme environment. *J. infect. Dis.* **130**, 300–306.

Moulder, J. W. 1979 The cell as an extreme environment. *Proc. R. Soc. Lond.* B **204**, 199–210 (this volume).

Muscatine, L. & Pool, R. R. 1979 Regulation of numbers of intracellular algae. *Proc. R. Soc. Lond.* B **204**, 131–139 (this volume).

Muscatine, L., Pool, R. R. & Cernichiari, E. 1972 Some factors affecting selective release of soluble organic material by zooxanthellae from reef corals. *Mar. Biol.* **13**, 298–308.

Oschman, J. L 1966 Development of the symbiosis of *Convoluta roscoffensis* Graff and *Platymonas* sp. *J. Phycol.* **2**, 105–111.

Preer, J. R., Preer, L. B. & Jurand, A. 1974 Kappa and other endosymbionts in *Paramecium aurelia*. *Bact. Rev.* **38**, 113–163.

Ruby, E. G. & Nealson, K. H. 1976 Symbiotic association of *Photobacterium fischeri* with the marine luminous fish *Monocentris japonica*: a model of symbiosis based on bacterial studies. *Biol. Bull.* **151**, 574–586.

Smith, D. C. 1974 Transport from symbiotic algae and symbiotic chloroplasts to host cells. In *Transport at the cellular level* (28th Symp. Soc. exp. Biol.) (ed. M. A. Sleigh & D. M. Jennings), pp. 485–520. Cambridge University Press.

Stanier, R. Y. & Cohen-Bazire, G. 1977 Phototrophic prokaryotes: the cyanobacteria. *A. Rev. Microbiol.* **31**, 225–74.

Strand, R. & Laetsch, W. M. 1977 Cell and endophyte structure of the nitrogen fixing root nodules of *Ceanothus integerrimus*. II. Progress of the endophyte into young cells of the growing nodule. *Protoplasma* **93**, 179–90.

Taylor, D. L. 1971 On the symbiosis between *Amphidinium klebsii* (Dinophyceae) and *Amphiscolops langerhansi* (Turbellaria: Acoela). *J. mar. biol. Ass. U.K.* **51**, 301–313.

Taylor, F. J. R. 1979 Symbionticism revisited: a discussion of the evolutionary impact of intracellular symbioses. *Proc. R. Soc. Lond.* B **204**, 267–286 (this volume).

Trench, R. K. 1971 The physiology and biochemistry of zooxanthellae symbiotic with marine coelenterates. II. Liberation of fixed ^{14}C by zooxanthellae *in vitro*. *Proc. R. Soc. Lond.* B **177**, 237–250.

Trench, R. K., Boyle, J. E. & Smith, D. C. 1973 The association between chloroplasts of

Codium fragile and the mollusc *Elysia viridis*. II. Chloroplast ultrastructure and photosynthetic carbon fixation in *E. viridis*. *Proc. R. Soc. Lond.* B **184**, 63–81.

Trench, R. K., Pool, R. R., Logan, M. & Engelland, A. 1978 Aspects of the relation between *Cyanophora paradoxa* (Korschikoff) and its endosymbiotic cyanelles *Cyanocyta korschikoffiana* (Hall & Claus). I. Growth, ultrastructure, photosynthesis and the obligate nature of the association. *Proc. R. Soc. Lond.* B **202**, 423–443.

Wilkinson, C. R. 1978 Microbial associations in sponges. III. Ultrastructure of the *in situ* associations of coral reef sponges. *Mar. Biol.* (In the press.)

BACTERIAL PLASMIDS INSTRUMENTAL IN THE ORIGIN OF EUKARYOTES?

SORIN SONEA

Département de microbiologie et d'immunologie
Faculté de médecine, Université de Montréal
Montréal, Québec, Canada

The possibility that eukaryotes originated from a heterotrophic, anaerobic prokaryotic ancestor by a series of symbioses with other prokaryotes seems to be generally accepted [5], although the prokaryotic cytoplasmic membrane presented a serious barrier to the penetration of the original endosymbionts [10]. One preliminary condition for this decisive evolutionary step had to be an increase in the size of the ancestral prokaryotic host so that future symbiotic cells could first be "phagocytized". The present-day average volume of eukaryotic cells is indeed thousands of times larger than that of bacteria. Many supplementary genes were also necessary for such a cell with an increased volume and had to be progressively accumulated : a contemporary nucleus in a eukaryotic cell contains far more DNA than a prokaryote. This considerable increase in the quantity of DNA in the ancestor cell differs entirely from the usual behaviour of the prokaryotes which keep their cellular size and intracellular DNA content small [9].

The present essay presents the hypothesis that bacterial small replicons (plasmids and genomes of temperate phages) have probably been instrumental in the realization of this important preliminary step in the origin of eukaryotes.

Plasmids, including prophages, are available for most bacteria and constitute a common genetic denominator for many different strains; their genetic information is occasionally beneficial, as supplementary and temporary genes are added to the basic information of the bacterial chromosome [8].

Under the selective pressure of the environment, intracellular plasmids can be replaced by others. As in present-day bacteria, in the ancestral strain from which eukaryotes derived, many such small replicons were probably "at home", although only as occasional "visitors". It is likely that the ancestor strain became progressively polylysogenic and/or it accumulated conjugation factors and non-transmissible plasmids and attachment sites for even more of the latter were provided for by a preliminary augmentation in size [6]. As a plasmid is more readily integrated into a resident plasmid than into the chromosome of a bacterium [8], each resident plasmid could combine with newly arrived ones to form progressively larger replicons; thus progress toward the much larger eukaryotic chromosomes probably started.

The total available genome of this ancestor cell would have been intracellular as it is in eukaryotes, and not mainly extracellular as it is in bacteria [9] and therefore the process was not one of "inventing" new genes, but probably an assemblage of "prefabricated" genes in many simultaneously duplicating small replicons. There are indications that vestiges of autonomous "replicons" may exist in the intranuclear genome of eukaryotes [4].

After the first symbiosis between the ancestors of eukaryotes and before the final change into a really eukaryotic cell, the genes of some plasmids including prophages, which instructed for their transmission and multiplication and which are generally grouped together [7], probably became the ancestors of the DNA viruses of the present eukaryotes. It is tempting to suggest that some eukaryotic viruses may be descendents of conjugation plasmids and that a few may be transmitted only as genomes by cell contact and not by virions.

The plasmids, in which generally most of the genes are highly repressed, might also have been at the origin of the "closed" chromosome [1] of the present eukaryotes, which contain mostly repressed genes, in contrast to the usual bacterial chromosome which is open [2].

The high sensitivity of eukaryotes to radiation, compared to that of bacteria is perhaps related to the fact that lysogenic bacteria are in general less resistant to U. V. than members of the same strain which have lost their prophage or which have not yet acquired one [3].

In contradistinction to the haploidy of the bacterial chromosome, plasmids may present a form of diploidy. For instance, two compatible non-transmissible penicillinase plasmids in *Staphylococcus* normally form dissociated diploids, recombination between them being relatively rare [8]. Conjugation plasmids also fall into incompatible isogenic groups, and those present in the enlarged ancestor cell might have finally realized paired, dissociated, equivalent replicons. Diploidy in eukaryotes may have such an origin.

The appearance and rapid settlement of eukaryotes on earth is easier to explain if most of their early supplementary genes were composed of recently assembled and captive preexisting plasmids and similar small replicons which were previously harboured only occasionally by the ancestor strain in the usual prokaryotic manner, and were already highly selected for their usefulness as supplementary genes for the ancestor of eukaryotes.

ABSTRACT — The hypothesis that eukaryotes originated by a series of symbioses between prokaryotes is generally accepted. It is suggested in the present essay that the increased DNA of the eukaryote nucleus may have started by a permanent accumulation of plasmids, including prophages. This hypothesis may also explain the presence of viruses in all animals and plants, and also the repressed genome of the eukaryotes, their diploidy and their chromosomes.

ABRÉGÉ — **Les plasmides et les prophages à l'origine des eukaryotes.** Si on accepte l'hypothèse que les eukaryotes dérivent d'une cellule prokaryote qui aurait réalisé une série de symbioses avec autres prokaryotes, il reste à expliquer l'accumulation initiale de l'acide désoxyribonucléique du futur noyau. Il est possible que ceci ait été réalisé chez l'ancêtre des eukaryotes par une accumulation de plasmides, les prophages inclus, devenus par la suite captifs. Cette hypothèse expliquerait en partie la diploïdie des eukaryotes, leur chromosome réprimé, la présence de chromosomes, et la présence de virus chez tous les animaux et les végétaux.

REFERENCES

1. BULLOUGH, W. S. : *The Evolution of Differentiation*. Academic Press, London and New York, 1967.

2. DAVIDSON, E. H. : *Gene Activity in Early Development*. Academic Press, London and New York, 1968.

3. MORSE, M. L. and J. W. LABELLE : Characteristics of a staphylococcal phage capable of transduction. *J. Bacteriol.*, **83** : 775, 1962.

4. HUBERMAN, J. A. and A. D. RIGGS : On the mechanism of DNA replication in mammalian chromosomes. *J. Mol. Biol.*, **32** : 327, 1968.

5. MARGULIS, L. : *Origin of Eukaryotic Cells*. Yale University Press, New Haven and London, 1970.

6. MARVIN, D. A. : Control of DNA replication by membrane. *Nature* (London), **219** : 485, 1968.

7. NOVICK, R. P. : Extrachromosomal inheritance in bacteria. *Bacteriol. Rev.*, **33** : 210, 1969.

8. RICHMOND, M. H. : In *Organization and Control in Prokaryotic and Eukaryotic Cells* (ed. by H. P. Charles and B. C. J. G. Knight). Cambridge University Press, Cambridge, 1970.

9. SONEA, S. : A Tentative unifying view of bacteria. *Rev. Can. Biol.*, **30** : 239, 1971.

10. STANIER, R. Y. : In *Organization and Control in Prokaryotic and Eukaryotic Cells* (ed. by H. P. Charles and B. C. J. G. Knight). Cambridge University Press, Cambridge, 1970.

Evidence for a Natural Gene Transfer from the Ponyfish to Its Bioluminescent Bacterial Symbiont *Photobacter leiognathi*

THE CLOSE RELATIONSHIP BETWEEN BACTERIOCUPREIN AND THE COPPER-ZINC SUPEROXIDE DISMUTASE OF TELEOST FISHES*

Joseph P. Martin, Jr.‡ and Irwin Fridovich§

From the Department of Biochemistry, Duke University Medical Center, Durham, North Carolina 27710

The copper and zinc-containing superoxide dismutases of six teleost fish species and of the bioluminescent bacterium *Photobacter leiognathi* were isolated and characterized, as was the iron-containing superoxide dismutase of *P. leiognathi*.

The amino acid composition of the CuZn enzyme from *P. leiognathi* was more closely related to the corresponding fish enzymes than it was to any other yet described CuZn superoxide dismutase. Rank order correlation analysis of amino acid difference indices, computed for all pairwise comparisons of bacteriocuprein and the ponyfish enzyme with other CuZn-superoxide dismutases, suggested that the bacterial and ponyfish enzymes occupy similar phylogenetic positions, being most closely related to the superoxide dismutases of teleost fish. Discriminant functions analysis of all available superoxide dismutase amino acid compositions demonstrated that correct group classification could be achieved on the sole basis of amino acid composition. Thus, Fe and Mn superoxide dismutase groups could be delineated and both groups differed dramatically from the CuZn-superoxide dismutases. Bacteriocuprein was classified with the CuZn-superoxide dismutase group. Moreover, this analytical method subdivided the CuZn group into three subgroups made up of the avian plus mammalian, plant plus fungal, and the fish enzymes, with the *P. leiognathi* enzyme placed in the fish subgroup. The physicochemical properties of the *P. leiognathi* CuZn-superoxide dismutases were very similar to those of the comparable CuZn enzymes from other sources, especially those of teleost fish.

In view of the long standing symbiosis between *P. leiognathi* and its teleost hosts, the leiognathid fishes, these results suggest that the gene specifying the CuZn-superoxide dismutase was transferred from the fish to the bacterium, during their coevolutionary history.

Three major types of superoxide dismutases have been described. The first of these contains Cu and Zn, the second contains Mn, and the third contains Fe. Comparisons of amino acid sequences indicate that the Fe and the Mn enzymes are derived from a common ancestral protein, whereas the CuZn enzymes represent an independent line of descent (1–9). The CuZn-superoxide dismutase[1] is characteristic of the cytosol of eukaryotic cells and the Fe-superoxide dismutase of prokaryotes, while the Mn-superoxide dismutase has been found in prokaryotes and in mitochondria (10–15), presumably a reflection of the symbiotic origin of these organelles (16–20).

Puget and Michelson (21) discovered that *Photobacter leiognathi* contains a CuZn-superoxide dismutase, in addition to the typically prokaryotic Fe-superoxide dismutase (22). Numerous marine and terrestrial enteric bacteria, which are closely related to *P. leiognathi*, contain only the Fe-superoxide dismutase (23). Since *P. leiognathi* is a symbiont of the ponyfish (24), being found in a specific gland (25, 26) and imparting a characteristic luminescence (27, 28), it was proposed that the gene for the CuZn-superoxide dismutase may have been transferred from the host fish to the symbiotic bacterium (14). Comparisons of superoxide dismutases, on the basis of physicochemical properties and amino acid compositions, support this proposal and are described below.

[*Editors' Note:* The Experimental Procedures and Results section has been omitted.]

---

* This work was supported by research grants from the National Institute of General Medical Sciences, National Institutes of Health (Bethesda, MD), from the United States Army Research Office (Research Triangle Park, NC), and from the Merck, Sharp and Dohme Research Laboratories (Rahway, NJ). The costs of publication of this article were defrayed in part by the payment of page charges. This article must therefore be hereby marked "*advertisement*" in accordance with 18 U.S.C. Section 1734 solely to indicate this fact.

‡ Recipient of a National Science Foundation postdoctoral fellowship.

§ To whom reprint requests should be addressed.

[1] The abbreviations used are: CuZn-superoxide dismutase, copper- and zinc-containing superoxide dismutase; Mn-superoxide dismutase, manganese-containing superoxide dismutase; Fe-superoxide dismutase, iron-containing superoxide dismutase; PMSF, phenylmethanesulfonyl fluoride; SDS, sodium dodecyl sulfate; SOD, superoxide dismutase.

Purity—All of the CuZn-superoxide dismutases prepared from teleosts and from *P. leiognathi* were homogeneous by several criteria. Thus: 1) the specific activities were in the range seen for such enzymes from a variety of sources; 2) sodium dodecyl sulfate-gel electrophoresis demonstrated a single band of protein; and 3) electrophoresis of native enzyme, followed by staining for activity and for protein, showed no inactive protein components. Since whole ponyfish were used, one might suspect contamination by the enzyme derived from the symbiotic *P. leiognathi*. Given the size of the *P. leiognathi* colony/ponyfish (26), we calculate that no more than 0.5% of the CuZn enzyme in whole ponyfish would be found in the bacterial symbionts. Furthermore, as shown in Fig. 2, immunodiffusion analysis demonstrated no detectable cross-reactivity between the isolated ponyfish and *P. leiognathi* CuZn-superoxide dismutases. Immunodiffusion analyses were also carried out with rabbit antibodies to human, bovine, chicken, and swordfish CuZn-superoxide dismutases. These antibodies were tested against the antigens given in the legend of Fig. 2. Anti-swordfish CuZn-superoxide dismutase gave a precipitin line with marlin CuZn-superoxide dismutase but reacted only weakly with the red snapper and the ponyfish enzymes and did not cross-react at all with the other CuZn-superoxide dismutases tested. Rabbit antibodies to chicken, human, and bovine CuZn-superoxide dismutases reacted with the homologous antigen but gave no precipitin line with any heterologous CuZn-superoxide dismutase.

Comparisons of Superoxide Dismutases from Teleosts and from P. leiognathi—The CuZn-superoxide dismutases from the teleosts and from *P. leiognathi* were all inhibited by 1.0 mM CN^- and were resistant to exposure to chloroform plus ethanol. In contrast, the Fe-superoxide dismutase from *P. leiognathi* was insensitive to CN^- but was inactivated by the Tsuchihashi fractionation.

Charge heterogeneity was characteristic of these enzymes and the electrophoretic variants co-purified. *P. leiognathi* thus contains two Fe-superoxide dismutases and three CuZn-superoxide dismutases and these were discernible in the purified enzymes or in the crude cell extracts. The method of Hedrick and Smith (40) demonstrated that these multiple forms were charge isomers of identical size. Charge heteroge-

TABLE V
Relative electrophoretic mobilities and isoelectric points of superoxide dismutases

Species	Enzyme	Relative mobility	Isoelectric point
P. leiognathi	CuZn-superoxide dismutase	0.26	8.4
	CuZn-superoxide dismutase	0.33	8.1
	CuZn-superoxide dismutase	0.39	7.9
	Fe-superoxide dismutase	0.86	4.43
	Fe-superoxide dismutase	0.90	4.37
Ponyfish	CuZn-superoxide dismutase	0.32	8.2
	CuZn-superoxide dismutase	0.39	8.0
	CuZn-superoxide dismutase	0.44	7.8
Snapper	CuZn-superoxide dismutase	0.21	8.7
	CuZn-superoxide dismutase	0.25	8.5
	CuZn-superoxide dismutase	0.30	8.3
	CuZn-superoxide dismutase	0.36	8.2
Sea bass	CuZn-superoxide dismutase	0.30	8.13
	CuZn-superoxide dismutase	0.35	8.08
Croaker	CuZn-superoxide dismutase	0.35	7.8
	CuZn-superoxide dismutase	0.40	7.6
Marlin	CuZn-superoxide dismutase	0.49	6.5
	CuZn-superoxide dismutase	0.54	6.4
	CuZn-superoxide dismutase	0.59	6.2
Trout	CuZn-superoxide dismutase	0.50	5.4
	CuZn-superoxide dismutase	0.56	4.9
	CuZn-superoxide dismutase	0.62	4.3
Human	CuZn-superoxide dismutase	0.50	
	CuZn-superoxide dismutase	0.55	
	CuZn-superoxide dismutase	0.60	

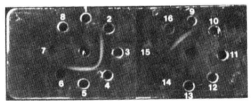

FIG. 2. **Ouchterlony double-diffusion analysis of purified CuZn-superoxide dismutases using antibodies prepared against bacteriocuprein.** The *central well* contained antisera to bacteriocuprein (≃100 µg of protein). The *peripheral wells* contained purified CuZn-superoxide dismutases (≃10 µg) of: *1*, white marlin; *2*, speckled trout; *3*, bacteriocuprein; *4*, inactive bacteriocuprein; *5*, bacteriocuprein; *6*, red snapper; *7*, ponyfish; *8*, black sea bass; *9*, wheat germ; *10*, pig; *11*, human; *12*, bovine; *13*, *E. coli* Fe-superoxide dismutase; *14*, *E. coli* Mn-superoxide dismutase; *15*, Fe-superoxide dismutase of *P. leiognathi*; *16*, bacteriocuprein.

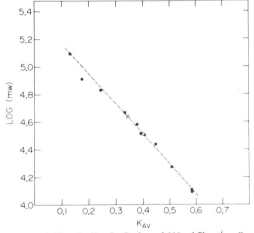

FIG. 3. **Calibration line for Sephacryl-200 gel filtration.** Conditions are as described in the text. The standard proteins, in order of decreasing molecular weight, were aldolase, human transferrin, bovine serum albumin, ovalbumin, *E. coli* Fe-superoxide dismutase, chicken liver CuZn-superoxide dismutase, bovine CuZn-superoxide dismutase, carbonic anhydrase, myoglobin, cytochrome *c* (two points) and, *P. leiognathi* Fe-superoxide dismutase (△). Ponyfish CuZn-superoxide dismutase (○) and *P. leiognathi* CuZn-superoxide dismutase (○) are also represented.

neity of superoxide dismutases has been noted previously (10). The isoelectric points and relative mobilities of the teleost, human and *P. leiognathi* CuZn-superoxide dismutases, and of the *P. leiognathi* Fe-superoxide dismutases are given in Table V.

Gel exclusion chromatography was used to estimate the molecular weights of the teleost and *P. leiognathi* superoxide dismutases (Fig. 3). The *P. leiognathi* Fe-superoxide dismutase exhibited a molecular weight of approximately 42,000 while all of the CuZn-superoxide dismutases showed molecular weights ranging from 30,000 to 34,000. SDS-gel electrophoresis of these enzymes (Fig. 4) demonstrated, in all cases, that the subunit weight was half of the native molecular weight (Table VI).

Optical Spectra—The ultraviolet absorption spectra of the teleost and the *P. leiognathi* CuZn-superoxide dismutases are shown in Fig. 5. For comparative purposes the spectra of the corresponding human and porcine enzymes are also given. From these spectra it appears that the *P. leiognathi* and some of the teleost enzymes do contain tryptophan, in contrast to the porcine enzyme, which seems not to.

Inactivation by Modification of Arginine Residues—The bovine erythrocyte CuZn-superoxide dismutase has been shown, by x-ray crystallography, to contain an arginine residue, number 141 in the amino acid sequence, which is posi-

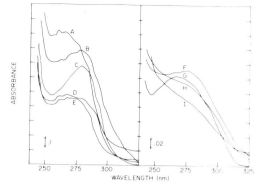

FIG. 5. **Ultraviolet spectra of purified CuZn-superoxide dismutase from various sources.** Spectra were measured in 0.05 M K_2HPO_4, pH 8.0, at 23 °C. The base-lines have been shifted in order to better juxtapose the spectra. *A*, speckled trout; *B*, white marlin; *C*, black sea bass; *D*, human; *E*, pig; *F*, *P. leiognathi*; *G*, ponyfish; *H*, red snapper; *I*, Atlantic croaker.

tioned close to the active-site Cu (117). Modification of this arginine residue, with $\alpha\beta$-diketones, caused a loss of activity (49). CuZn-superoxide dismutases from bovine liver, porcine liver, chicken liver, and wheat germ were comparably inactivated by arginine-specific reagents, whereas bacterial Mn-superoxide dismutase and Fe-superoxide dismutase were unaffected. The teleost and *P. leiognathi* superoxide dismutases were incubated in 50 mM sodium borate, pH 9.0, with either 30 mM 2,3-butanedione or of 1,2-cyclohexanedione. Borate buffer, *per se*, had no effect on activity but, as shown in Fig. 6, butanedione plus borate caused a loss of activity of these CuZn-superoxide dismutases and had no effect on the *P. leiognathi* Fe-superoxide dismutase. 1,2-Cyclohexane dione gave results similar to those obtained with the butanedione.

Relationships Derived from Amino Acid Compositions— Results of quantitative analyses of the teleost and the *P. leiognathi* superoxide dismutases are given in Table VII. The teleost and the *P. leiognathi* CuZn-superoxide dismutase do contain tryptophan, which is a distinguishing characteristic, inasmuch as several other CuZn-superoxide dismutases have been found to lack this amino acid. On the basis of visual comparisons of these compositions it is clear that the *P. leiognathi* and the ponyfish CuZn-superoxide dismutases are very similar while the *P. leiognathi* Fe-superoxide dismutase is very unlike any of the CuZn-superoxide dismutases.

Quantitative comparisons were computed as described under "Experimental Procedures" and the results are shown in Table VIII. The phylogenetic relationships are presented in Fig. 7 and the D^2 values[3] for each enzyme, which are measures of difference in composition, are tabulated in the order of increasing phylogenetic distance. It is clear from the results that the *P. leiognathi* CuZn-superoxide dismutase is more closely related to the ponyfish CuZn-superoxide dismutase than it is to any other CuZn-superoxide dismutase.

If the *P. leiognathi* CuZn-superoxide dismutase is derived from the corresponding ponyfish enzyme, then a positive correlation should exist between the D^2 values for the *P. leiognathi*-eukaryotic CuZn-superoxide dismutase pairs and the phylogenetic distance of these various pairs. In contrast, if there was no gene transfer from the ponyfish to *P. leiognathi*, then the CuZn-superoxide dismutase in this bacterium

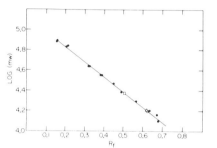

FIG. 4. **The calibration line for sodium dodecyl sulfate-polyacrylamide gel electrophoresis.** Experimental conditions are as described in the text. The standard proteins in order of decreasing molecular weight are human transferrin (two points), human serum albumin, bovine serum albumin, ovalbumin (two points), pepsin (two points), carbonic anhydrase, trypsin, myoglobin, bovine CuZn-superoxide dismutase, lysozyme, and cytochrome *c*. The *line* is a composite of two standard runs. Also presented are *P. leiognathi* Fe-superoxide dismutase (□), ponyfish CuZn-superoxide dismutase (△), and *P. leiognathi* CuZn-superoxide dismutase (○).

TABLE VI
Native and subunit molecular weights of superoxide dismutases

Source	Enzyme	Molecular weight	
		Native	Subunit
Escherichia coli	Mn-superoxide dismutase	40,000	
Streptococcus faecalis	Mn-superoxide dismutase	42,000	
P. leiognathi	Fe-superoxide dismutase	42,000	23,000
E. coli	Fe-superoxide dismutase	39,000	21,000
P. leiognathi	CuZn-superoxide dismutase	32,000	16,000
Bovine liver	CuZn-superoxide dismutase	30,000	15,000
Human liver	CuZn-superoxide dismutase	34,000	
Chicken liver	CuZn-superoxide dismutase	34,000	16,000
Ponyfish	CuZn-superoxide dismutase	32,000	16,000
Croaker	CuZn-superoxide dismutase	31,000	16,000
Marlin	CuZn-superoxide dismutase	31,000	16,000
Sea bass	CuZn-superoxide dismutase	31,000	17,000
Red snapper	CuZn-superoxide dismutase	32,000	16,000
Speckled trout	CuZn-superoxide dismutase	31,000	17,000

[3] $D^2 = (\Sigma (X_{ij} - X_{ik})^2)$ where X_{ij} = the *i*th amino acid residue in the *j*th composition and X_{ik} = the *i*th amino acid residue in the *k*th composition.

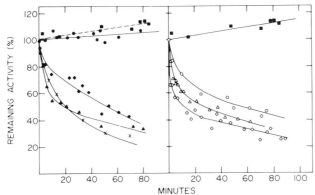

FIG. 6. **The effect of 2,3-butanedione on purified superoxide dismutase from fish and bacteria.** The enzymes were incubated in 0.05 M borate buffer, pH 9.0, in the presence of 30 mM 2,3-butanedione. The enzyme concentration was ≃1.5 µM in each incubation mixture. Aliquots were removed at various time intervals and assayed in the standard assay mixture + 0.1 M DL-mannitol. (■- - -■), bovine CuZn control, no butanedione; (●———●), *P. leiognathi* Fe-superoxide dismutase + 50 mM butanedione; (♦), bovine CuZn-superoxide dismutase + butanedione; ▲, ponyfish CuZn-superoxide dismutase + butanedione; (×, *P. leiognathi* CuZn-superoxide dismutase + butanedione; ○, speckled trout CuZn-superoxide dismutase + butanedione; △, red snapper CuZn-superoxide dismutase + butanedione; ◇, white marlin CuZn-superoxide dismutase + butanedione.

TABLE VII
Amino acid compositions of fish and bacterial superoxide dismutases

Residue[a]	P. leiognathi Fe-superoxide dismutase[c]	P. leiognathi CuZn-superoxide dismutase	Ponyfish	Red snapper	Atlantic croaker	Speckled trout	Black sea bass	White Marlin	Swordfish[b]
Lysine	17	22	22	20	18	20	19	18	20
Histidine	10	12	13	13	13	14	14	14	16
Arginine	5	6	6	9	11	7	8	10	14
Aspartic acid or asparagine	37	37	38	39	42	37	44	36	37
Threonine	20	17	20	21	16	23	25	24	23
Serine	17	18	19	18	16	24	25	22	9
Glutamic acid or glutamine	38	29	28	26	39	26	31	29	28
Proline	14	14	12	14	14	16	10	13	11
Glycine	25	45	46	52	45	46	50	41	46
Alanine	37	28	33	28	27	25	25	26	28
Valine	21	20	20	23	20	24	14	26	26
Methionine	2	4	5	6	5	3	3	4	2
Isoleucine	13	11	13	13	13	15	12	15	20
Leucine	28	29	21	19	18	20	18	19	17
Tyrosine	13	7	3	4	5	3	5	4	5
Phenylalanine	19	11	11	7	9	7	7	9	9
Tryptophan[d]	12	2	2	2	0	2	2	2	0
Total residues/dimer	328	312	312	312	312	312	312	312	312

[a] Residue values are expressed as mole of residue/mol of native enzyme and are estimated to the nearest integral value. Values for cysteine were not determined in this study.
[b] Taken from Ref. 37 and determined by hydrolysis in 6 N HCl.
[c] Taken from Ref. 76 and determined by hydrolysis in 6 N HCl.
[d] Determined by the method of Edelhoch (47).

should be only distantly related to that in any of the eukaryotes. In that case no particular relationship between D^2 values and the *P. leiognathi*-eukaryote phylogenetic distances would be expected. The order of the D^2 values in Table VIII was correlated with the ranked order of phylogenetic distances by Spearman's rank order correlation coefficient. A highly significant r value was obtained ($r = 0.830$, $t_{23} = 7.136$, $p < 0.001$). The positive correlation shown by the data in Table VIII is thus in direct accord with the expectations of the gene transfer hypothesis. Another similarity between the ponyfish and the *P. leiognathi* CuZn-superoxide dismutases is observed when either one or the other is compared with a range of eukaryotic CuZn-superoxide dismutases and the D^2 values correlated with phylogenetic distance (Table VIII). Thus, the ponyfish enzyme exhibits a significant rank ordering of D^2 values with phylogenetic distance when compared with other CuZn-superoxide dismutases.

Discriminant functions analysis, described under "Experimental Procedures," was applied to the compositions of the teleost and the *P. leiognathi* superoxide dismutases, as well as to all other superoxide dismutases for which amino acid compositions have been reported. Our first goal was to see whether such an analysis of amino acid compositions could distinguish among the known groups of superoxide dismutases, *i.e.* among CuZn-superoxide dismutases, Mn-superoxide dismutases, and Fe-superoxide dismutases, and where it would

Evidence for a Eukaryote to Prokaryote Gene Transfer

TABLE VIII

Values of D^2, the euclidian distance, for comparisons of P. leiognathi CuZn-superoxide dismutase and ponyfish CuZn-superoxide dismutase, with eucaryote CuZn-superoxide dismutases

Position in phylogeny	Species	D^2 P. leiognathi[a]	Ponyfish[b]
2	Ponyfish	0.0100	
3	Red snapper	0.0113	0.0083
4	Atlantic croaker	0.0218	0.0205
5	Speckled trout	0.0135	0.0105
6	Black bass	0.0161	0.0101
7	White marlin	0.0113	0.0128
8	Swordfish	0.0301	0.0317
9	Chicken	0.0255	0.0238
10	Bovine	0.0383	0.0340
10	Human	0.0493	0.0397
10	Pig	0.0345	0.0327
10	Horse	0.0459	0.0368
10	Rat	0.0407	0.0333
11	Yeast	0.0348	0.0304
11	Neurospora	0.0481	0.0441
12	Spinach	0.0536	0.0484
12	Wheat germ I	0.0598	0.0525
12	Wheat germ II	0.0742	0.0774
12	Peas	0.0731	0.0567

[a] Spearman's rank order correlation coefficient, $r = 0.83$; $t_{23} = 7.136$, $p < 0.001$.

[b] Spearman's rank order correlation coefficient, $r = 0.78$; $t_{22} = 5.712$, $p < 0.001$.

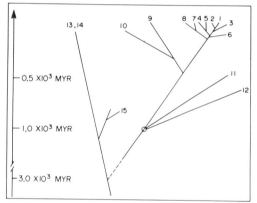

FIG. 7. **Two hypotheses on the relationship of bacterial and eukaryotic CuZn-superoxide dismutase illustrated by the phylogenetic relationships among organisms.** The ordinate is calibrated in millions of years (*MYR*). Uncertainties in divergence time of ancestral lines are depicted as nodules (○). *Position 1* is that occupied by *P. leiognathi* CuZn-superoxide dismutase if acquired from the ponyfish by gene transfer. *Position 15* represents the phylogenetic position of the marine enterobacteria, the expected position of *P. leiognathi* CuZn-superoxide dismutase if only distantly related to eukaryotic CuZn-superoxide dismutases. The other numbered positions represent: *2*, ponyfish; *3*, red snapper; *4*, Atlantic croaker; *5*, speckled trout; *6*, black sea bass; *7*, white marlin; *8*, swordfish; *9*, all mammals; *10*, birds; *11*, plants; *12*, fungi; *13* and *14*, prokaryotes containing Fe and/or Mn-superoxide dismutase.

place the *P. leiognathi* CuZn-superoxide dismutase in this array of superoxide dismutases. Table IX lists the two discriminant functions F_I and F_{II} which best separated the three groups and the amino acid residues found to be most useful in the separation of the groups. Function I maximized the distance between the CuZn-superoxide dismutases on the one hand and the Fe-superoxide dismutases and Mn-superoxide dismutases on the other. Function II distinguished Mn-superoxide dismutases from Fe-superoxide dismutases. Amino acid residues were emphasized in the functions to the degree in which they systematically varied among the groups of superoxide dismutases. The F values indicate the relative importance of the amino acids in differentiating one group of superoxide dismutases from another. The CuZn-superoxide dismutase differed from Mn-superoxide dismutases and Fe-superoxide dismutases in possessing relatively more glycine, arginine, and lysine, and less of proline, threonine, and the aromatic residues, while the Mn-superoxide dismutases differ from the Fe-superoxide dismutases primarily with respect to phenylalanine, lysine, tryptophan, leucine, isoleucine, and threonine.

Fig. 8 presents a map of the individual superoxide dismutase compositions. The position of each superoxide dismutase is determined by the values assumed by its two canonical variables. All of the CuZn-superoxide dismutases cluster into a distinct group, well separated from the Mn-superoxide dismutases and the Fe-superoxide dismutases, which, in turn, form separate but slightly overlapping groups. Since the Fe-

TABLE IX

Coefficients of discriminant functions I and II, F values for included amino acids, and the cumulative proportions of the total sample dispersion explained by the functions

Amino acid	Coefficient function I (λ)	Coefficient function II (λ)	F	DF[a]
Glycine	19.317	6.770	147.07[b]	2, 52
Phenylalanine	−17.742	24.767	18.168[b]	2, 51
Tryptophan	−19.728	7.388	13.006[b]	2, 50
Tyrosine	−12.461	−3.613	5.703[c]	2, 49
Lysine	8.748	−12.970	4.479[d]	2, 48
Leucine	−3.723	−23.526	3.057[e]	2, 47
Isoleucine	−9.701	−17.855	3.157[f]	2, 46
Proline	−21.580	−3.277	2.680[e]	2, 45
Arginine	10.507	−6.461	1.861	2, 44
Threonine	−19.728	7.388	1.800	2, 43
Constant[g] (λ₀)	2.494	0.716		
Cumulative proportion of sample variance explained by functions I and II.	0.947	1.00		

[a] DF, degrees of freedom.
[b] $p < 0.001$.
[c] $p < 0.01$.
[d] $p < 0.025$.
[e] $p < 0.1$.
[f] $p < 0.05$.
[g] λ₀ is chosen such that the mean of the canonical variables is 0.

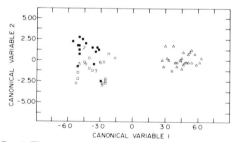

FIG. 8. **The separation of the three superoxide dismutase groups (CuZn, Fe, Mn) in two-dimensional discriminant space.** Each canonical variable is a linear composite of the amino acids that best discriminate among the groups. The positions of the group means and individual samples are indicated by: *1*, CuZn-superoxide dismutase (△); *2*, Fe-superoxide dismutase (■); *3*, Mn-superoxide dismutase (□); *4*, bacterial CuZn-superoxide dismutase; *5*, *M. bryantii*; *6*, *B. campestris*.

TABLE X

Classification matrix for amino acid compositions in Groups CuZn-superoxide dismutase, Fe-superoxide dismutase, and Mn-superoxide dismutase

Source	Number of compositions classified into group[a]			Correct classification
	CuZn	Fe	Mn	
				%
CuZn	27	0	0	100
Fe	0	12	2	85.7
Mn	0	4	12	75
Bacteriocuprein	1	0	0	
Methanogen	0	0	1	
Mustard	0	0	1	

[a] The first three group means differed significantly from one another, $F_{9,45} > 5.17, p < 0.001$.

TABLE XI

Coefficients of discriminant functions I and II, F values for included amino acids, and the cumulative proportions of the total sample dispersion explained by the functions, derived from the CuZn-superoxide dismutase subgroups

Amino acid	Coefficient Function I (λ)	Coefficient Function II (λ)	F	DF[a]
Alanine	−57.216	19.738	18.557[b]	2, 23
Lysine	−18.643	47.838	14.508[b]	2, 22
Valine	−7.848	−49.629	8.410[c]	2, 21
Glycine	−40.146	−31.025	7.288[c]	2, 20
Proline	51.579	13.142	6.856[d]	2, 19
Arginine	−18.643	47.838	4.136[e]	2, 18
Leucine	−23.528	15.355	2.712[f]	2, 17
Aspartic acid	44.394	−8.892	3.443[f]	2, 16
Constant (λ_0)	−0.482	4.610		
Cumulative proportion of total variance explained by functions I and II	0.625	1.000		

[a] DF, degrees of freedom.
[b] $p < 0.001$.
[c] $p < 0.005$.
[d] $p < 0.01$.
[e] $p < 0.05$.
[f] $p < 0.1$.

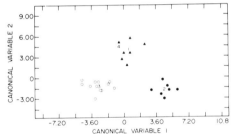

FIG. 9. **The subdivision of the CuZn-superoxide dismutase group into three categories in discriminant space:** *1,* fish (▲); *2,* plant and fungi (●); *3,* mammals and birds (○); *4,* bacterial CuZn-superoxide dismutase.

TABLE XII

Classification matrix for amino acid compositions of CuZn-superoxide dismutases in groups fish, birds and mammals, and plants and fungi

Source	Number of compositions classified into groups[a]			Correct classification
	Fish	Birds and mammals	Plants and fungi	
				%
Fish	7	0	0	100
Birds and mammals	0	12	0	100
Plants and fungi	0	0	7	100
Bacteriocuprein	1	0	0	

[a] The first three group means differed significantly, $F_{8,16} > 14.88, p < 0.001$ for all pairwise comparisons.

ation from the original data set. The Mahalanobis distances of the sample from the adjusted group means were computed and the sample was classified as a member of the closest group. Table X presents the resultant classification matrix. The CuZn, Fe, and Mn group means were significantly different. There were only a few sample misclassifications and these occurred between the closely related Mn-superoxide dismutases and Fe-superoxide dismutases. In at least one case, what appeared to be a misclassification was actually a correction of an erroneous report in the literature. Thus, the superoxide dismutase of *Bacillus megaterium* was reported to be a Fe-superoxide dismutase (78), while the computer analysis of its amino acid composition classified it with the Mn-superoxide dismutases. Re-examination of this enzyme by metal removal and replacement demonstrated it to be a Mn-superoxide dismutase (119), in agreement with the discriminant analysis. Bacteriocuprein was properly classified as a typical CuZn-superoxide dismutase, while the *P. leiognathi* Fe-superoxide dismutase was positively identified with the Fe-superoxide dismutase group.

The locations of two unusual and recently characterized Fe-superoxide dismutases, in this discriminant space, were also calculated. These were the Fe-superoxide dismutases from the higher plant *Brassica campestris* (89) and from the archaebacterium *Methanobacterium bryantii*.[4] These Fe-superoxide dismutases appeared between the Fe-superoxide dismutase and the Mn-superoxide dismutase groups and slightly closer to the Mn-superoxide dismutase group mean. It appears, from phylogenetic data (11–13), that Fe-superoxide dismutases predated Mn-superoxide dismutases and that the *M. bryantii* and the mustard Fe-superoxide dismutases may be more closely related to that Fe-superoxide dismutase which was ancestral to both more modern Fe-superoxide dismutases and Mn-superoxide dismutases. Consequently one might expect that these two superoxide dismutases would be approximately intermediate to the Mn-superoxide dismutase and Fe-superoxide dismutase groups.

Discriminant analysis was next applied to the set of CuZn-superoxide dismutases. Table XI presents the sets of two discriminant functions which were computed, while Fig. 9 presents the map of these compositions in discriminant space. The CuZn-superoxide dismutases can be separated into three distinct groups based on their compositions: 1) fish, 2) plant-fungal, and 3) bird-mammals. When classified as previously described, the *P. leiognathi* CuZn-superoxide dismutase fell into the fish group (Table XII).

superoxide dismutases and Mn-superoxide dismutases are closely related to each other, while being unrelated to the CuZn-superoxide dismutases, on the basis of amino acid sequence data, these results demonstrate the validity and utility of the discriminant functions analysis applied to amino acid compositions.

The group membership of each individual superoxide dismutase was calculated by a jackknife procedure (118). The positions of the group means in discriminant space were recalculated after removing the composition under consider-

DISCUSSION

Gene transfer among prokaryotes is a well documented phenomenon and is the basis of transformation, transduction, and sexduction. Gene transfer from prokaryotes to eukaryotes

[4] T. Kirby (1980), unpublished observations.

is also known, a clear case being that between *Agrobacter tumifaciens* and host plants in the etiology of crown gall tumor (120, 121). In contrast, gene transfer from eukaryotes to prokaryotes has been less certain as a natural process, although achievable by recombinant DNA technology. One possible case relates to the microorganism *Progenitor cryptocides* which is found in close association with certain human tumors and which, during *in vitro* culture, secretes a substance similar to human chorionic gonadotropin (122-124).

The very close relationship between the ponyfish, other teleosts, and the *P. leiognathi* CuZn-superoxide dismutases, as deduced by physicochemical analysis and quantitative analysis of amino acid compositions, now suggests gene transfer from the host fish to its symbiotic bacterium. This particular case was closely examined because the very presence of a CuZn-superoxide dismutase in a bacterium is an anomaly, it being characteristic of the cytosol of eukaryotes. Although prokaryotes may contain both Fe-superoxide dismutase and Mn-superoxide dismutase, these molecules show extensive amino acid sequence homology and are clearly related. In contrast the *P. leiognathi* CuZn-superoxide dismutase differs dramatically and appears unrelated to the Fe-superoxide dismutase also found in these bacteria. The existence of unrelated superoxide dismutases in eukaryotic cells is well known, but since eukaryotic cells are of chimeric origin, the appearance of two unrelated forms of superoxide dismutase is understandable.

The lack of immunological cross reactivity between the ponyfish and the *P. leiognathi* CuZn-superoxide dismutases is not an indication of unrelatedness. Thus, the available amino acid sequence data indicate that human, bovine and horse CuZn-superoxide dismutases are closely related, exhibiting at least 83% absolute homology (3, 4, 125, 126). Yet, human, bovine, and porcine enzymes do not cross react in our hands. Shields *et al.* (127) similarly noted that rabbit anti-human CuZn-superoxide dismutase did not react with the corresponding enzyme from cow, rabbit, or chicken. Furthermore, we have noted that rabbit anti-swordfish CuZn-superoxide dismutase fails to cross-react with CuZn-superoxide dismutases from several other fish. It is clear that serological cross-reactivity is not a reliable indication of relatedness among the slowly evolving CuZn-superoxide dismutases from animals.

It might be supposed that the CuZn-superoxide dismutase evolved prior to the prokaryote-eukaryote divergence and that its appearance in *P. leiognathi* is a retained primitive characteristic. This supposition is not supported by the available data on the phylogenetic distribution of this enzyme. Thus CuZn-superoxide dismutases are ubiquitous among higher eukaryotes, while not being found in protozoa, eukaryotic algae, and in numerous classes of bacteria (11-13, 22, 23, 29, 34, 96). Moreover, numerous species of marine photobacteria, closely related to *P. leiognathi*, have been examined and found to lack detectable CuZn-superoxide dismutase (23). If bacteriocuprein were a retained primitive characteristic, why was it retained by *P. leiognathi* and not by other bacteria and why is it found in all higher eukaryotes?

Alternately, one might postulate that the CuZn-superoxide dismutase arose in *P. leiognathi* in response to some selection pressure peculiar to its symbiotic lifestyle. The fossil record indicates the age of the family Leiognathidae to be approximately 30×10^6 years. Since most of the fishes in closely related families do not harbor symbiotic photobacteria, we must take this to be a relatively recent adaptation by the Leiognathidae. Hence, an independent evolution of a CuZn-superoxide dismutase in *P. leiognathi* would have, of necessity, to have been accomplished in less than 30×10^6. Is it likely that over 10^9 years of protein evolution could be duplicated in 10^7 years? That some precursor protein, already present in *P. leiognathi*, could be brought to the same molecular weight, subunit composition, amino acid composition, active site conformation, and catalytic function as the ponyfish enzyme in less than 30×10^6 years would be very unlikely. We are left with gene transfer as the most parsimonious explanation of the data.

It was easy to imagine that *P. leiognathi* could have gotten the CuZn-superoxide dismutase gene from the host ponyfish since their symbiotic relationship has persisted long enough for the fish to have evolved a special gland in which to house the bacterium. Furthermore, as mentioned above, *P. leiognathi* is the only photobacterium found in the light gland of the ponyfish (128) and it is the only one, among more than 93 strains of marine and terrestrial enterobacteria thus far examined, which has been found to contain a CuZn-superoxide dismutase (23). Given the occurrence of mechanisms for the uptake of DNA by prokaryotes (129), it is not difficult to imagine how the CuZn-superoxide dismutase gene could have been passed from the host ponyfish to *P. leiognathi*. It is, however, difficult to imagine what selective advantage could have caused this gene to be retained by the bacterium. Stronger support for the gene-transfer hypothesis could be gained from knowledge of the amino acid sequences of the ponyfish and of the *P. leiognathi* CuZn-superoxide dismutases. Hopefully such data will be forthcoming.

Acknowledgments—We would like to thank Nikos Cerletti, John L. Abernethy, and Ralph Wylie for help in conducting amino acid analyses.

REFERENCES

1. Steinman, H. M. (1978) *J. Biol. Chem.* **253**, 8708-8720
2. Brock, C. J., and Walker, J. E. (1980) *Biochemistry* **19**, 2873-2882
3. Steinman, H. M., Naik, V. R., Abernethy, J. L., and Hill, R. L. (1974) *J. Biol. Chem.* **249**, 7326-7338
4. Abernethy, J. L., Steinman, H. M., and Hill, R. L. (1974) *J. Biol. Chem.* **249**, 7339-7347
5. Johansen, J. T., Overballe-Peterson, C., Martin, B., Hasemann, V., and Svendsen, I. B. (1979) *Carlsberg Res. Commun.* **44**, 201-217
6. Steinman, H. M. (1980) *J. Biol. Chem.* **255**, 6758-6765
7. Harris, J. I., Auffret, A. D., Northrop, F. D., and Walker, J. E. (1980) *Eur. J. Biochem.* **106**, 297-303
8. Harris, J. J., and Steinman, H. (1977) in *Superoxide and Superoxide Dismutases* (Michelson, A. M., McCord, J. M., and Fridovich, I., eds) pp. 225-231, Academic Press, New York
9. Steinman, H. M., and Hill, R. L. (1973) *Proc. Natl. Acad. Sci. U. S. A.* **70**, 3725-3729
10. Weisiger, R. A., and Fridovich, I. (1973) *J. Biol. Chem.* **248**, 3582-3592
11. Asada, K., Kanematsu, S., Takahashi, M., and Kono, Y. (1975) in *Advances in Experimental Medicine and Biology in Iron and Copper Proteins* (Yasunobu, K. T., Mower, H. F., and Hayaishi, O., eds) Vol. 74, pp. 551-565, Plenum Press, New York
12. Asada, K., Kanematsu, S., and Uchida, K. (1977) *Arch. Biochem. Biophys.* **179**, 243-256
13. Asada, K., and Kanematsu, S. (1978) in *Evolution of Protein Molecules* (Matsubara, H., and Yamanaka, T., eds) pp. 361-372, Japan Scientific Society Press, Tokyo
14. Fridovich, I. (1975) *Annu. Rev. Biochem.* **44**, 147-159
15. Lumsden, J., Henry, L., and Hall, D. O. (1977) in *Superoxide and Superoxide Dismutases* (Michelson, A. M., McCord, J., and Fridovich, I., eds) pp. 437-451, Academic Press, New York
16. Fridovich, I. (1974) *Life Sci.* **14**, 819-826
17. Margulis, L. (1970) *Origin of Eucaryotic Cells*, Yale University Press, New Haven
18. Schwartz, R. M., and Dayhoff, M. O. (1978) *Science* **199**, 395-403
19. Doolittle, W. F. (1980) *Trends Biochem. Sci.* 146-151

20. Fox, G. E., Stackebrandt, E., Hespell, R. B., Gibson, J., Maniloff, J., Dyer, T. A., Wolfe, R. S., Balch, W. E., Tanner, R. S., Magrum, L. J., Zablen, L. B., Blakemore, R., Gupta, R., Bonen, L., Lewis, B. J., Stahl, D. A., Luehrsen, K. R., Chen, K. N., and Woese, C. R. (1980) *Science* **209**, 457–463
21. Puget, K., and Michelson, A. M. (1974) *Biochem. Biophys. Res. Commun.* **58**, 830–838
22. Puget, K., Lavelle, F., and Michelson, A. M. (1977) in *Superoxide and Superoxide Dismutases* (Michelson, A. M., McCord, J., and Fridovich, I., eds) pp. 139–151, Academic Press, New York
23. Bang, S. S., Woolkalis, M. J., and Baumann, P. (1978) *Curr. Microbiol.* **1**, 371–376
24. Haneda, Y. (1950) *Pac. Sci.* **4**, 214–227
25. Boisvert, H., Chatelain, R., and Bassot, J. M. (1967) *Ann. Instit. Pasteur (Paris)* **112**, 520–525
26. Hastings, J. W., and Mitchell, G. (1971) *Biol. Bull.* **141**, 261–268
27. Hastings, J. W. (1971) *Science* **173**, 1016–1017
28. Hastings, J. W., and Nealson, K. (1977) *Annu. Rev. Microbiol.* **31**, 549–595
29. Puget, K., and Michelson, A. M. (1974) *Biochimie* **56**, 1255–1267
30. Dahlberg, M. D. (1975) *Guide to the Coastal Fishes of Georgia and Nearby States*, University of Georgia Press, Athens
31. McCord, J. M., and Fridovich, I. (1969) *J. Biol. Chem.* **244**, 6049–6055
32. Beauchamp, C. O., and Fridovich, I. (1971) *Anal. Biochem.* **44**, 276–287
33. Keele, B. B., Jr., McCord, J. M., and Fridovich, I. (1971) *J. Biol. Chem.* **246**, 2875–2880
34. Waud, W. R., Brady, F. O., Wiley, R. D., and Rajagopalan, K. V. (1975) *Arch. Biochem. Biophys.* **169**, 695–701
35. Beauchamp, C. O. (1973) Ph.D. dissertation, Duke University, Durham
36. Lowry, O. H., Rosebrough, N. J., Farr, A. L., and Randall, R. J. (1951) *J. Biol. Chem.* **193**, 265–275
37. Bannister, J. V., Anastasi, A., and Bannister, W. H. (1977) *Comp. Biochem. Physiol.* **56B**, 235–238
38. Davis, B. J. (1964) *Ann. N. Y. Acad. Sci.* **121**, 404–427
39. Ornstein, L. (1964) *Ann. N. Y. Acad. Sci.* **121**, 321–349
40. Hedrick, J. L., and Smith, A. J. (1968) *Arch. Biochem. Biophys.* **126**, 155–164
41. Weber, K., and Osborn, M. (1969) *J. Biol. Chem.* **244**, 4406–4412
42. Andrews, P. (1964) *Biochem. J.* **91**, 222–233
43. Righetti, P., and Drysdale, J. W. (1971) *Biochim. Biophys. Acta* **236**, 17–28
44. Blakesly, R. W., and Boezi, J. A. (1977) *Anal. Biochem.* **82**, 580–582
45. Ouchterlony, O. (1968) *Handbook of Immunodiffusion and Immunoelectrophoresis*, Ann Arbor Scientific Publishers, Ann Arbor
46. Simpson, R. J., Neuberger, M. R., and Liu, T.-Y. (1976) *J. Biol. Chem.* **251**, 1936–1940
47. Edelhoch, H. (1967) *Biochemistry* **6**, 1948–1954
48. Riordan, J. F., McElvany, K. D., and Borders, C. L. (1977) *Science* **195**, 884–886
49. Malinowski, D. P., and Fridovich, I. (1979) *Biochemistry* **18**, 5909–5917
50. Harris, C. E., and Teller, D. C. (1973) *J. Theor. Biol.* **38**, 347–362
51. Kanematsu, S., and Asada, K. (1979) *Arch. Biochem. Biophys.* **195**, 535–545
52. Marchalonis, J. J., and Weltman, J. K. (1971) *Comp. Biochem. Physiol.* **38B**, 609–625
53. Sullivan, J. B., Bonaventura, J., Bonaventura, C., Pennell, L., Boyum, R., and Lambie, W. (1975) *J. Mol. Evol.* **5**, 103–115
54. Black, J. A., and Harkins, R. N. (1977) *J. Theor. Biol.* **66**, 281–295
55. Cornish-Bowden, A. (1977) *J. Theor. Biol.* **65**, 735–742
56. Cornish-Bowden, A. (1979) *J. Theor. Biol.* **76**, 369–386
57. Cornish-Bowden, A. (1980) *Anal. Biochem.* **105**, 233–238
58. Helwig, J. T., and Council, K. A., eds (1979) *SAS Users Guide*, SAS Institute, Inc., Raleigh
59. Johnson, S. C. (1967) *Psychometrika* **32**, 251–254
60. Sokal, R. R., and Rohlf, F. J. (1969) *Biometry*, W. H. Freeman and Co., San Francisco
61. Siegal, S. (1956) *Nonparametric Statistics for the Behavioral Sciences*, McGraw-Hill, New York
62. Rohlf, F. J., and Sokal, R. R. (1969) *Statistical Tables*, John Freeman and Co., San Francisco
63. Sharoyan, S. G., Shalijain, A. A., Nalbandyan, R. M., and Buniatian, H. C. (1977) *Biochim. Biophys. Acta* **493**, 478–487
64. Abu-Erreish, G., Magnes, L., and Li, T.-K. (1978) *Biol. Reprod.* **18**, 554–560
65. Albergoni, V., and Cassini, A. (1974) *Comp. Biochem. Physiol.* **47B**, 767–777
66. Barikowiak, A., Lpyko, W., and Fried, R. (1975) *Comp. Biochem. Physiol.* **62B**, 61–66
67. Bannister, W. H., Anastasi, A., and Bannister, J. V. (1977) in *Superoxide and Superoxide Dismutases* (Michelson, A., McCord, J., and Fridovich, I., eds) pp. 107–129, Academic Press, New York
68. Carrico, R. J., and Deutsch, H. F. (1969) *J. Biol. Chem.* **244**, 6087–6093
69. Marklund, S., Beckman, G., and Stigbrand, T. (1976) *Eur. J. Biochem.* **65**, 415–422
70. Crapo, J. D., and McCord, J. M. (1976) *Am. J. Physiol.* **231**, 1196–1203
71. Goscin, S. A., and Fridovich, I. (1972) *Biochim. Biophys. Acta* **289**, 276–283
72. Misra, H. P., and Fridovich, I. (1972) *J. Biol. Chem.* **247**, 3410–3414
73. Sawada, Y., Ohyama, T., and Yamazaki, I. (1972) *Biochim. Biophys. Acta* **268**, 305–312
74. Asada, K., Urano, M., and Takahashi, M. (1973) *Eur. J. Biochem.* **36**, 257–266
75. Beauchamp, C. O., and Fridovich, I. (1973) *Biochim. Biophys. Acta* **317**, 50–64
76. Vanopdenbosch, B., Crichton, R. R., and Puget, K. (1977) in *Superoxide and Superoxide Dismutases* (Michelson, A. M., McCord, J., and Fridovich, I., eds) pp. 199–207, Academic Press, New York
77. Yost, F. J., Jr., and Fridovich, I. (1973) *J. Biol. Chem.* **248**, 4905–4908
78. Anastasi, A., Bannister, J. V., and Bannister, W. H. (1979) *Int. J. Biochem.* **7**, 541–546
79. Yamakura, F. (1976) *Biochim. Biophys. Acta* **422**, 280–294
80. Hatchikian, C. E., Bell, G. R., and LeGall, J. (1977) in *Superoxide and Superoxide Dismutases* (Michelson, A. M., McCord, J., and Fridovich, I., eds) pp. 159–173, Academic Press, New York
81. Hatchikian, E. C., and Henry, Y. A. (1977) *Biochimie* **59**, 153–161
82. Baldensperger, J. B. (1978) *Arch. Microbiol.* **119**, 237–244
83. Kanematsu, S., and Asada, K. (1978) *FEBS Lett.* **91**, 94–98
84. Kanematsu, S., and Asada, K. (1978) *Arch. Biochem. Biophys.* **185**, 473–482
85. Asada, K., Yoshikawa, K., Takahashi, M., Maeda, Y., and Enmanji, K. (1975) *J. Biol. Chem.* **250**, 2801–2807
86. Misra, H. P., and Keele, B. B., Jr. (1975) *Biochim. Biophys. Acta* **379**, 418–425
87. Lumsden, J., and Hall, D. O. (1974) *Biochim. Biophys. Res. Commun.* **58**, 35–41
88. Kusunose, E., Ichihara, K., Noda, Y., and Kusunose, M. (1976) *J. Biochem. (Tokyo)* **80**, 1343–1352
89. Salin, M. L., and Bridges, S. M. (1980) *Arch. Biochem. Biophys.* **201**, 369–374
90. Salin, M. L., Day, E. D., Jr., and Crapo, J. D. (1978) *Arch. Biochem. Biophys.* **187**, 223–228
91. McCord, J., Boyle, J. A., Day, E. D., Rizzolo, L. J., and Salin, M. (1977) in *Superoxide and Superoxide Dismutases* (Michelson, A., McCord, J., and Fridovich, I., eds) pp. 129–139, Academic Press, New York
92. Marklund, S. (1978) *Int. J. Biochem.* **9**, 299–306
93. Ravindranath, S. D., and Fridovich, I. (1975) *J. Biol. Chem.* **250**, 6107–6112
94. Lavelle, F., Durasay, P., and Michelson, A. M. (1974) *Biochemie* **56**, 451–458
95. Keele, B. B., Jr., McCord, J. M., and Fridovich, I. (1970) *J. Biol. Chem.* **245**, 6176–6181
96. Britton, L., Malinowski, D. P., and Fridovich, I. (1978) *J. Bacteriol.* **134**, 229–236
97. Vance, P. G., Keele, B. B., Jr., and Rajagopalan, K. V. (1972) *J. Biol. Chem.* **247**, 4782–4786
98. Sato, S., and Harris, J. I. (1977) *Eur. J. Biochem.* **73**, 373–381
99. Sato, S., and Nakayaza, K. (1978) *J. Biochem. (Tokyo)* **83**, 1165–1171
100. Harris, J. I. (1977) in *Superoxide and Superoxide Dismutases* (Michelson, A. M., McCord, J., and Fridovich, I., eds) pp. 151–

159, Academic Press, New York
101. Kusunose, M., Noda, Y., Ichihara, K., and Kusunose, E. (1976) *Arch. Microbiol.* **108,** 65–73
102. Ichihara, K., Kusunose, E., Kusunose, M., and Mori, T. (1977) *J. Biochem.* **81,** 1427–1433
103. Misra, H. P., and Fridovich, I. (1977) *J. Biol. Chem.* **252,** 6421–6423
104. Lumsden, J., Cammack, R., and Hall, D. O. (1976) *Biochim. Biophys. Acta* **438,** 380–392
105. Dayhoff, M. O. (1978) *Atlas of Protein Sequence and Structure,* Vol. 5, Suppl. 3, National Biomedical Research Foundation, Washington, D. C.
106. Romer, A. S. (1966) *Vertebrate Paleontology,* University of Chicago Press, Chicago
107. Nelson, J. S. (1976) *Fishes of the World,* Wiley-Interscience, New York
108. Greenwood, P. H., Rosen, D. E., Weitzman, S. H., and Myers, G. S. (1966) *Bull. Am. Mus. Nat. Hist.* **131,** 339–456
109. Anderson, T. W. (1958) *Introduction to Multivariate Statistical Analysis,* John Wiley & Sons, New York
110. Rao, C. R. (1965) *Linear Statistical Inference and Its Applications,* John Wiley & Sons, New York
111. Snedecor, G. W., and Cochran, W. G. (1967) *Statistical Methods,* Iowa State University Press, Ames
112. Gould, S. J., and Johnson, R. F. (1972) *Annu. Rev. Ecol. Syst.* **3,** 1–30
113. Gould, S. J. Woodruff, D., and Martin, J. (1974) *Syst. Zool.* **23,** 518–536
114. Calhoon, R. E., and Aaronson, S. (1979) *J. Theor. Biol.* **78,** 225–239
115. Dixon, W. J. (1977) *Biomedical Computer Programs,* University of California Press, Berkeley
116. Tsuchihashi, M. (1923) *Biochem. Z.* **140,** 65–112
117. Richardson, D. C. (1977) in *Superoxide and Superoxide Dismutases* (Michelson, A. M., McCord, J., and Fridovich, I., eds) pp. 217–225, Academic Press, New York
118. Lackenbruch, P., and Mickey, R. M. (1968) *Technometrics* **10,** 1–11
119. Kirby, T., Blum, J., Kahane, I., and Fridovich, I. (1980) *Arch. Biochem. Biophys.* **201,** 551–555
120. Drummond, M. H., Gordon, M. P., Nester, E. W., and Chilton, M.-D. (1977) *Nature* **269,** 535–536
121. Drummond, M. (1979) *Nature* **281,** 343–347
122. Cohen, H., and Strampp, A. (1976) *Proc. Soc. Exp. Biol. Med.* **152,** 408–411
123. Livingston, V., and Livingston, A. (1974) *Trans. N. Y. Acad. Sci.* **36,** 569–582
124. Livingston, V., and Alexander-Johnson, E. (1970) *Ann. N. Y. Acad. Sci.* **174,** 636–653
125. Ammer, D., and Lerch, K. (1980) in *Chemical and Biological Aspects of Superoxide and Superoxide Dismutase* (Bannister, J. V., and Hill, H. A. O., eds) pp. 230–236, Elsevier/North Holland, Amsterdam
126. Barra, D., Martini, F., Bannister, J. V., Schinina, M. E., Rotilio, G., Bannister, W. H., and Bossa, F. (1980) *FEBS Lett.* **120,** 53–56
127. Shields, G. S., Markowitz, H., Lassen, W. H. K., Cartwright, G. E., and Wintrobe, M. M. (1961) *J. Clin. Invest.* **40,** 2007–2019
128. Reichelt, J. L., Nealson, K., and Hastings, J. W. (1977) *Arch. Microbiol.* **112,** 157–161
129. Lacks, S. A. (1977) in *Genetic Interaction and Gene Transfer—Brookhaven Symposia in Biology* (Anderson, C., ed) Vol. 29, pp. 147–161, Brookhaven National Laboratory, Upton

MALARIA PARASITES ADOPT HOST CELL SUPEROXIDE DISMUTASE

A. S. Fairfield and S. R. Meshnick

Abstract. *Aerobic organisms depend on superoxide dismutase to suppress the formation of dangerous species of activated oxygen. Intraerythrocytic stages of the malaria parasite exist within a highly aerobic environment and cause the generation of increased amounts of activated oxygen.* Plasmodium berghei *in mice was found to derive a substantial amount of superoxide dismutase activity from the host cell cytoplasm. Plasmodia isolated from mouse red cells contained mouse superoxide dismutase, whereas rat-derived parasites contained the rat enzyme. This is believed to be the first example of the acquisition of a host cell enzyme by an intracellular parasite.*

The superoxide anion (O_2^-), a partially reduced form of molecular oxygen, is generated spontaneously and metabolically within biological organisms. Because O_2^- can react to form highly destructive hydroxyl radical ($\cdot OH$), aero-tolerant organisms may need protection against O_2^-. Indeed, almost all known aerobic organisms contain the enzyme superoxide dismutase (SOD) [1]:

$$2\ O_2^- + 2\ H^+ \xrightarrow{SOD} H_2O_2 + O_2$$

This enzyme may be particularly important for mammalian erythrocytes because activated oxygen species (O_2^-, H_2O_2, and $\cdot OH$) form spontaneously within these cells [2]. Abnormally large amounts of activated oxygen (especially H_2O_2) occur within malaria-infected murine red cells [3]. One would therefore expect intraerythrocytic forms of malaria to have particularly effective oxidant defense mchanisms.

In view of the probable dependence of the malaria parasite on antioxidant enzymes, we investigated malarial SOD with a view to identifying a new target for the development of chemotherapeutic agents. Accordingly, we measured SOD [4] in normal mouse erythrocytes, mouse erythrocytes infected with *Plasmodium berghei*, and isolated parasites [5]. Contrary to our expectation, malaria-infected erythrocytes showed less SOD activity than normal mouse red cells when expressed on a per cell basis (Table 1). A portion of this decrement undoubtedly reflected parasite-mediated digestion of host cell cytoplasmic contents. Thus, if SOD activity was expressed per unit of protein or hemoglobin, infected cells appeared to have nor-

Table 1. Superoxide dismutase activities of normal and *P. berghei*–infected murine red blood cells (RBC's) and of isolated *P. berghei* [5]. The results are expressed as means ± standard deviation.

Sample	SOD activity		Percentage inhibition by cyanide ($10^{-3}M$)
	$U/10^9$ cells	U/mg protein	
Normal RBC's	89.7 ± 12 ($N = 8$)	4.4 ± 0.5 ($N = 8$)	85.2 ± 15.5 ($N = 5$)
Infected RBC's	63.8 ± 16 ($N = 10$)	6.1 ± 1.5 ($N = 10$)	
Isolated parasites		12.3 ± 4.1 ($N = 10$)	82.6 ± 1.1 ($N = 3$)

nal or increased amounts of activity (Table 1 and Fig. 1).

Despite the decreased SOD within infected cells, isolated parasites contained substantial SOD activity (Table 1) (6). Indeed, in red cells from animals with various percentages of infected cells, there was a concomitant decline in host cell SOD with an increase in parasite-associated SOD activities. At very high parasitemias (90 to 100 percent of cells infected) approximately 20 percent of the total SOD activity was parasite-associated, whereas red cell SOD activity was 20 percent lower (not shown). These results suggested that some host SOD might be taken up by the parasite. Furthermore, although isolated parasites are unavoidably contaminated with host cell cytoplasm, such contamination could not account for the substantial parasite-associated SOD activity because the ratio of SOD to hemoglobin was much higher in the parasite than the host cell (Fig. 1). Indeed, this ratio was more than ten times higher in crude parasite lysates than in uninfected red cells and almost 70 times higher when membrane-associated and particulate hemoglobin was removed by ultracentrifugation (Fig. 1).

To characterize further this parasite-associated SOD, we assessed the inhibitory effects of cyanide. Whereas the mammalian copper- and zinc-containing SOD is cyanide-sensitive (1), those of protozoan parasites—such as Leishmania tropica, Trypanosoma brucei, and Trypanosoma cruzi—have iron at the center and are cyanide-insensitive (7). Contrary to our expectation, the SOD associated with malaria parasites was as readily inhibited by cyanide as that of the host cell (Table 1). This cyanide inhibition of plasmodial SOD suggests that the enzyme contains copper and zinc and, therefore, resembles host SOD more than it does protozoan SOD's.

Plasmodial and mouse enzymes are indistinguishable electrophoretically. Isoelectric focussing (IEF) (7) and polyacrylamide gel electrophoresis (PAGE) (8) were performed on extracts of mouse red cells and of isolated parasites (9) and stained for SOD activity (10). The mobilities of host and parasite SOD's were identical on PAGE (Fig. 2A) and IEF (Fig. 2B) ($pI \cong 5.1$).

These results suggest that the malarial SOD might be entirely of host origin. Indeed, we have obtained direct evidence that this is the case. If the plasmodial SOD originates from the host cell, the characteristics of this enzyme should depend on the species of infected animal. To test this, we infected weanling rats

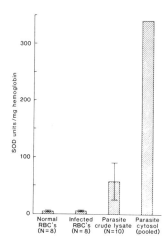

Fig. 1. Ratio of SOD activity to hemoglobin content of normal and P. berghei–infected murine red blood cells (RBC's), and of isolated P. berghei (5) (mean ± standard deviation). The parasite cytosol was prepared by ultracentrifugation (100,000g) of homogenized parasites (pooled samples from 12 mice) for 1 hour (12). Mean SOD activities per milligram of protein in these preparations (see Table 1) were 4.4 ± 0.5 (red cells), 6.1 ± 1.5 (infected red cells), 12.3 ± 4.1 (crude parasite lysates) and 16.8 (parasite cytosol). Note the much greater ratio of SOD to hemoglobin in parasite-associated material.

with the same strain of P. berghei. Examination by PAGE (Fig. 2A) and IEF (Fig. 2B) of parasite cell free extracts isolated from these animals revealed bands of activity identical to those from extracts of uninfected rat erythrocytes. The isoelectric point of the parasite-associated SOD ($pI \cong 5.7$) was characteristic of rat red cell SOD and quite distinct from that of parasites isolated from infected mice.

Thus several lines of evidence indicate that SOD activity associated with isolated P. berghei originates from the host cell. First, although the total cellular SOD activity declines within infected erythrocytes, substantial SOD is present within isolated P. berghei. In progressively more heavily infected animals, the host cytoplasm SOD activity decreases whereas that associated with isolated parasites increases reciprocally. Second, the SOD activity of isolated P. berghei is as sensitive to cyanide inhibition as is that of the host cell. In this regard, malarial SOD is unlike those reported for any other parasitic protozoan. Third, control studies (not shown) indicate no detectable passive adsorption of exogenously added (bovine) SOD by isolated P. berghei. Finally, the electrophoretic properties of P. berghei–associated SOD depend on whether the parasite is raised in mice or rats and, in both cases, the parasite SOD is electrophoretically indistinguishable from the host cell enzymes.

These observations indicate that P. berghei lacks an endogenous SOD and may depend on the host cell as a source of active enzyme. The absence of detectable SOD synthesized by the parasite itself is especially surprising in an organism living in such an oxygen-rich environment. The fact that it does not produce SOD is probably due to adoption of the active host cell enzyme. To our knowledge, this is the first example of direct parasitic incorporation of a host cell enzyme. The uptake of nutrients and cofactors by malarial parasites is well established (11). Malarial parasites rapidly endocytose and digest host cell hemoglobin (11) and, probably, other cytoplasmic constituents. SOD that is taken up along with the hemoglobin may es-

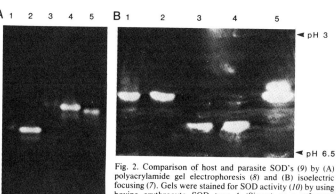

Fig. 2. Comparison of host and parasite SOD's (9) by (A) polyacrylamide gel electrophoresis (8) and (B) isoelectric focusing (7). Gels were stained for SOD activity (10) by using bovine erythrocyte SOD type 1 (Sigma) as a reference [$pI = 4.95$ (13)]. Lane 1, SOD of mouse RBC's; lane 2, mouse-derived P. berghei SOD; lane 3, SOD from rat RBC's; lane 4, rat-derived P. berghei SOD; lane 5, bovine SOD.

cape digestion and either remain within the food vacuole or emigrate into the parasite cytoplasm. These results serve to emphasize the complexity and intricacy of parasite dependence on host resources. Knowledge of such dependence may, in the future, suggest alternative approaches to the control of parasitic diseases.

References and Notes

1. I. Fridovich, *Annu. Rev. Biochem.* **44**, 147 (1975); J. M. McCord, B. B. Kaele, Jr., I. Fridovich, *Proc. Natl. Acad. Sci. U.S.A.* **68**, 1024 (1971).
2. R. P. Hebbel, J. W. Eaton, M. Balasingam, M. H. Steinberg, *J. Clin. Invest.* **70**, 1253 (1982).
3. N. L. Etkin and J. W. Eaton, in *Erythrocyte Structure and Metabolism*, G. J. Brewer, Ed. (Liss, New York, 1975), p. 219; M. J. Friedman, *Nature (London)* **280**, 245 (1979).
4. M. L. Salin and J. M. McCord, *J. Clin. Invest.* **54**, 1005 (1974).
5. *Plasmodium berghei* (NYU-2) was maintained in NCS or BALB/c mice by intraperitoneal injection of infected blood to yield a maximum parasitemia ($\geq$ 75 percent) in 4 days with no increase in reticulocyte number. Mice were bled intracardially with heparin, and Hanks balanced salt solution (HBSS) was used to wash the red blood cells. White cells and platelets were removed by using a Sigmacell/Cellulose column [E. Beutler, *Red Cell Metabolism: A Manual of Biochemical Methods* (Grune & Stratton, New York, 1975), p. 10]. Erythrocytes were washed three times in HBSS after which parasites were isolated by saponin lysis [A. Zuckerman, *Bull. WHO* **37**, 431 (1967)]. Parasites were sedimented by centrifugation at 4500g for 5 minutes and washed four times in HBSS. The initial parasite-free supernatant was used to represent the host red blood cell cytoplasm. All fractions were frozen and thawed and Triton X-100 was added to 0.05 percent before the SOD, hemoglobin, and protein determinations were done. Hemoglobin was measured with Drabkin's reagent as described in N. W. Teitz *et al.*, *Fundamentals of Clinical Chemistry* (Saunders, Philadelphia, 1976), p. 412. Total protein was determined by the method of M. M. Bradford [*Anal. Biochem.* **72**, 248 (1976)].
6. U. Suthipark, J. Krungkrai, A. Jearnpipatkul, Y. Yuthavong, B. Panijpan, *J. Parasitol.* **68**, 337 (1982).
7. S. R. Meshnick and J. W. Eaton, *Biochem. Biophys. Res. Commun.* **102**, 970 (1981); S. R. Meshnick, N. L. Trang, K. Kitchener, A. Cerami, J. W. Eaton, in *Oxy Radicals and Their Scavenger Systems: Molecular Aspects*, G. Cohen and R. Greenwald, Eds. (Elsevier, New York, in press). N. L. Trang, S. R. Meshnick, K. Kitchener, J. W. Eaton, A. Cerami, *J. Biol. Chem.* **258**, 125 (1983).
8. B. J. Davis, *Ann. N.Y. Acad. Sci.* **121**, 404 (1964).
9. Erythrocyte SOD-containing extracts were prepared by chloroform-ethanol treatment and acetone precipitation of normal red blood cell lysates [J. M. McCord and I. Fridovich, *J. Biol. Chem.* **244**, 6049 (1969)]. Parasite extracts were prepared by sonication (three times for 20 seconds in 25 mM tris buffer, pH 6.8) of washed isolated parasites followed by centrifugation (27,000g for 1 hour at 4°C). Both extracts were dialyzed at 4°C in three changes of 25 mM tris containing 1 mM EDTA, pH 6.8, before they were used.
10. C. O. Beauchamp and I. Fridovich, *Anal. Biochem.* **44**, 276 (1971).
11. I. W. Sherman, *Microbiol. Rev.* **43**, 453 (1979).
12. For these experiments, isolated parasites were homogenized in a motorized Potter-Elvehjem-type homogenizer (40 strokes, 1200 rev/min) in sucrose-TKM buffer [I. W. Sherman and L. A. Jones, *J. Protozool.* **26**, 489 (1979)]. Homogenates were centrifuged at 100,000g for 1 hour and the supernatants were dialyzed and assayed for protein, hemoglobin, and SOD activity.
13. J. V. Bannister, W. H. Bannister, E. Wood, *Eur. J. Biochem.* **18**, 178 (1971).
14. We thank A. Cerami and J. Haldane for discussions and K. Kitchener, C. Schneider, and S. Swiecicki for technical assistance. Supported by NIH grants AI 16975 and HL 16833, by Rockefeller Foundation grant RF 81060, and by grants from the WHO/UNDP/World Bank Special Programme for Research and Training in Tropical Diseases. J.W.E. is a recipient of an NIH Research Career Development Award and S.R.M. is a recipient of an Irma T. Hirschl Trust Career Scientist Award.

Part III
ORIGIN OF MITOCHONDRIA

Editors' Comments
on Papers 13 through 18

13 THOMAS and WILKIE
Recombination of Mitochondrial Drug-Resistance Factors in Saccharomyces Cerevisiae

14 BARATH and KÜNTZEL
Cooperation of Mitochondrial and Nuclear Genes Specifying the Mitochondrial Genetic Apparatus in Neurospora crassa

15 JOHN and WHATLEY
Paracoccus denitrificans *and the Evolutionary Origin of the Mitochondrion*

16 GRAY et al.
Organization and Evolution of Ribosomal RNA Genes in Wheat Mitochondria

17 VAN DEN BOOGAART, SAMALLO, and AGSTERIBBE
Similar Genes for a Mitochondrial ATPase Subunit in the Nuclear and Mitochondrial Genomes of Neurospora crassa

18 OBAR and GREEN
Excerpt from *Molecular Archaeology of the Mitochondrial Genome*

The diversity in morphology, genetics, and biochemistry of the mitochondria found in the four kingdoms of eukaryotes suggests a polyphyletic origin for these organelles.

Paracoccus bacteria are possibly descendants of one original line of symbiotic bacteria that became mitochondria. The similarities between the modern *Paracoccus* bacteria and modern mitochondria are reviewed by John and Whatley in Paper 15. The research described in Paper 16 is an example of the accelerated gathering of evidence for similarities between mitochondria and bacteria that followed the advent of new techniques of restriction mapping and nucleotide sequencing in the 1970s. The concept of mitochondria as independent genetic entities with their own sexual system is discussed by Thomas and Wilkie in Paper 13.

The mitochondrial genome is greatly reduced. Control over

Editors' Comments on Papers 13 through 18

several important functions is shared between the mitochondria and the nucleus (Paper 14). It is clear that genes from mitochondria have been transferred directly or indirectly to the nucleus (Paper 17) as well as to the plastid (Paper 23). Also, mitochondria have somewhat different genetic codes than do prokaryotes and the nuclei of eukaryotes. The complexity of mitochondrial biochemistries leads to considerable difficulty in reconstructing the events that led to the greatly modified mitochondrial genome, a topic reviewed and discussed by Obar and Green in Paper 18.

RECOMBINATION OF MITOCHONDRIAL DRUG-RESISTANCE
FACTORS IN SACCHAROMYCES CEREVISIAE

D.Y. Thomas and D. Wilkie

Department of Botany, University College London.

Received January 17, 1968

Re-assortment of mitochondrially-located drug-resistance factors in the vegetative progeny of individual zygotes will be described and provides evidence of a recombinational process involving the genetic material of the organelle.

The specific inhibition of respiratory enzyme synthesis in yeast by certain protein inhibitors of bacteria was first reported by Linnane and his collaborators (see Clark-Walker and Linnane, 1967) and the isolation and characterization of spontaneous resistant mutants to some of these antibiotics is described in Wilkie et al (1967). A detailed account of a series of mutants resistant to erythromycin is given in Roodyn and Wilkie (1967) and Thomas and Wilkie (1968) in which the procedure for locating genetic factors for resistance in the so-called mitochondrial rho factor (generally assumed to be mitochondrial DNA (MDNA)) is fully described. Briefly this involves isolating the cytoplasmic petite mutant, in which there is effective loss of MDNA (see Mounolou et al, 1966), from those strains exhibiting non-chromosomal inheritance of resistant factors and demonstrating concomitant loss of the resistance factor in subsequent crosses to sensitive strains. In this way a number of rho factor mutants resistant to various antibiotics have now been obtained and the present report concerns those resistant to the macrolide antibiotics erythromycin and spiramycin and the aminoglycoside paromomycin respectively. It may be emphasized that these drugs affect the protein-synthesizing system of the mitochondrion only and cause inhibition of cell growth only in media containing non-fermentable substrate.

METHODS

Genetically marked yeast strains of this laboratory were used. General techniques for crossing, zygote manipulation and determination of drug tolerance levels are detailed in Thomas and Wilkie (1968).

Since mitochondria are not synthesized and are not seen in cells of S. cerevisiae grown under anaerobic conditions (Lukins et al, 1967), although all genetic information for re-synthesis continues to be transmitted, it was thought that crossing of strains under anaerobic conditions would facilitate any recombinational process involving MDNA. Thus crosses were set up between anaerobic cultures of differently auxotrophic haploid strains in liquid minimal medium (Wickerham's) containing 4% glucose under a nitrogen atmosphere in a Fildes cylinder. Electron microscope studies show no evidence of mitochondrial structures in cells grown under our anaerobic conditions (unpublished results of Drs. R. Marchant and D. Smith of this laboratory).

The zygotes produced were usually in comparatively high numbers and were plated on solid minimal medium containing 2% glucose incubated aerobically for 2 days at $30^\circ C$, and the resulting prototrophic diploid colonies to which each zygote gave rise were picked off and suspended in water. Each clone of cells was then tested for drug resistance by dropping out onto the appropriate drug series of plates. Some of the multiple-resistant recombinant clones were confirmed as such by testing cells grown on one antibiotic for ability to grow on the other drug or drugs. Confirmatory tests were also made on multiple-drug plates. This ruled out the possibility of clones of this type being comprised of mixtures of mitochondrial types either at the cell level or at the population level.

RESULTS AND DISCUSSION

The drop-out method of analysis of clones derived from individual zygotes enables an estimate to be made of the frequency of resistant and sensitive cells provided the cell density is not greater than about 10^3 per drop. In nearly all cases all cells within a clone derived from a zygote formed under anaerobic conditions seemed to contain mitochondria of one type only, and the segregation of types of clones is given in Table 1. It must be emphasized that these segregations and recombinations have taken place in diploid cells testifying to the non-chromosomal nature of the resistance factors.

If it is assumed that recombination involves crossing-over between MDNA strands carrying their respective resistance factors, it would appear that under anaerobic conditions only one MDNA strand survives in the zygote. However, the significant excess of multiple-sensitive recombinant clones in crosses 1, 3 and 4 must be accounted for. If it is further assumed that MDNA is circular, it will be appreciated that a single cross-over between two circles would give a single large loop while two or an even number of cross-overs would allow separation of two small circles.

This is not unlikely in view of Avers (1967) claim to have isolated circular MDNA filaments from yeast of different lengths. Thus a cross-over (or odd number of cross-overs) at any point

Table 1

Analysis of clones from individual
zygotes in various crosses

Strains and crosses		Genotypes*	Clone types and number				Total clones
1-617		$E^s\ P^r$	$E^s\ P^r$ 25		$E^s\ P^s$ 31		80
X	(1)		$E^r\ P^s$ 19		$E^r\ P^r$ 5		
6-82		$E^r\ P^s$					
1-617		$E^s\ S^s\ P^r$	$E^s\ S^s\ P^r$ 49		$E^s\ S^r\ P^s$ 3		119
X	(2)		$E^r\ S^r\ P^s$ 25		$E^s\ S^r\ P^r$ 4		
			$E^s\ S^s\ P^s$ 14		$E^r\ S^s\ P^s$ 1		
6-82		$E^r\ S^r\ P^s$	$E^r\ S^r\ P^r$ 23				
1-601		$E^s\ S^s\ P^{r1}$	$E^s\ S^s\ P^r$ 49		$E^r\ S^r\ P^r$ 10		156
X	(3)		$E^r\ S^r\ P^r$ 47		$E^s\ S^r\ P^s$ 1		
6-82		$E^r\ S^r\ P^s$	$E^s\ S^s\ P^s$ 21				
1-619		$E^s\ S^{r1}$	$E^s\ S^r$ 2		$E^s\ S^s$ 152		195
X	(4)		$E^r\ S^r$ 40		$E^r\ S^s$ 1		
6-82		$E^r\ S^r$					

* E^s, sensitivity to 10 μg/ml erythromycin
  E^r, resistance to >3 mg/ml erythromycin
  S^s, sensitivity to 50 μg/ml spiramycin
  S^r, resistance to 2 mg/ml spiramycin
  P^s, sensitivity to 50 μg/ml paromycin
  P^r, resistance to 1 mg/ml paromycin

would result in "diploid" MDNA heterozygous for resistance factors which would be expected to be recessive if they make altered mitochondrial ribosomes (discussed in Roodyn and Wilkie, 1967), the probable sites of action of these antibiotics. A conjoined configuration could be relatively stable and give only occasional resistant segregants within a clone as has been seen in some cases.

On the other hand, recombination may result from mixtures of the different MDNA strands within individual mitochondria, but if this generally applied one would expect much less distinction between clones and less uniformity within a clone regarding mitochondrial type. These points may be resolved by further tests particularly tetrad analysis.

Crosses carried out aerobically yielded clones of mixed cells showing various proportions of the two parental types and plating on multiple-drug plates revealed very few recombinant cells of the multiple-resistant type.

It is perhaps premature to consider mapping resistance factors but it could be inferred from the data that the "genes" E and S for macrolide resistance are closely linked while these and the P "gene" are comparatively far apart on the mitochondrial genome.

The main features of this communication were given in a paper by Wilkie and Thomas read at the meeting of the British Genetical Society on 10 November 1967.

REFERENCES

Avers, C.J., Proc.Nat.Acad.Sci., 58, 620 (1967).

Clark-Walker, G.D. and Linnane, A.W., Biochem.Biophys.Res.Commun., 25, 8 (1967).

Lukins, H.B., Tham, S.H., Wallace, P.G. and Linnane, A.W., Biochem.Biophys.Res.Commun., 23, 363 (1967).

Mounolou, J., Jakob, H. and Slonimski, P.P., Biochem.Biophys.Res.Commun., 24, 218 (1966).

Roodyn, D.B. and Wilkie, D., The Biogenesis of Mitochondria, Methuen, London (1967).

Thomas, D.Y. and Wilkie, D., Genet.Res.Camb., 11 (1968). In the press.

Wilkie, D., Saunders, G. and Linnane, A.W., Genet.Res.Camb., 10, 199 (1967).

Cooperation of Mitochondrial and Nuclear Genes Specifying the Mitochondrial Genetic Apparatus in *Neurospora crassa*

(chloramphenicol/ethidium bromide/repressor control/model)

ZOLTAN BARATH* AND HANS KÜNTZEL

Max-Planck-Institut für experimentelle Medizin, Abteilung Chemie, Göttingen, Hermann-Rein-Str. 3, Germany

Communicated by David E. Green, March 22, 1972

ABSTRACT Enzymes involved in the expression of the mitochondrial genome in *Neurospora crassa* are induced by chloramphenicol and ethidium bromide, which block transcription and translation of mitochondrial DNA. It is concluded that most, if not all, proteins of the mitochondrial genetic apparatus are coded by nuclear genes, synthesized on cytoplasmic ribosomes, and controlled by a repressor-like mitochondrial gene product. A model explaining the coordination of nuclear and mitochondrial division cycles by repressor control is discussed.

In some aspects mitochondria behave like endosymbiotic bacterial cells, because they proliferate by division, growth, and distribution to the progeny (1–3) and because they contain a circular genome that is expressed by enzymes of bacterial specificity (4). On the other hand, mitochondria can also be considered as giant multienzyme complexes that are almost completely built up by the nuclear-cytoplasmic protein synthesizing system. Less than 10% of the mitochondrial proteins are synthesized on mitochondrial ribosomes; this fraction consists of probably not more than eight species of hydrophobic proteins that are incorporated into the inner membrane where they assemble cytochromes, cytochrome oxidase, and ATPase (4). A mitochondrial genetic origin of these proteins is highly probable, but has not been proven.

The only mitochondrial gene products that have positively been identified are ribosomal RNA and transfer RNA (4). Although the coding capacity of mitochondrial DNA would be large enough to code for at least an additional 20 proteins, there is no evidence that enzymes involved in the expression of the mitochondrial genome are coded by this genome, with the possible exception of a replication factor (5, 6). On the other hand, most workers agree that not only all proteins of the outer mitochondrial membrane and most proteins of the inner membrane, but also the proteins of the mitochondrial genetic apparatus are synthesized on cytoplasmic ribosomes (4, 7–9). Furthermore, a few enzymes involved in mitochondrial protein synthesis have been suggested to be coded by nuclear DNA, namely a mitochondrial leucyl-tRNA synthetase from *Neurospora* (10) and the two mitochondrial peptide chain elongation factors from yeast (11). We have studied the biosynthesis of mitochondrial RNA and ribosomes and of two enzymes of bacterial specificity involved in mitochondrial protein synthesis in *Neurospora*. The finding that specific inhibitors of mitochondrial transcription and translation stimulate the biosynthesis of mitochondrial enzymes strongly suggests a nuclear origin of these proteins and a control by mitochondrial gene products.

METHODS

Growth Conditions. *Neurospora crassa* (wild type, Em 5256) was grown at 30° for 14 hr. The inoculum was 2×10^6 conidia per ml. Chloramphenicol (2 mg/ml) was added together with conidia; 75 μM ethidium bromide was added 4 hr after inoculation. The hyphae were homogenized in a carborundum mill (12, 13). The methods for cell fractionation (7) and for isolation of mitochondrial RNA (14), and mitochondrial and cytoplasmic ribosomes (15) have been described.

Assay of Elongation Factors G (EF-G). Hyphae were washed with 20 mM Tris·HCl (pH 7.8)–5 mM MgCl$_2$–14 mM 2-mercaptoethanol–10 mM KCl, ground with sea sand, and extracted with half their wet weight of the above buffer. Debris were removed by centrifugation for 5 min at 3000 rpm in a Christ centrifuge. The supernatant was sonified four times in ice with a Branson sonifier for 15 sec at full power and centrifuged for 20 min at 17,000 rpm in a Sorvall centrifuge. The supernatant was centrifuged for 1 hr at 65,000 rpm in a Spinco centrifuge. Peptide chain elongation factors were isolated from the high-speed supernatant by (NH$_4$)$_2$SO$_4$ precipitation and analyzed by filtration on Sephadex G-150 as described (16).

The fractions were assayed in two different reaction mixtures. Both mixtures contained 96 mM Tris·HCl (pH 7.8), 13 mM KCl, 5 mM phosphoenolpyruvate, 11 mM 2-mercaptoethanol, 0.8 mM GTP, 16 μg/ml of pyruvate kinase, 100 μg/ml poly(U), and 100,000 cpm/ml [^{3}H]phenylalanyl-tRNA. Assay mixture A (G$_{70}$-activity) was supplemented with 70S ribosomes (28 A units/ml) and 33 μg/ml of elongation factor T (EF-T) (19), both from *Escherichia coli*. Assay mixture B (G$_{80}$-activity) was supplemented with cytoplasmic 77S ribosomes from *Neurospora* (35 A units/ml). After incubation for 30 min at 32°, the radioactivity in the hot Cl$_3$CCOOH-insoluble fraction was determined.

Assay of N^{10}-Formyltetrahydrofolate: Methionyl-tRNA transformylase. Hyphae were washed with 10 mM Tris·HCl (pH 7.5)–10 mM MgCl$_2$–10 mM 2-mercaptoethanol, ground with sea sand, and extracted with the same buffer. High-speed supernatants were prepared from sonified crude extracts as de-

Abbreviations: EF-G, elongation factor G; EF-T, elongation factor T.

* Permanent address: Biological Institute of Slovak Akademy of Sciences, Bratislava, Czechoslovakia.

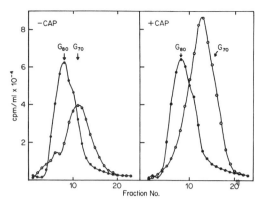

FIG. 1. Filtration on Sephadex G-150 of crude peptide chain elongation factors from whole cell extracts of *Neurospora* grown in the absence and presence of chloramphenicol. For details see *Methods* and ref. 16. 10-μl aliquots of the 1.5-ml fractions were assayed in 50-μl mixtures containing either 70S ribosomes from *E. coli* (O——O) or 77S ribosomes from *Neurospora* (●——●).

scribed above and dialyzed overnight against the extraction buffer. [^{3}H]Methionyl-tRNA (from *E. coli*) was prepared as described (17), except that the formyl donor was omitted.

Transformylase was tested in a mixture containing 50 mM Tris·HCl (pH 7.5), 10 mM MgCl$_2$, 6 mM 2-mercaptoethanol, 0.01 mM N^{10}-formyltetrahydrofolic acid, and 1.2 mg/ml [^{3}H]methionyl-tRNA (600 cpm/μg). After incubation at 37°, aliquots were treated with 0.1 volume of 25% ammonia for 5 min at room temperature and subjected to paper electrophoresis for 2 hr at 1500 V on Whatman 3 MM sheets. The dried electropherograms were cut into strips and counted in a liquid scintillation counter in toluene–2,5-diphenyloxazole–1,4-bis-(5-phenyloxazolyl-2)-benzene.

RESULTS

Table 1 shows that the amount of mitochondrial RNA in relation to postmitochondrial supernatant protein has about doubled in cells grown in the presence of 2 mg/ml of chloramphenicol. The mass of mitochondrial protein relative to nonmitochondrial protein has increased 1.3-fold; this explains why the RNA content of mitochondria has increased only 1.5-fold, but also suggests that the increased formation of mitochondrial proteins by cytoplasmic ribosomes may compensate for the loss of cytochrome assembling proteins and the reduction of respiratory efficiency. We have previously shown that the biosynthesis of mitochondrial RNA polymerase in *Neurospora* is stimulated by chloramphenicol and ethidium bromide (18). This suggests that not the rate of transcription, but rather the formation of the mitochondrial transcription complex is stimulated by the antibiotic.

Table 1 also demonstrates that the amount of mitochondrial ribosomes has increased more than 2-fold in comparison to cytoplasmic ribosomes. The ribosome preparations from cells grown in the presence and absence of chloramphenicol did not differ in their activity in poly(U)-dependent cell-free systems and in their specific response to fusidic acid (16). This result confirms the previous finding that all ribosomal proteins essential for the function of mitochondrial ribosomes are synthesized on cytoplasmic ribosomes (7–9). In addition, it demonstrates that the biosynthesis of mitochondrial, but not of cytoplasmic, ribosomal proteins is stimulated by blockage of mitochondrial protein synthesis, suggesting a control of nuclear genes coding for mitochondrial ribosomal proteins by mitochondrial gene products (see *Discussion*).

Fig. 1 shows the activity pattern of mitochondrial and cytoplasmic peptide chain elongation factors after filtration through a Sephadex G-150 column (16). The enzyme fractions have been isolated from high-speed supernatants containing both mitochondrial soluble matrix proteins and cytosolic proteins. The two complementary elongation factors, EF-G and EF-T, are not separated by this procedure; however, because of the extreme lability of the mitochondrial EF-T (19), the mitochondrial EF-G was complemented with EF-T from *E. coli* for assay. As can be seen from Fig. 1, the enzyme fractions from cells grown in the absence and presence of chloramphenicol contain about the same activity of cytoplasmic elongation factors, whereas the activity of the mitochondrial EF-G is about doubled in the fraction from cells treated with chloramphenicol.

The finding that the biosynthesis of mitochondrial proteins is unaffected, or even stimulated, by chloramphenicol excludes an intramitochondrial site of synthesis but does not exclude a mitochondrial origin of messenger RNA coding for these proteins. The transcriptional origin of mitochondrial enzymes can be determined with ethidium bromide, a drug known to induce mitochondrial "petite" mutants in yeast and to inhibit repair, transcription, and, indirectly, translation of mitochondrial, but not of nuclear, DNA (4). This method cannot be used to study the origin of mitochondrial ribosomal proteins because these proteins are only detectable after assembly of the mitochondrial ribosome, a step that is directly

TABLE 1. *Stimulation of mitochondrial RNA and ribosome synthesis by chloramphenicol*

	Mitochondrial RNA (μg)		Mitochondrial ribosomes (μg/mg cytoplasmic ribosomes)	Mitochondrial protein (mg/mg postmitochondrial supernatant protein)
Growth conditions	per mg Mitochondrial protein	per mg Postmitochondrial supernatant protein		
Without chloramphenicol	31	3.5	39	0.11
With chloramphenicol	48	7.3	90	0.14
Stimulation factor	1.5	2.1	2.3	1.3

Protein was determined by the method of Lowry *et al.* (33). The mitochondrial ribosomes had a A_{260}/A_{280} ratio of 1.8, the cytoplasmic ribosomes a ratio of 2.0. 1 mg of ribosomes was taken as 10 A_{260} units.

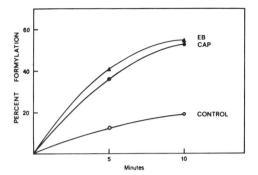

FIG. 2. Methionyl-tRNA transformylase activity of high-speed supernatant fractions from whole cells of *Neurospora* grown in the absence and presence of chloramphenicol (CAP) or ethidium bromide (EB). For details see *Methods*. The amount of protein added per ml assay mixture was 0.63 mg (*Control*), 0.57 mg (*CAP*), and 0.76 mg (*EB*). Previous experiments have shown that the initial formylation rate increased linearly with the protein concentration.

controlled by ethidium bromide-sensitive transcription of mitochondrial ribosomal RNA genes.

We have, therefore, studied the effect of ethidium bromide on the biosynthesis of methionyl-tRNA transformylase a mitochondrial enzyme of bacterial specificity that does not depend on assembly steps to be tested and that is normally absent from the cytoplasm (20–23). The enzyme was assayed by N-formylation of [³H]methionyl-tRNA from *E. coli*, followed by alkaline hydrolysis and electrophoretic separation of methionine and N-formylmethionine. Fig. 2 demonstrates that the transformylase activity of high-speed supernatant fractions obtained from whole cells is increased more than 2-fold if cells have been grown in the presence of chloramphenicol or ethidium bromide.

DISCUSSION

Chloramphenicol is known to suppress *in vivo* the synthesis of inner membrane proteins contributed by intramitochondrial protein synthesis and, thus, to prevent the proper assembly of cytochromes a, a_3, and b and of cytochrome oxidase, resulting in the formation of respiratory-deficient mitochondria characteristic for mitochondrial mutants of yeast (24) and *Neurospora* (25). However, mitochondria isolated from chloramphenicol-treated yeast cells still incorporate amino acids into protein, indicating that the proteins of the mitochondrial translational apparatus in yeast are synthesized on cytoplasmic ribosomes. This agrees with our finding that chloramphenicol does not inhibit biosynthesis of mitochondrial RNA, mitochondrial ribosomes, and mitochondrial enzymes like RNA polymerase (18), ribosomal translocase, and methionyl-tRNA transformylase. Moreover, there is increasing evidence that most, if not all, proteins involved in the expression of mitochondrial DNA are coded by nuclear DNA. A nuclear origin has been suggested for mitochondrial enzymes like DNA polymerase (26), RNA polymerase (18), EF-G (11), and leucyl-tRNA synthetase (10). The results shown in Fig. 2 allow us to add methionyl-tRNA transformylase to this list.

The unexpected finding that the biosynthesis of these enzymes is stimulated by agents blocking transcription and translation of mitochondrial DNA suggests that the nuclear cistrons coding for these proteins are controlled by one or several repressor-like proteins which, in turn, are coded by mitochondrial DNA and synthesized on mitochondrial ribosomes. A similar repressor mechanism has been suggested by Williamson (5) to explain the high efficiency of ethidium bromide in inducing "petite" mutants in yeast. However, in contrast to Williamson who postulates an irreversible repression, we explain the ethidium effect in *Neurospora* by derepression of nuclear genes coding for mitochondrial proteins.

Fig. 3 illustrates the cooperation of nuclear and mitochondrial genes responsible for mitochondrial biogenesis. We have

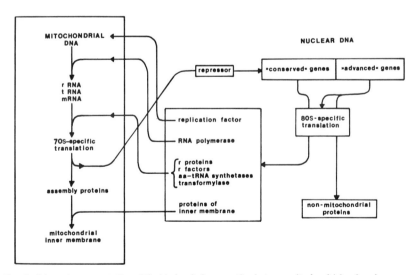

FIG. 3. Schematic representation of the biochemical cooperation between mitochondrial and nuclear genes.

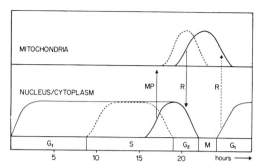

FIG. 4. Possible coordination of nuclear and mitochondrial division cycles by repressor control. The time scale reflects the mitotic cycle of synchronized HeLa cells (31). - - -, DNA synthesis; ———, synthesis of nonmitochondrial proteins; ———, synthesis of mitochondrial proteins. R = repressor; MP = mitochondrial proteins.

to assume that the nuclear genome contains a set of genes that code for mitochondrial proteins and that have been conserved since the origin of the eukaryotic cell, probably because they had to replace functionally the lost genes of the mitochondrial genome. A part of these gene products indeed exhibit a strict specificity for prokaryotic protein synthesis (4). Another set of genes, which probably evolved later by duplication of the early genome, is responsible for the synthesis of the nuclear/cytoplasmic genetic system, of the mitochondrial outer membrane, nuclear membrane, and outer cell membranes, and of other nonmitochondrial proteins. It is suggested that the conserved part of the nuclear genome responsible for mitochondrial biogenesis is controlled by its "partner," the mitochondrial genome, which produces repressor-like proteins binding to the nuclear partner genes.

There is good evidence that in *Neurospora* (2), yeast (3, 27, 28), and HeLa cells (29, 30) all mitochondria divide and grow synchronously and that the replication periods of nuclear and mitochondrial DNA alternate periodically. The possible role of mitochondrial repressor(s) in maintaining the rhythm of nuclear and mitochondrial division is depicted in Fig. 4, showing the mitotic cycle of HeLa cells (31). In HeLa cells, mitochondrial DNA synthesis starts in the S phase and reaches its maximum in the G_2 phase (29), whereas mitochondrial transcription starts in the G_2 phase (30) and proceeds into the metaphase (32). Consequently mitochondrial protein synthesis would start in the G_2 phase and produce a repressor that terminates translation of nuclear genes coding for mitochondrial proteins. After metaphase only the unrepressed nuclear genes would start to be expressed (G_1 phase), whereas expression of the repressed "conserved" genes would be possible only after derepression by a new round of nuclear replication (late S phase).

This model predicts three periods of protein synthesis during the cell cycle: A first period, starting after metaphase, in which nonmitochondrial proteins are produced; a second period, starting in the late S phase, in which the enzymes of the mitochondrial genetic apparatus and most of the inner membrane proteins are synthesized; and a third period of intramitochondrial protein synthesis during the G_2 phase, in which the previously formed "soluble" proteins of the respiratory chain and oxidative phosphorylation are assembled by the insoluble mitochondrial gene products to enlarge the inner membrane during mitochondrial growth. This relatively short period of mitochondrial replication, division, and growth could conceivably be initiated by the nuclear production of a mitochondrial replication factor (5, 6) and be terminated either by a repressor produced at the onset of nuclear gene expression after metaphase or, more likely, by the exhaustion of the pool of soluble mitochondrial proteins leading to a halt of assembly-controlled intramitochondrial protein synthesis.

This investigation was supported by the Deutsche Forschungsgemeinschaft. Z. B. holds a fellowship of the Deutsche Akademische Austauschdienst. We thank Dr. F. Cramer for generous support and Miss A. Helms for expert technical cooperation.

1. Luck, D. J. L. (1963) *J. Cell Biol.* **16**, 483–499.
2. Hawley, E. S. & Wagner, R. P. (1967) *J. Cell Biol.* **35**, 489–499.
3. Osumi, M. & Sando, N. (1969) *J. Electronmicros.* **18**, 47–55.
4. For references see: Schatz, G. (1969) in *Membranes of Mitochondria and Chloroplasts*, ed. Racker, E. Van Nostrand Reinhold Corp., New York, pp. 251–314; Küntzel, H. (1971) *Curr. Top. Microbiol. Immunol.* **54**, 94–118; Borst, P., *Ann. Rev. Biochem.*, in press.
5. Williamson, D. H., Maroudas, N. G. & Wilkie, D. (1971) *Mol. Gen. Genet.* **111**, 209–228.
6. Weislogel, P. O. & Butow, R. A. (1970) *Proc. Nat. Acad. Sci. USA* **67**, 52–58.
7. Küntzel, H. (1969) *Nature* **222**, 142–146.
8. Neupert, W., Sebald, W., Schwab, A. J., Massinger, P. & Bücher, T. (1969) *Eur. J. Biochem.* **10**, 589–591.
9. Davey, P. J., Yu, R. & Linnane, A. W. (1969) *Biochem. Biophys. Res. Commun.* **36**, 30–34.
10. Gross, S. R., McCoy, M. T. & Gilmore, E. B. (1968) *Proc. Nat. Acad. Sci. USA* **61**, 253–260.
11. Richter, D. (1971) *Biochemistry*, **10**, 4422–4424.
12. Weiss, H., von Jagow, G., Klingenberg, M. & Bücher, T. (1970) *Eur. J. Biochem.* **14**, 75–82.
13. Schäfer, K. P., Bugge, G., Grandi, M. & Küntzel, H. (1971) *Eur. J. Biochem.* **21**, 478–488.
14. Schäfer, K. P. & Küntzel, H. (1972) *Biochem. Biophys. Res. Commun.*, **46**, 1312–1319.
15. Küntzel, H. & Noll, H. (1967) *Nature* **215**, 1340–1345.
16. Grandi, M., Helms, A. & Küntzel, H. (1971) *Biochem. Biophys. Res. Commun.* **44**, 864–871.
17. Sala, F. & Küntzel, H. (1970) *Eur. J. Biochem.* **15**, 280–286.
18. Küntzel, H. & Barath, Z., *Hoppe-Seyler's Z. Physiol. Chem.*, in press.
19. Grandi, M. & Küntzel, H. (1970) *FEBS Lett.* **10**, 25–28.
20. Smith, A. E. & Marcker, K. A. (1968) *J. Mol. Biol.* **38**, 241–243.
21. Galper, J. B. & Darnell, J. E. (1969) *Biochem. Biophys. Res. Commun.* **34**, 205–214.
22. Küntzel, H. & Sala, F. (1969) *Hoppe-Seyler's Z. Physiol. Chem.* **350**, 1158–1159.
23. Epler, J. L., Shugart, L. R. & Barnett, W. E. (1970) *Biochemistry* **9**, 3575–3579.
24. Linnane, A. W. & Haslam, J. M. (1970) *Curr. Top. Cell. Regul.* **2**, 101–172.
25. Haskins, F. A., Tissieres, A., Mitchell, H. K. & Mitchell, M. B. (1953) *J. Biol. Chem.* **200**, 819–826.
26. Westergaard, O., Marcker, K. A. & Keiding, J. (1970) *Nature* **227**, 708–710.
27. Smith, D., Tauro, P., Schweizer, E. & Halvorson, H. O. (1968) *Proc. Nat. Acad. Sci. USA* **60**, 936–942.
28. Cottrell, S. T. & Avers, C. J. (1970) *Biochem. Biophys. Res. Commun.* **38**, 973–980.
29. Pica-Mattoccia, L. & Attardi, G., *J. Mol. Biol.*, in press.
30. Pica-Mattoccia, L. & Attardi, G. (1971) *J. Mol. Biol.* **57**, 615–621.
31. Terasima, T. & Tolmach, L. J. (1963) *Exp. Cell Res.* **30**, 344–362.
32. Fan, H. & Penman, S. (1970) *J. Mol. Biol.* **50**, 655–670.
33. Lowry, O. H., Rosebrough, N. J., Farr, A. L. & Randall, R. J. (1951) *J. Biol. Chem.* **193**, 265–275.

Paracoccus denitrificans and the evolutionary origin of the mitochondrion

Philip John & F. R. Whatley

Botany School, South Parks Road, Oxford OX1 3RA, UK

It is demonstrated that Paracoccus denitrificans *resembles a mitochondrion more closely than do other bacteria, in that it effectively assembles in a single organism those features of the mitochondrial respiratory chain and oxidative phosphorylation which are otherwise distributed at random among most other aerobic bacteria. A feasible evolutionary transition from the plasma membrane of an ancestral aerobic bacterium resembling* P. denitrificans *to the inner mitochondrial membrane is suggested.*

ACCORDING to the endosymbiotic theory of the evolutionary origin of the eukaryotic cell, as developed by Margulis[1], the mitochondrion has evolved from a free-living prokaryote resembling a present-day aerobic bacterium, which had been taken up by a plastid-free, amoeboid cell, the protoeukaryote, that depended on fermentation for the production of ATP. In time the prokaryote suffered a progressive loss of autonomy: its proliferation kept pace with the cell division of its host; many of its biosynthetic capabilities were lost or taken over by its host; and its metabolism became integrated with that of its host.

The endosymbiotic theory implies that the outer mitochondrial membrane is derived from the protoeukaryote, and that the inner mitochondrial membrane and its invaginations (the cristae) are homologous with the plasma membrane of present-day bacteria. Here we describe the modifications of the bacterial plasma membrane which would be necessitated by the evolutionary transition from a free-living, aerobic bacterium resembling *Paracoccus denitrificans* (previously *Micrococcus denitrificans*[2]) to a mitochondrion. In this transition the ATPase and the constitutive components of the respiratory chain are retained; the adaptive components of the respiratory chain are lost; and the transport properties of the plasma membrane are altered so that, in its new role as the inner mitochondrial membrane, it is able to integrate the metabolism of the organelle with that of the surrounding cell. We chose *P. denitrificans* as a plausible ancestor because, when all possible biochemical parameters are compared, it resembles a mitochondrion much more closely than do other aerobic bacteria. This comparison is summarised below and will be presented in detail elsewhere[3].

Comparison

Those mitochondrial features which *P. denitrificans* possesses, and which have a widespread (or perhaps universal) distribution among those bacteria which can make a respiratory chain, include nicotinamide dinucleotide transhydrogenase[4,5], NADH and succinate dehydrogenases[6-8], and the tricarboxylic acid cycle[9,10].

Those mitochondrial features which *P. denitrificans* possesses, but which have a limited distribution among other bacteria, include phosphatidyl choline as a major constituent of the membrane phospholipid fraction[11,12], straight-chain saturated and unsaturated fatty acids accounting for nearly all the membrane fatty acids[11,13,14], ubiquinone-10 as the sole functional quinone of the respiratory chain[6,15,16], two b-type and two c-type cytochromes as easily distinguishable components of the respiratory chain of aerobically grown cells[16-19], cytochrome aa_1 as the cytochrome oxidase[6,16,20], and a sensitivity to low concentrations of antimycin A and rotenone, both of which are inhibitors of mitochondrial electron transport[16,21] (see Fig. 1). Furthermore, mitochondrial cytochrome *c* more closely resembles the soluble cytochrome *c* isolated from *P. denitrificans* than the c-type cytochromes isolated from other bacteria, in both physical and structural properties[22-24]. The soluble cytochrome *c* of *P. denitrificans* is unusual among bacterial c-type cytochromes in that it is to a large extent interchangeable with mitochondrial cytochrome *c* in its ability to react with mitochondrial cytochrome oxidase[22,25-27] and with the two cytochrome oxidases of *P. denitrificans*, that is, the constitutive cytochrome aa_1[27,28] and the inducible cytochrome *cd*, which functions *in vivo* as a nitrite reductase[28,29].

Furthermore, *P. denitrificans*, like *Hydrogenomonas eutropha*[31] but unlike *Escherichia coli*[32] and *Bacillus megaterium*[33], resembles a mitochondrion[34,35] in the stoichiometry of oxidative phosphorylation as indicated by measuring H$^+$/O ratios. Phosphorylating particles prepared from the plasma membrane of *P. denitrificans* show a mitochondrial type of respiratory control[16,37] that is rarely observed in other phosphorylating bacterial preparations[8,38]. As in mitochondria the respiratory rate is increased by the addition of ADP (respiratory control), or by the addition of carbonylcyanide-*p*-trifluoromethoxyphenylhydrazone (FCCP) (an uncoupler of oxidative phosphorylation) and decreased by the addition of venturicidin which, like oligomycin, inhibits ATPase[39] (Fig. 2).

We should like to emphasise that although none of these mitochondrial features is unique to *P. denitrificans* no other species of bacterium possesses as many[3]. *Rhodopseudomonas spheroides*, however, when grown aerobically in the dark has a mitochondrial type of respiratory chain[19], but the stoichiometry of oxidative phosphorylation and the presence of respiratory control are not known. Future research will probably reveal that *P. denitrificans* is not unique in the degree to which it resembles a mitochondrion, and it will then be viewed as a representative of a small group of aerobic bacteria (probably including *Rps. spheroides*) all having an obvious affinity with the mitochondrion.

In addition to the mitochondrial features found in *P.*

Fig. 1 The respiratory chain of mitochondrion and of *P. denitrificans*. Components of the mitochondrial respiratory chain, and of the constitutive respiratory chain of *P. denitrificans* are in heavy print; the additional adaptive components of the respiratory chain of *P. denitrificans* are in lighter print. Cytochrome *o* has been found in the mitochondria of only a few organisms[68], and its significance in the constitutive respiratory chain of *P. denitrificans* is not known. From ref. 3.

denitrificans described above, there seems to be no significant feature of the mitochondrial respiratory chain or ATPase which is present in another bacterium, but known to be absent from *P. denitrificans*.

Most of the differences between the respiratory chains of a mitochondrion and *P. denitrificans* can be accounted for by the presence in *P. denitrificans* of adaptive components which are absent from mitochondria. These adaptive components may be viewed as *ad hoc* additions to the constitutive portion of the respiratory chain, which is itself essentially similar to the mitochondrial respiratory chain (Fig. 1). The constitutive components of the respiratory chain of *P. denitrificans* are nicotinamide nucleotide transhydrogenase, NADH and succinate dehydrogenases, flavoprotein, iron-sulphur proteins, ubiquinone-10, cytochromes of the *b*-type and *c*-type, and cytochrome aa_3. The adaptive components are hydrogenase, formate and lactate dehydrogenases, nitrate reductase, and nitrite reductase (cytochrome *cd*) (Fig. 1).

In their criticism of the endosymbiotic theory, Raff and Mahler[40] pointed out that, although aerobic bacteria possess cytochrome respiratory chains similar in function to mitochondrial cytochromes, "there are significant differences which suggest a considerable evolutionary divergence between mitochondrial and bacterial cytochromes". The differences cited are (1) the insensitivity of bacterial electron transport systems to some of the generally used inhibitors of mitochondrial electron transport; (2) the greater variety of bacterial cytochromes; (3) the poor cross reactivity of mitochondrial cytochrome *c* with most bacterial cytochrome oxidases, and vice versa; and (4) the dissimilarities in the primary sequence, redox potential and isoelectric point of bacterial and mitochondrial *c*-type cytochromes. We have noted above that while differences (1), (3) and (4) may apply to bacteria in general, these differences do not apply to *P. denitrificans* and *Rps. spheroides* since in these respects these bacteria resemble the mitochondrion while most other aerobic bacteria do not. Furthermore, it is shown below that the presence in *P. denitrificans* of "bacterial" cytochromes in addition to "mitochondrial" cytochromes raises no serious objection to an evolutionary transition from an aerobic bacterium resembling *P. denitrificans* to a mitochondrion, since these "bacterial" cytochromes are readily dispensed with under the appropriate environmental conditions.

Hypothetical evolutionary transition

A hypothetical transition from an aerobic bacterium resembling *P. denitrificans* to a mitochondrion, as envisaged by the endosymbiotic theory[1], would involve the loss of the adaptive components and the retention of the constitutive components of the respiratory chain of *P. denitrificans* (Fig. 1). The synthesis of the adaptive components in *P. denitrificans* is readily suppressed in the presence of alternative electron donors or electron acceptors that are "more acceptable". Hydrogenase, for example, is present only when cells are grown in the presence of hydrogen and in the absence of organic compounds such as glucose[41], and nitrate reductase is absent from cells grown aerobically in the presence of nitrate[42]. Similarly cells of *P. denitrificans* grown anaerobically in the presence of nitrate can use either nitrate or oxygen as terminal electron acceptors but, when both are supplied, use oxygen in preference to nitrate[3,43].

The postulated evolutionary transition does not require the adoption of new features by the constitutive respiratory chain of *P. denitrificans*, although modification of some components may occur. The limited extent of such changes is clear from X-ray structure analyses[24] of the readily solubilised *c*-type cytochromes of *P. denitrificans* and of the mitochondrion. The indications are that there is a very considerable similarity in the amino acid sequences of the two cytochromes[24]. The exact extent of this similarity will be revealed by the detailed amino acid sequencing now being carried out (E. Margoliash, personal communication).

The similarity in the H^+/O ratios observed with *P. denitrificans* and with the mitochondrion imply that the respiratory chain is arranged in a similar way in the plasma membrane of *P. denitrificans* and in the inner mitochondrial membrane. The ATPases are also essentially similar in their mode of operation in *P. denitrificans* and in a mitochondrion[44], suggesting that the process of oxidative phosphorylation need not be altered during the evolution of the mitochondrion.

So far, the proposed evolutionary transition from an aerobic bacterium resembling *P. denitrificans* has not necessitated the acquisition by its plasma membrane of any entirely new features, nor has it involved a significant alteration in the mode of operation of those features already present. When the respective transport systems of *P. denitrificans* and a mitochondrion are compared, however, it becomes apparent that there are mitochondrial features not found in bacteria, such as *P. denitrificans*, as well as bacterial features not found in mitochondria. Furthermore, some carriers operate in a different way in bacteria and mitochondria.

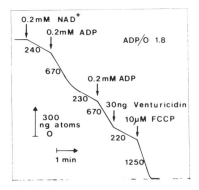

Fig. 2 Mitochondrial type of respiratory control in particles prepared from *P. denitrificans*. Membrane particles were prepared by a slight modification of the procedure of John and Hamilton[37]. The reaction mixture contained, in a total volume of 3 ml: 30 μmol Tris-acetate (*p*H 7.3), 15 μmol magnesium acetate, 30 μl ethanol, 0.1 mg alcohol dehydrogenase (Sigma A7011), and 0.54 mg particle protein. Further additions were made as indicated. Oxygen uptake was measured in a Clark-type oxygen electrode at 30 °C. The ADP/O ratio was calculated as described for mitochondria by Chance and Williams[69]. The numbers alongside the traces refer to the rates of oxygen uptake in ng atoms per min per mg protein.

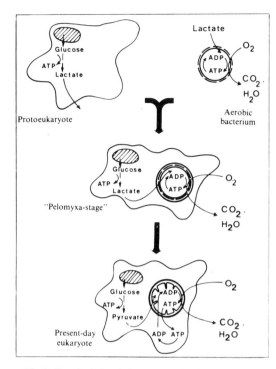

Fig. 3 Hypothetical evolutionary transition from a free-living aerobic bacterium to a mitochondrion. A fermenting protoeukaryote takes up a respiring bacterium to the advantage of both partners (the "Pelomyxa-stage", where lactate, the end product of fermentation, serves as the respiratory substrate for the bacterium). By the acquisition of an adenine nucleotide carrier the ATP synthesising potential of the bacterium is made available to the host cell (present-day eukaryote, containing mitochondria).

As with those features of the respiratory system, which are present in *P. denitrificans* but are absent from mitochondria, so also the transport systems which are uniquely bacterial can be related to the potentially more varied and unstable environment of bacteria compared with that of mitochondria. These bacterial transport systems, which would be lost in the transition to a mitochondrion, include carriers responsible for sugar transport, the periplasmic binding proteins and the extracellular iron chelators[45,46].

Both the plasma membrane of *P. denitrificans* (J. Burnell, P. J., and F. R. W., in preparation) and the inner mitochondrial membrane[47] contain a sulphydryl-sensitive, phosphate carrier which mediates the uptake of P_i coupled to the simultaneous uptake of protons by an electroneutral process equivalent to the proton-symport of Mitchell[48]. Thus this carrier, already present in the plasma membrane of *P. denitrificans*, would require no modification to operate in the mitochondrion.

On the other hand, the sulphate carriers in *P. denitrificans* and in a mitochondrion seem to operate by different mechanisms. In *P. denitrificans* sulphate is accumulated by an electroneutral proton-symport (J. Burnell, P. J., and F. R. W., in preparation) whereas, in mitochondria, sulphate permeates the inner mitochondrial membrane principally in exchange for succinate or malate[49]. Similarly, while carboxylate carriers in a number of aerobic bacteria[50–53] probably bring about the uptake of carboxylates coupled to the simultaneous uptake of protons[46], mitochondrial di- and tricarboxylate carriers act as exchange diffusion carriers, mediating the electroneutral, equimolar P_i-dicarboxylate, dicarboxylate–tricarboxylate, and heterologous dicarboxylate–dicarboxylate exchanges[47]. The carboxylate exchange carriers of the inner mitochondrial membrane clearly function to integrate the operation of the tricarboxylic acid cycle in the mitochondrial matrix with extramitochondrial metabolism[54]. On the other hand, the carboxylate carriers of the bacterial membrane function simply in the accumulative uptake of carboxylates from the bacterial environment. Thus in an evolutionary transition from an aerobic bacterium resembling *P. denitrificans* to a mitochondrion, modification would be necessary to the mechanism of the sulphate and carboxylate carriers from a proton-symport (equivalent to an hydroxyl-exchange) to a sulphate–carboxylate, carboxylate–carboxylate, and carboxylate–P_i exchange.

The only new component necessary for an evolutionary transition from the plasma membrane of *P. denitrificans* to the inner mitochondrial membrane is the adenine nucleotide carrier[55], which seems to be absent from all bacteria and present in all mitochondria. This carrier makes ATP, synthesised on the matrix (inner) side of the mitochondrial membrane[56], available to the ATP-utilising sites in the rest of the cell.

If it be assumed that the protomitochondrion was tolerated by its host for a sufficient length of time for a stable symbiotic relationship to have developed before the acquisition of the adenine nucleotide carrier by the protomitochondrion, then it is necessary to consider what advantage this protomitochondrion may have conferred on its host for the symbiotic relationship to have survived the selective pressures to which it was exposed. We suggest that the protomitochondrion, by virtue of its tricarboxylic acid cycle and respiratory chain, was able to completely oxidise fermentation products such as lactic acid and ethanol, which may have accumulated to potentially toxic levels in a relatively large and undifferentiated protoeukaryote dependent on the Emden-Meyerhof pathway for the production of ATP (Fig. 3). In support of this conjecture we may note that the giant amoeba *Pelomyxa palustris*, which lacks mitochondria but contains respiratory, endosymbiotic bacteria[57–59], uses the Emden-Meyerhof pathway to supply its energy needs, produces lactic acid but has a requirement for oxygen[60], presumably for its endosymbiotic partners[58]. *P. denitrificans* could serve in such a role, since it can utilise both lactic acid and ethanol as sources of energy and carbon[61].

The most conspicuous morphological features of the mitochondrion are the invaginations of the inner membrane, termed cristae. By contrast the plasma membrane of *P. denitrificans* does not project into the interior of the cell, but remains applied to the internal surface of the cell wall[62,63]. Presumably the development of cristae would be associated with the progressive specialisation of the aerobic bacterium resembling *P. denitrificans* essentially into a generator of ATP by oxidative phosphorylation. As this bacterium assumed its new role as a mitochondrion, the plasma membrane, as the site of oxidative phosphorylation, would increase in importance relative to the bacterial cytoplasm, which, as the site of redundant metabolic activities, would decrease in importance. Reflecting these changes, we could expect an expansion in the surface area of the plasma membrane, and a contraction in the volume of the cytoplasm, thus leading to the development of cristae.

Integration

By endowing the ancestor of the mitochondrion with the properties of the present-day bacterium *P. denitrificans*, we have provided a plausible account of the evolutionary origin of the inner mitochondrial membrane from the bacterial plasma membrane. The major changes involved in this transition are: first, the loss of redundant components of the bacterial respiratory chain and, second, a modification in the transport properties of the bacterial plasma membrane, so that effective biochemical communication is attained between the mitochondrial matrix and the rest of the cell.

Many of the components of the mitochondrion that we believe to have come from the ancestral bacterium are now coded for in the nucleus and not in the mitochondrion[64,65]. This implies that gene transfer has occurred during the evolution of the mitochondrion from a *Paracoccus*-like ancestor. Support for this conclusion may well come from comparing the amino acid sequence of the soluble cytochrome cytochrome c from *P. denitrificans* (when it is published) with the amino acid sequence of mitochondrial cytochrome c.

The evidence provided here lends support to the endosymbiotic theory, firstly, by removing some of the objections raised by previous authors[40], concerning the affinity of bacterial and mitochondrial respiratory chains and, secondly, by offering a feasible evolutionary transition from a bacterial plasma membrane to the inner mitochondrial membrane. This evidence on its own is, however, not incompatible with the non-symbiotic evolutionary origin of the mitochondrion proposed by Raff and Mahler[40], since their theory considers features common to bacteria and mitochondria to be "retained primitive states"[66]. Furthermore, the changes which we have described for an evolutionary origin of the inner mitochondrial membrane from the plasma membrane of an aerobic bacterium resembling *P. denitrificans* would also apply to an evolutionary origin of the inner mitochondrial membrane from the plasma membrane of an aerobic protoeukaryote possessing a respiratory system similar to that of *P. denitrificans*. Having noted these points, however, we agree with Taylor[67] that it is "unrealistic to discuss the origin of mitochondria independently from that of chloroplasts". Thus it is in the context of the greater applicability of the endosymbiotic theory to explain the evolutionary origin of the eukaryotic cell as a whole[67] that we have chosen to compare *P. denitrificans* with the mitochondrion and to show how a relationship of limited advantage to both symbiotic partners could readily have developed into the present close relationship between a mitochondrion and its "host" cell.

We thank the Science Research Council for a research grant.

Received December 11, 1974; revised March 3, 1975.

[1] Margulis, L., *Origin of Eukaryote Cells* (Yale University Press, New Haven, 1970).
[2] Davis, D. H., Doudoroff, M., Stanier, R. Y., and Mandel, M., *Int. J. syst. Bact.*, 19, 375–390 (1969).
[3] John, P., and Whatley, F. R., *Adv. bot. Res.* (in the press).
[4] Asano, A., Imai, K., and Sato, R., *Biochim. biophys. Acta*, 143, 477–486 (1967).
[5] Murthy, P. S., and Brodie, A. F., *J. biol. Chem.*, 239, 4292–4297 (1964).
[6] Imai, K., Asano, A., and Sato, R., *Biochim. biophys. Acta*, 143, 462–476 (1967).
[7] Gel'man, N. S., Lukoyana, M. A., and Ostrovskii, N. D., *Respiration and Phosphorylation of Bacteria* (Plenum, New York, 1967).
[8] Smith, L., in *Biological Oxidations* (edit. by Singer, T. P.), 55–122 (Interscience, New York, 1968).
[9] Forget, P., and Pichnoty, F., *Annls Inst. Pasteur, Paris*, 108, 364–377 (1965).
[10] Doelle, H. W., *Bacterial Metabolism* (Academic, New York, 1969).
[11] Wilkinson, B. J., Morman, M. R., and White, D. C., *J. Bact.*, 112, 1288–1294 (1972).
[12] Parsons, D. F., Williams, G. R., Thompson, W., Wilson, D., and Chance, B., in *Mitochondrial Structure and Compartmentation* (edit. by Quagliariello, E., Papa, S., Slater, E. C., and Tager, J. M.), 29–70 (Adriatica Editrice, Bari, 1967).
[13] Girard, A. E., *Can. J. Microbiol.*, 17, 1503–1508 (1971).
[14] Getz, G. S., Bartley, W., Stirpe, F., Notton, B. M., and Renshaw, A., *Biochem. J.*, 83, 181–194 (1962).
[15] Crane, F. L., in *Biochemistry of Quinones* (edit. by Morton, R. A.), 183–206 (Academic, New York, 1965).
[16] Scholes, P. B., and Smith, L., *Biochim. biophys. Acta*, 153, 363–375 (1968).
[17] Shipp, W. S., *Archs Biochem. Biophys.*, 150, 482–488 (1972).
[18] Shipp, W. S., *Archs Biochem. Biophys.*, 150, 459–472 (1972).
[19] Dutton, P. L., and Wilson, D. F., *Biochim. biophys. Acta*, 346, 165–212 (1974).
[20] Kamen, M. D., and Horio, T., *A. Rev. Biochem.*, 39, 673–700 (1970).
[21] Erickson, S. K., *Biochim. biophys. Acta*, 245, 63–69 (1971).
[22] Kamen, M. D., and Vernon, L. P., *Biochim. biophys. Acta*, 17, 10–22 (1955).
[23] Scholes, P. B., McLain, G., and Smith, L., *Biochemistry*, 10, 2072–2075 (1971).
[24] Timkovich, R., and Dickerson, R. E., *J. molec. Biol.*, 79, 39–56 (1973).
[25] Yamanaka, T., and Okunuki, K., *J. biol. Chem.*, 239, 1813–1817 (1964).
[26] Yamanaka, T., *Space Life Sci.*, 4, 490–504 (1973).
[27] Smith, L., Newton, N., and Scholes, P. B., in *Hemes and Hemoproteins* (edit. by Chance, B., Estabrook, R. W., and Yonetani, T.), 395–403 (Academic, New York, 1966).
[28] Lam, Y., and Nicholas, D. J. D., *Biochim. biophys. Acta*, 180, 459–472 (1969).
[29] Newton, N., *Biochim. biophys. Acta*, 185, 316–331 (1969).
[30] Scholes, P. B., and Mitchell, P., *J. Bioenerg.*, 1, 309–323 (1970).
[31] Beatrice, M. C., and Chappell, J. B., *Biochem. Soc. Trans.*, 2, 151–153 (1973).
[32] Lawford, H. G., and Haddock, B. A., *Biochem. J.*, 136, 217–220 (1973).
[33] Downs, A. J., and Jones, C. W., *Biochem. Soc. Trans.*, 2, 526–529 (1974).
[34] Mitchell, P., *FEBS Symp.*, 28, 353–370 (1972).
[35] Moyle, J., and Mitchell, P., *Biochem. J.*, 105, 1147–1162 (1973).
[36] John, P., and Hamilton, W. A., *FEBS Lett.*, 10, 246–248 (1970).
[37] John, P., and Hamilton, W. A., *Eur. J. Biochem.*, 23, 528–532 (1971).
[38] Jones, C. W., Erickson, S. K., and Ackrell, B. A. C., *FEBS Lett.*, 13, 33–35 (1971).
[39] Walter, P., Lardy, H. A., and Johnson, D., *J. biol. Chem.*, 242, 5014–5018 (1967).
[40] Raff, R. A., and Mahler, H. R., *Science*, 177, 575–582 (1972).
[41] Fewson, C. A., and Nicholas, D. J. D., *Biochim. biophys. Acta*, 48, 208–210 (1961).
[42] Pichinoty, F., *Annls. Inst. Pasteur, Paris*, 109, 248–255 (1965).
[43] Pichinoty, F., and D'Ornano, L., *Biochim. biophys. Acta*, 52, 386–389 (1961).
[44] Ferguson, S. J., John, P., Lloyd, W. J., Radda, G. K., and Whatley, F. R., *Biochim. biophys. Acta*, 357, 457–461 (1974).
[45] Tait, G. H., *Biochem. Soc. Trans.*, 2, 657–658 (1974).
[46] Hamilton, W. A., *Adv. mic. Physiol.*, 12, 1–53 (1975).
[47] Chappell, J. B., and Haarhoff, K. N., in *Biochemistry of Mitochondria* (edit. by Slater, E. C., Kaniuga, Z., and Wojtczak, L.), 77–91 (Academic, New York, 1967).
[48] Mitchell, P., in *Organization and Control in Prokaryotic and Eukaryotic Cells* (edit. by Charles, H. P., and Knight, B. C. J. G.), 121–166 (Cambridge University Press, Cambridge, 1970).
[49] Crompton, M., Palmieri, F., Capano, M., and Quagliariello, E., *Biochem. J.*, 142, 127–137 (1974).
[50] Postma, P. W., and Van Dam, K., *Biochim. biophys. Acta*, 249, 515–527 (1971).
[51] Visser, A. S., and Postma, P. W., *Biochim. biophys. Acta*, 298, 333–340 (1973).
[52] Ghei, O. K., and Kay, W. W., *FEBS Lett.*, 20, 137–140 (1972).
[53] Lawford, H. G., and Williams, G. R., *Biochem. J.*, 123, 571–577 (1971).
[54] Chappell, J. B., *Br. med. Bull.*, 24, 150–157 (1968).
[55] Klingenberg, M., *Essays in Biochemistry*, 6, 119–159 (1970).
[56] Lehninger, A. L., *The Mitochondrion* (Benjamin, New York, 1964).
[57] Daniels, E. W., Breyer, E. P., and Kudo, R. R., *Z. Zellforsch. mikrosk. Anat.*, 73, 367–383 (1966).
[58] Leine, M., Schweikhardt, F., Blaschke, G., Konig, K., and Fischer, M., *Biol. Zbl.*, 87, 567–591 (1968).
[59] Chapman-Andresen, C., *A. Rev. Microbiol.*, 25, 27–48 (1971).
[60] Andresen, N., Chapman-Andresen, C., and Nilsson, J. R., *C. r. Trav. Lab. Carlsberg*, 36, 285–317 (1968).
[61] Vogt, M., *Archs Mikrobiol.*, 50, 256–281 (1965).
[62] Kocur, M., Martinec, T., and Mazanec, K., *Antonie van Leeuwenhoek*, 34, 19–26 (1968).
[63] Scholes, P. B., and Smith, L., *Biochim. biophys. Acta*, 153, 350–362 (1968).
[64] Sherman, F., and Stewart, J. W., *A. Rev. Genet.*, 5, 257–296 (1971).
[65] Tzagoloff, A., Rubin, M. S., and Sierra, M. F., *Biochim. biophys. Acta*, 301, 71–104 (1973).
[66] Uzzell, T., and Spolsky, C., *Science*, 180, 516–517 (1973).
[67] Taylor, J. F. R., *Taxon*, 23, 229–258 (1974).
[68] Kronick, P., and Hill, G. C., *Biochim. biophys. Acta*, 368, 173–180 (1974).
[69] Chance, B., and Williams, G. R., *J. biol. Chem.*, 217, 383–393 (1955).

ERRATUM

On page 496, lines 4 and 5 of the caption to Figure 2 should read "The reaction mixture contained, in a total volume of 3 ml: 30 μmol Tris-P$_i$ (pH 7.3), ..."

ORGANIZATION AND EVOLUTION OF RIBOSOMAL RNA GENES IN WHEAT MITOCHONDRIA

Michael W. Gray, Tai Y. Huh, Murray N. Schnare,
David F. Spencer, and Denis Falconet*

Department of Biochemistry
Dalhousie University
Halifax, Nova Scotia B3H 4H7 (CANADA)

INTRODUCTION

Mitochondrial DNA (mtDNA) is known to encode the distinctive ribosomal RNA (rRNA) species found in the organelle-specific ribosomes that function in mitochondrial protein synthesis [1]. Qualitatively and quantitatively, rRNA production is a major genetic function of mtDNA, and the factors controlling mitochondrial rRNA (mt-rRNA) synthesis are therefore of great interest. Moreover, since homologous rRNA species are found in all systems (eukaryotic cytosol, prokaryotic, chloroplast, mitochondrial) having a ribosome-based translation system, comparative studies of these rRNAs and their genes are an important means of illuminating the evolutionary history of the eukaryotic cell and its organelles [2]. For these reasons, we have been investigating the organization and evolution of the rRNA genes in plant mitochondria, particularly those of wheat, <u>Triticum aestivum.</u>

Novel Arrangement of rRNA Genes in Plant mtDNA

T_1 Oligonucleotide fingerprinting [3] and cataloguing [4] studies and RNA sequence analysis [5-7] have shown that wheat 26S, 18S, and 5S mt-rRNAs are distinct from their cytosol counterparts, with which they share no homology detectable in rRNA-DNA hybridization experiments [8]. Identification of restriction fragments encoding the individual wheat mt-rRNAs provided the first direct demonstration of specific genes encoded by plant mtDNA [8] and led us to infer that:

---

*Present address: Laboratoire de Biologie Moléculaire Végétale, Universite de Paris-Sud, Centre d'Orsay, 91405 Orsay Cedex (FRANCE)

(1) Genes for wheat 26S and 18S mt-rRNAs must be far apart in the wheat mitochondrial genome, since they are always found on different fragments in a given restriction digest.

(2) Genes for wheat 5S and 18S mt-rRNAs must be closely linked in wheat mtDNA, since they are found on the same restriction fragments in a given digest. This localization of 5S rRNA genes differs from that in all other genomes (nuclear, prokaryotic, and chloroplast) that encode a 5S rRNA species.

(3) Since each individual rRNA probe usually hybridizes with multiple restriction fragments in each digest, there must be multiple distinct rRNA cistrons in wheat mtDNA. The 5S mt-rRNA gene (if it is colinear with the RNA sequence; [6]) should contain no internal EcoRI or SalI sites, yet 5S mt-rRNA hybridizes with at least four distinct EcoRI and three different SalI restriction fragments. On the other hand, the 5S rRNA hybridizes with only a single XhoI fragment. Taken together, these data imply that there is a basic structural unit encoding the 5S rRNA, but that sequences flanking this unit are not identical. The multiplicity of rRNA restriction fragments hybridizing with the same rRNA cannot be ascribed to incomplete endonuclease cleavage of wheat mtDNA arising from partial methylation of restriction sites, since identical restriction and rRNA hybridization patterns are observed when wheat mtDNA is digested with either MspI or HpaII [9].

More recently, we have examined other plant mtDNAs (rye, corn, broad bean, pea, cucumber) in order to determine whether this novel arrangement of mt-rRNA genes is peculiar to wheat or is more widespread. The results [10] suggest that this same pattern of rRNA gene organization (18S and 5S genes close together but remote from 26S genes) is a general (perhaps universal) feature of angiosperm mtDNA.

rRNA Genes in Cloned Fragments of Wheat mtDNA

We have collaborated with F. Quetier's group (Orsay) in examining cloned restriction fragments carrying wheat mt-rRNA genes. SalI fragments of wheat mtDNA were inserted into pBR322 by the Orsay group, and selected recombinant plasmids hybridizing with wheat 18S and 5S mt-rRNAs were characterized in our laboratory. Restriction maps of Sal fragments S21 and S19 [8] (5.5 and 6.2 kbp, respectively) are shown in FIG. 1. It can be seen that the two fragments share identical restriction sites over most of their lengths, starting at the same leftward SalI site (S_L) and proceeding at least as far as a common ClaI site 4760 bp away. The remaining lengths of S21 (740 bp) and S19 (1440 bp) bear no homology detectable by restriction analysis, and the fragments terminate at distinct rightward SalI sites (S_R).

Initial Southern hybridization experiments localized the coding regions for the 18S and 5S mt-rRNAs within a PstI-XhoI subfragment 2750 bp long and common to both Sal inserts. The 3'-end of the 18S

MITOCHONDRIAL RIBOSOMAL RNA GENES

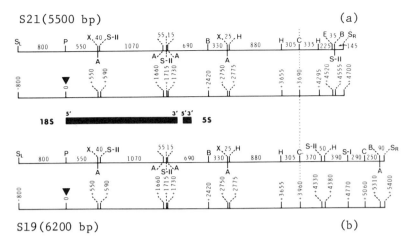

Fig. 1. Restriction maps of fragments S21 (a) and S19 (b), cloned from a SalI digest of wheat mtDNA. S_L and S_R = leftward and rightward SalI sites, respectively; A = AvaI; B = BamHI; C = ClaI; E = EcoRI; H = HindIII; P = PstI; S-I = SstI; S-II = SstII; X = XhoI. Restriction site coordinates (in bp) are calculated relative to the unique PstI site (solid triangle). The dotted line denotes the point at which S21 and S19 diverge in restriction sites. Solid bars mark the 18S and 5S rRNA coding regions.

gene was placed close to the 5S gene by hybridization experiments which employed a 3'-specific probe derived from [5'-^{32}P]pCp-labeled wheat 18S mt-rRNA [7]. Coding boundaries have since been refined by sequence analysis of cloned rDNA. Based on 5'- [unpublished] and 3'- [7] nucleotide sequences directly determined from the 18S mt-rRNA, the 5' end of the 18S gene is 25 bp downstream from the unique PstI site (which is 800 bp from S_L), while the 3'-terminus is 200 bp downstream from the +1730 AvaI site. By comparison of the DNA sequence determined for this region of cloned rDNA with the primary sequence of wheat 5S mt-rRNA [6], the 5'-end of the 5S gene has been positioned about 105 bp downstream from the 3'-end of the 18S gene, on the same strand. This arrangement raises the possibility that the 18S and 5S mt-rRNAs are co-transcribed from a single promoter.

Multiple Distinct Cistrons for 18S and 5S rRNA Genes in Wheat mtDNA

In addition to bands corresponding to S21 and S19, a third band of hybridization (>16 kbp) was evident when SalI digests of wheat mtDNA were probed with either 18S or 5S mt-rRNAs [8]. In screening for clones which might account for this latter hybridization, we uncovered two additional, high-molecular-weight fragments carrying 18S and 5S mt-rRNA genes. Detailed restriction analysis has shown that these fragments, designated S5,6a (18.4 kbp) and S5,6b (19.1 kbp), share the same leftward SalI site, different from that in S21 and S19, and are identical over most of their lengths, diverging close to their rightward ends. We also found (somewhat unexpectedly) that in the region of divergence, each of the larger fragments is identical to one of the smaller fragments (either S21 or S19). These relationships are illustrated in FIG. 2, which shows the XhoI restriction maps of the four SalI fragments. Note that each possesses the same basic 18S-5S rRNA coding unit, largely contained within the common 2200 bp XhoI subfragment that is the main site of 18S hybridization, and exclusive site of 5S hybridization, in XhoI digests of wheat mtDNA [8].

It seems likely, although not yet definitively proven, that there is one genome equivalent of each of the rDNA fragments identified in this study. Based on analysis of densitometer tracings [8], S19 and S21 are present in amounts roughly equimolar with each other and with neighboring fragments in gels of SalI digests of wheat mtDNA. As shown in FIG. 3, this implies that the same rRNA coding sequence (R) is repeated four times in the wheat mitochondrial genome, and is flanked by different arrays of other repeated sequences (U, V, W, W'). How these SalI fragments are organized within wheat mtDNA remains to be investigated.

Assuming a stoichiometry of one for S21 in SalI digests of wheat mtDNA, and based on its known size (5.5 kbp) and the mass proportion it occupies in a SalI digest (determined from densitometer tracings of

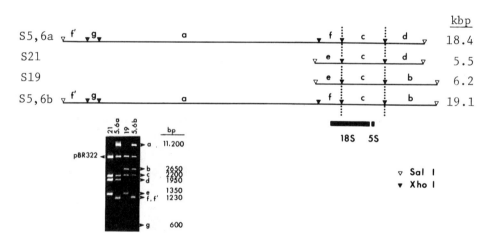

Fig. 2. XhoI restriction maps showing the structural relationships among SalI rDNA fragments cloned from wheat mtDNA. Dotted lines delineate the common XhoI fragment (c, 2200 bp) that carries most of the 18S, and all of the 5S, gene (Fig. 1). The inset shows restriction profiles visualized after agarose gel electrophoresis of double digests (XhoI + SalI) of recombinant plasmids carrying the SalI rDNA inserts.

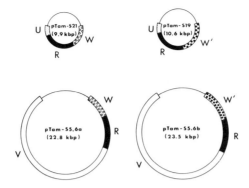

Fig. 3. Structural relationships among recombinant plasmids carrying wheat mt-rRNA genes. Mitochondrial DNA inserts represent rDNA-containing SalI fragments cloned into the SalI site of pBR322 (thin line). R = common rRNA coding unit corresponding to the 2.75 kbp PstI-XhoI fragment (0 to +2750) shown in Fig. 1. The sizes (in kbp) of the flanking regions are: U = 0.8; V = 13.7; W = 1.95; W' = 2.65. W and W' show extensive restriction site homology (Fig. 1).

a SalI restriction pattern; see Fig. 2 of [8]), we estimate that the physical size of wheat mtDNA is minimally about 225 kbp (150 MDal), which would make it some 14 times as large as animal mtDNA. The four cloned SalI fragments described here total 49 kbp, and so would account for about 20% of the wheat mitochondrial genome, if they are each present in unimolar amounts.

Evolutionary Origin of Plant Mitochondrial rRNAs

Ribosomal RNA sequence data have been instrumental in solidifying the view that chloroplasts are of endosymbiotic, specifically eubacterial, origin [2]. In contrast, conclusions about the evolutionary origin of mitochondria, drawn from similar data, are complicated by the pronounced diversity in primary sequence, base composition, and size of the mitochondrial rRNAs [1,2]. This may be due in part to a variable (and in some cases, quite rapid) rate of evolution of mtDNA and its contained rRNA genes [1]. Plant mt-rRNAs, however, are not "atypical" in size and base composition [11] and strong evidence in support of an endosymbiotic, eubacterial, origin of mitochondria has come from studies of the primary structure of plant mt-rRNAs.

We showed a number of years ago that the T_1 oligonucleotide catalogue of wheat 18S mt-rRNA strongly resembles T_1 oligonucleotide catalogues of eubacterial and chloroplast 16S rRNAs, but lacks detectable homology with the T_1 catalogue of wheat cytosol 18S rRNA

[4]. In addition, among those T_1 oligonucleotides which the 18S mt-rRNA shares with eubacterial and chloroplast 16S rRNAs, there is an especially high proportion of ones identified [12] as conserved in the evolution of eubacterial 16S but not eukaryotic cytosol 18S rRNA species [13]. More recently, direct determination of the sequences of the 3'-terminal ca. 100 nucleotides of wheat mitochondrial and cytosol 18S and E. coli 16S rRNAs [7] has demonstrated that:

(1) There is a substantially greater degree of primary sequence homology between wheat mitochondrial 18S and E. coli 16S rRNAs (72%) than between either wheat mitochondrial and cytosol 18S rRNAs (53%) or wheat cytosol 18S and E. coli 16S rRNAs (60%). In fact, except that it lacks the "Shine-Dalgarno" sequence [14] present in E. coli and chloroplast 16S rRNAs, wheat mitochondrial 18S rRNA is as homologous to E. coli 16S rRNA in this region as is maize chloroplast 16S rRNA.

(2) At a position occupied by 3-methyluridine (m^3U) in E. coli 16S rRNA, the same (or a very similar) modified nucleoside is present in wheat mitochondrial 18S rRNA but not in wheat cytosol 18S rRNA. Interestingly, m^3U is present in a "universal" T_1 oligonucleotide (U*AACAAGp of [12]) in E. coli 16S rRNA. Since two other eubacterial 16S universals, CCm^7GCGp and $m^4CmCCGp$, also appear in the wheat 18S mt-rRNA T_1 catalogue [4,13], the latter rRNA shows evidence of being highly homologous to eubacterial 16S rRNAs not only in primary sequence, but in post-trancriptional modification pattern, as well.

Extensive primary sequence homology between E. coli and wheat mitochondrial rRNAs is also indicated by the fact that the bacterial 16S and 23S rRNAs hybridize specifically to those restriction fragments containing wheat mitochondrial 18S and 26S rRNA genes, respectively [7]. The high degree of primary sequence homology found within the 3'-terminal 100 nucleotides of E. coli 16S and wheat mitochondrial 18S rRNAs has now been confirmed by DNA sequence analysis of the coding region of the latter. From the primary sequence data we have accumulated to date, overall homology to E. coli 16S rRNA appears to be 70-75%, with some impressively long stretches (in excess of 20 nucleotides) of complete identity.

Although we initially thought that the 18S mt-rRNA coding sequence was about 1750 bp long [15], precise determination of the ends of the gene now raises this estimate to about 1900 bp, making the wheat 18S mt-rRNA gene some 360 bp longer than that of E. coli 16S rRNA, and even longer than those of S. cerevisiae (1789 bp) and X. laevis (1825 bp) 18S rRNAs! We should therefore expect to find regions of non-homology between E. coli 16S and wheat mitochondrial 18S rRNA genes, corresponding to "inserts" in the mitochondrial gene sequence. One such example occurs within the 3'-terminal 200 nucleotides of the 18S mt-rRNA gene. Comparison with secondary structure models of E. coli and other small subunit rRNAs [16-18] shows that 47 nucleotides of the mitochondrial sequence have no counterpart in the bacterial sequence. The extra nucleotides in this part of wheat 18S mt-rRNA are localized within a helical region that is highly variable in size (ranging from 64-130 nucleotides) among small subunit rRNAs [18 & unpublished]. This secondary structure element is flanked by single-stranded segments (14 and 16 nucleotides long) that are absolutely conserved in E. coli and Z. mays chloroplast 16S rRNAs and wheat mitochondrial 18S rRNA.

In view of the pronounced eubacterial phylogenetic affinity of wheat 18S mt-rRNA, it has been somewhat surprising to find that wheat 5S mt-rRNA is not obviously eubacterial or eukaryotic in primary sequence or potential secondary structure, but in these parameters shows characteristics of both classes of 5S rRNA, as well as some unique features [6,19,20]. Nevertheless, the secondary structure model proposed by us for wheat 5S mt-rRNA [6] can be accommodated within a uniform model that has recently been shown to fit all known 5S rRNA sequences [21]. When scored in terms of positions within this uniform model that are diagnostic for different classes of 5S rRNA, wheat 5S mt-rRNA shares a somewhat higher number (7) of such positions with eubacterial 5S rRNA than with either eukaryotic (4 positions) or archaebacterial (2 positions) 5S rRNAs. The fact that wheat 18S mt-rRNA is obviously eubacterial in nature, whereas wheat 5S mt-rRNA is decidedly less so, may indicate that functional contraints on primary sequence divergence are more pronounced in 16S/18S rRNA than in 5S rRNA.

CONCLUDING REMARKS

Technical advances centered on recombinant DNA and nucleotide sequence analysis have confirmed a novel arrangement of rRNA genes in plant mitochondria, led to the identification of multiple rRNA cistrons in wheat mtDNA, and provided strong additional support for an endosymbiotic, specifically eubacterial, origin of plant mitochondria. Extension of these techniques should soon allow us to determine how these multiple copies of rRNA genes are organized with respect to each other and to other genes in the wheat mitochondrial genome, and the manner in which remotely located 26S and 18S genes, and closely linked 18S and 5S genes, are coordinately expressed.

ACKNOWLEDGMENTS

Continuing financial support from the Medical Research Council of Canada (grant MT-4124) is gratefully acknowledged.

REFERENCES

[1] Gray, M.W. (1982) Can. J. Biochem. 60:157-171.
[2] Gray, M.W. and Doolittle, W.F. (1982) Microbiol. Rev. 46:1-42.
[3] Cunningham, R.S., Bonen, L., Doolittle, W.F. and Gray, M.W. (1976) FEBS Lett. 69:116-122.

[4] Bonen, L., Cunningham, R.S., Gray, M.W. and Doolittle, W.F. (1977) Nucleic Acids Res. 4:663-671.
[5] MacKay, R.M., Spencer, D.F., Doolittle, W.F. and Gray, M.W. (1980) Eur. J. Biochem. 112:561-576.
[6] Spencer, D.F., Bonen, L. and Gray, M.W. (1981) Biochemistry 20:4022-4029.
[7] Schnare, M.N. and Gray, M.W. (1982) Nucleic Acids Res. 10:3921-3932
[8] Bonen, L. and Gray, M.W. (1980) Nucleic Acids Res. 8:319-335.
[9] Bonen, L., Huh, T.Y. and Gray, M.W. (1980) FEBS Lett. 111:340-346.
[10] Huh, T.Y. and Gray, M.W. (1982) Plant Mol. Biol. (in press).
[11] Leaver, C.J. and Gray, M.W. (1982) Annu. Rev. Plant Physiol. 33:373-402.
[12] Woese, C.R., Fox, G.E., Zablen, L., Uchida, T., Bonen, L., Pechman, K., Lewis, B.J. and Stahl, D. (1975) Nature 254:83-86.
[13] Cunningham, R.S., Gray, M.W., Doolittle, W.F. and Bonen, L. (1977) In Acides Nucléiques et Synthèse des Protéines Chez les Végétaux (Bogorad, L. and Weil, J.H., eds), pp. 243-248, Centre National de la Recherche Scientifique, Paris.
[14] Shine, J. and Dalgarno, L. (1974) Proc. Natl. Acad. Sci. U.S.A. 71:1342-1346.
[15] Gray, M.W., Bonen, L., Falconet, D., Huh, T.Y., Schnare, M.N. and Spencer, D.F. (1982) In Mitochondrial Genes (Slonimski, P., Borst, P. and Attardi, G., eds), pp. 483-488, Cold Spring Harbor Laboratory, Cold Spring Harbor, N.Y.
[16] Woese, C.R., Magrum, L.J., Gupta, R., Siegel, R.B., Stahl, D.A., Kop, J., Crawford, N., Brosius, J., Gutell, R., Hogan, J.J. and Noller, H.F. (1980) Nucleic Acids Res. 8:2275-2293.
[17] Stiegler, P., Carbon, P., Zuker, M., Ebel, J.-P. and Ehresmann, C. (1981) Nucleic Acids Res. 9:2153-2172.
[18] Zwieb, C., Glotz, C. and Brimacombe, R. (1981) Nucleic Acids Res. 9:3621-3640.
[19] Gray, M.W. and Spencer, D.F. (1981) Nucleic Acids Res. 9:3523-3529.
[20] MacKay, R.M., Spencer, D.F., Schnare, M.N., Doolittle, W.F. and Gray, M.W. (1982) Can. J. Biochem. 60:480-489.
[21] De Wachter, R., Chen, M.-W. and Vandenberghe, A. (1982) Biochimie 64:311-329.

Similar genes for a mitochondrial ATPase subunit in the nuclear and mitochondrial genomes of *Neurospora crassa*

Paul van den Boogaart, John Samallo & Etienne Agsteribbe

Laboratory of Physiological Chemistry, State University, Bloemsingel 10, 9712 KZ Groningen, The Netherlands

The proton-translocating ATPases from mitochondria, chloroplasts and bacteria consist of ~10 different polypeptide subunits[1]. The smallest subunit, a proteolipid of molecular weight ~8,000, is present in a stoichiometric amount of six—these six polypeptides are thought to form a proton channel through the membrane[2]. The proteolipid is affected by the ATPase inhibitors oligomycin and dicyclohexylcarbodiimide (DCCD). As DCCD is covalently bound to it, the proteolipid is commonly referred to as the DCCD-binding protein. Because of the spatial arrangement of the proteolipids in the membrane, binding is probably restricted to one specific glutamic acid residue[2,3]. In the yeast *Saccharomyces cerevisiae* the DCCD-binding protein is encoded by the mitochondrial DNA and synthesized inside the mitochondria[4], whereas in *Neurospora crassa*, another ascomycete, the gene for the DCCD-binding protein lies in the nucleus, and the protein is synthesized outside the mitochondria[5]. Here we report the presence on the mitochondrial DNA of *N. crassa* of a nucleotide sequence which potentially encodes another DCCD-binding protein. Although a translation product has not yet been found, we suggest a possible function of this gene and speculate on its evolutionary origin.

We have shown previously by heterologous hybridization that *Neurospora* mitochondrial DNA contains a sequence homologous to the yeast mitochondrial gene for the DCCD-binding protein[6]. We recently completed the nucleotide sequencing of this fragment (Fig. 1). It contains one open reading frame per strand. The longest reading frame, 225 nucleotides, is transcribed from the same strand as the ribosomal RNA (rRNA) genes and the genes for subunits 1 and 2 of cytochrome aa_3. This reading frame shows 65% homology with the yeast gene for the DCCD-binding protein of the mitochondrial ATPase[7]. The other reading frame is 183 nucleotides long and codes for an extremely basic polypeptide (Fig. 1). It is unknown whether this sequence encodes a protein.

Comparison of the amino acid sequence encoded by the former reading frame with the primary structures of DCCD-binding proteins from mitochondria, chloroplasts and bacteria suggests that the *Neurospora* mitochondrial gene codes for a typical DCCD-binding protein[8] (Fig. 2). This conclusion is based on the following observations: (1) The conservation of glycine at positions 27, 31 and 42, arginine at 45, proline at 47 and alanine at 66. These residues are thought to be indispensable for structure and function, as they have been conserved in all DCCD-binding proteins analysed so far. (2) The conservation of the DCCD-binding glutamic acid at position 65. Aspartic acid is found at the homologous position only in *Escherichia coli*. (3) The absence of histidine and tryptophan. (4) The distribution of the hydrophobic and hydrophilic amino acid residues along the polypeptide chain. Most of the few polar amino acids—all proteins have an extreme low polarity of ~20%—are clustered in the middle of the sequence. This polar segment is flanked on both sides by a long hydrophobic sequence of ~25 residues.

Sebald and co-workers[8] have demonstrated that in exponentially growing cells of *N. crassa*, the DCCD-binding protein of the mitochondrial ATPase is encoded by a nuclear gene and synthesized in the cytoplasm. Their evidence was based on the finding that cycloheximide, an inhibitor of cytoplasmic protein synthesis, blocks synthesis of the DCCD-binding protein; chloramphenicol, an inhibitor of mitochondrial protein synthesis, had no effect. Furthermore, oligomycin resistance in mutants of *N. crassa* segregates 4:4 at meiosis and not 8:0 as would be expected for a mitochondrial mutation. Sequence studies of the proteolipid from several mutants identified different amino acid substitutions. As no similar mitochondrially synthesized protein has been found in *Neurospora*, we are faced with the enigma of a gene having an as yet unknown

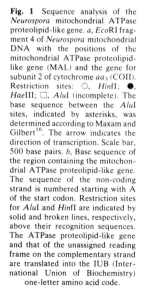

Fig. 1 Sequence analysis of the *Neurospora* mitochondrial ATPase proteolipid-like gene. *a*, *Eco*RI fragment 4 of *Neurospora* mitochondrial DNA with the positions of the mitochondrial ATPase proteolipid-like gene (MAL) and the gene for subunit 2 of cytochrome aa_3 (COII). Restriction sites: ○, *Hin*fI; ●, *Hae*III; □, *Alu*I (incomplete). The base sequence between the *Alu*I sites, indicated by asterisks, was determined according to Maxam and Gilbert[16]. The arrow indicates the direction of transcription. Scale bar, 500 base pairs. *b*, Base sequence of the region containing the mitochondrial ATPase proteolipid-like gene. The sequence of the non-coding strand is numbered starting with A of the start codon. Restriction sites for *Alu*I and *Hin*fI are indicated by solid and broken lines, respectively, above their recognition sequences. The ATPase proteolipid-like gene and that of the unassigned reading frame on the complementary strand are translated into the IUB (International Union of Biochemistry) one-letter amino acid code.

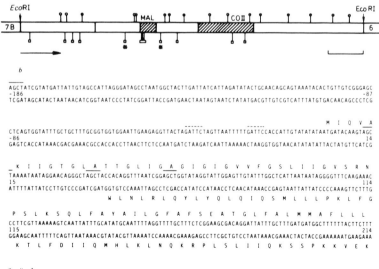

153

Fig. 2 Amino acid sequences of mitochondrial DCCD-binding proteins from: (1) *N. crassa*, as predicted from the mitochondrial gene; (2) *N. crassa*; (3) *S. cerevisiae*; and (4) bovine heart. The sequences are numbered starting with the first amino acid, tyrosine, of the *Neurospora* protein. Identical amino acid residues in the four sequences are indicated by an asterisk. The amino acid residues which are found to be conserved in all DCCD-binding proteins analysed so far, including those from chloroplasts and bacteria, are boxed. The glutamic acid residue, responsible for the binding of DCCD, is at position 65.

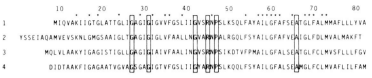

```
          10        20        30        40        50        60        70        80
          .....     .....     .....     .....     .....     .....     .....     .....
1   MIQVAKIIGTGLATTGLIGAGIGIGVVFGSLIIGVSRNPSLKSQLFAYAILGFAFSEATGLFALMMAFLLLYVA
2   YSSEIAQAMVEVSKNLGMGSAAIGLTGAGIGIGLVFAALLNGVARNPALRGQLFSYAILGFAFVEAIGLFDLMVALMAKFT
3   MQLVLAAKYIGAGISTIGLLGAGIGIAIVFAALINGVSRNPSIKDTVFPMAILGFALSEATGLFCLMVSFLLLFGV
4   DIDTAAKFIGAGAATVGVAGSGAGIGTVFGSLIIGYARNPSLKQQLFSYAILGFALSEAMGLFCLMVAFLILFAM
```

function. An obvious possibility is that the mitochondrial gene is silent. Analysis of the transcription products (Fig. 3) demonstrates that the gene lacks a small, unique, mRNA-like transcript, in contrast to the neighbouring gene for subunit 2 of cytochrome aa_3. This indicates that the gene possibly does not encode a functional protein. However, we believe that the gene is active because of the striking conservation of the primary structure of the polypeptide. Of course, a very recent inactivation of the gene, as apparently has occurred with the δ-globin gene in Old World monkeys[9], cannot yet be excluded.

That the translation product of the *Neurospora* mitochondrial gene has gone undetected so far, may be due to its low abundance. However, the DCCD-binding protein occurs as an oligomer of six subunits in each ATPase complex[2]. Therefore, if only one of these subunits were of mitochondrial origin, this would still constitute some 16% of the total amount of proteolipid of the ATPase complex. Such a high percentage would not have gone undetected in the experiments of Jackl and Sebald[5]. Moreover, our failure to detect a mRNA-like transcript from the gene suggests that in exponentially growing cells of *Neurospora* the gene product, if present, is only a minor mitochondrial translation product. A lower percentage than the 16% mentioned above would indicate functional heterogeneity of the ATPase complex, a phenomenon which, as far as we know, has never been observed. From these considerations we can almost exclude the possibility that the gene is expressed in exponentially growing cells of *Neurospora*. However, expression of the gene might be restricted to another stage of the life cycle of *Neurospora*, for example, to asexual or sexual reproduction. Differential expression of closely related genes is well documented in higher eukaryotes. The occurrence of fetal as opposed to adult haemoglobin chains in mammals[10], and the presence of oocyte-type and somatic cell-type 5S rRNA sequences in *Xenopus*[11] are good examples of this phenomenon.

The presence on *Neurospora* mitochondrial DNA of a gene coding for a protein homologous to a nucleus-encoded protein raises the question about the origins of these genes. According to the endosymbiont theory[12] most, if not all, of the nuclear genes encoding organelle-specific proteins have their origin on the endosymbiont genome. In *Neurospora*, the steps that could have led to the occurrence of similar genes on two different genomes are gene duplication, followed by transfer of one of the genes to the nucleus. An evolutionarily recent transfer implies a relatively high degree of homology between the two *Neurospora* proteins. Surprisingly, however, a comparison of the primary structures of the DCCD-binding proteins from mitochondria (Table 1) reveals that the degree of homology is almost the same, and that the similarity between the two *Neurospora* proteins is even lower than the similarities between the mitochondrial gene product from *Neurospora* and the proteins from yeast and bovine heart. This would mean that the divergence of both *Neurospora* genes has occurred in an early period of the evolution of the metazoa, implying that both genes existed in *Neurospora* for a long time. This reinforces our view that the mitochondrial gene is active; inactivation would have resulted in a pseudogene or disappearance of the gene. An alternative explanation for the relatively low homology between the two *Neurospora* proteins is the acquisition of the nuclear gene at a relatively late stage of evolution by uptake of foreign DNA. Such a horizontal gene transfer has similarly been invoked to explain the close homology between vertebrate haemoglobin genes and the leghaemoglobin gene from leguminous plants[13].

One might ask whether organisms closely related to *Neurospora*, like *Aspergillus* and *Botrydiplodia*[14], which also carry the genes for the DCCD-binding proteins on the nuclear genome, have a mitochondrial gene similar to that of *Neurospora*. If so, nucleotide sequences analysis of both nuclear and mitochondrial genes will not only bear on the origin of these genes, but will also elucidate the function of the mitochondrial gene product.

Table 1 Number of identical amino acids in the primary structures of DCCD-binding proteins from mitochondria of *N. crassa*, *S. cerevisiae* and bovine heart

	a	b	c	d
a	74			
b	43	81		
c	49	40	75	
d	46	40	45	76

a, *N. crassa* (primary structure predicted from the mitochondrial gene); *b*, *N. crassa* (nucleus-encoded); *c*, bovine heart; *d*, *S. cerevisiae*.

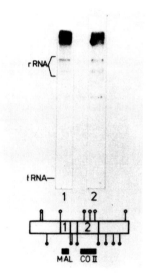

Fig. 3 Transcripts of the mitochondrial ATPase proteolipid-like gene (MAL) and the gene for subunit 2 of cytochrome aa_3 (COII). Preparation of blots of mitochondrial RNA on nitrocellulose filters and hybridization conditions were as described by Thomas[17]. Probes used for hybridization were *Hae*III and *Hinf*I subfragments (1 and 2) of *Eco*RI fragment 4 of *Neurospora* mitochondrial DNA. ●, ○, The *Hae*III and *Hinf*I restriction sites, respectively. The positions of the rRNAs and tRNAs which contain 3,200, 2,000 and ~80 nucleotides respectively, are indicated in the autoradiograph.

We thank Drs J. J. Beintema, P. Borst, A. M. Kroon, P. van 't Sant and H. de Vries for helpful discussions. This work was supported in part by the Netherlands Foundation for Chemical Research (SON) with financial aid from the Netherlands Organization for the Advancement of Pure Research (ZWO). Part of these results have been presented in preliminary form elsewhere[15].

Received 8 March; accepted 27 April 1982.

1. Boyer, P. D. et al. A. Rev. Biochem. **46**, 955–1026 (1977).
2. Sebald, W., Graf, T. & Lukins, H. B. Eur. J. Biochem. **93**, 587–599 (1979).
3. Sebald, W., Machleidt, W. & Wachter, E. Proc. natn. Acad. Sci. U.S.A. **77**, 785–789 (1980).
4. Orian, J. M., Murphy, M. & Marzuki, S. Biochim. biophys. Acta **652**, 234–239 (1981).
5. Jackl, G. & Sebald, W. Eur. J. Biochem. **54**, 97–106 (1975).
6. Agsteribbe, E., Samallo, J., De Vries, H., Hensgens, L. A. M. & Grivell, L. A. in *The Organization and Expression of Mitochondrial Genome* (eds Kroon, A. M. & Saccone, C.) 51–60 (Elsevier, Amsterdam, 1980).
7. Hensgens, L. A. M., Grivell, L. A., Borst, P. & Bos, J. L. Proc. natn. Acad. Sci. U.S.A. **76**, 1663–1667 (1979).
8. Sebald, W., Hoppe, J. & Wachter, E. in *Function and Molecular Aspects of Biomembrane Transport* (ed. Quagliariello, E.) 63–74 (Elsevier, Amsterdam, 1979).
9. Martin, S. L., Zimmer, E. A., Kan, Y. W. & Wilson, A. C. Proc. natn. Acad. Sci. U.S.A. **77**, 3563–3566 (1980).
10. Nienhaus, A. W. & Stamatoyannopoulos, G. Cell **15**, 307–315 (1978).
11. Ford, P. J. & Brown, R. D. Cell **8**, 485–493 (1976).
12. Margulis, L. *Origin of Eukaryotic Cells* (Yale University Press, New Haven, 1970).
13. Lewin, R. Science **214**, 426–429 (1981).
14. Wenzler, H. & Brambl, R. J. biol. Chem. **256**, 7166–7173 (1981).
15. van den Boogaart, P., Samallo, J. & Agsteribbe, E. *Cold Spring Harb. Conf. Mitochondrial Genes* (Cold Spring Harbor Laboratories, New York, 1982).
16. Maxam, A. M. & Gilbert, W. Meth. Enzym. **65**, 499–560 (1980).
17. Thomas, P. S. Proc. natn. Acad. Sci. U.S.A. **77**, 5201–5205 (1980).

18

Copyright © 1986 by Springer-Verlag, New York
Reprinted from manuscript accepted for publication in *J. Mol. Evol.*, vol. 22, issue 3 (1985)

MOLECULAR ARCHAEOLOGY OF THE MITOCHONDRIAL GENOME

Robert Obar and James Green

[*Editor's Note:* In the original, material precedes this excerpt.]

A MODEL FOR NET GENE TRANSFER

A minimal mechanism for the net transfer of a genetic element from a "donor" genome to a "recipient" genome is described by the following series of evolutionary steps:

Step A- DUPLICATION) A genetic element within the linear order of the donor genome is duplicated.

Step B- USE) Over time, the recipient becomes competent to regularly utilize some genetic or metabolic product of the genetic element. Note: Donor and recipient need not utilize the same product of this gene.

Step C- TRANSMIGRATION) At least one copy of the (donor) genetic element moves to and integrates into the recipient genome.

Step D- ACTIVATION) The newly acquired genetic element is transcribed and, if a protein is encoded, translated by the recipient.

Step E- REGULATION) The recipient fine-tunes its regulation of the genetic element and its product(s), and synthesis and transfer of each product to its functional site. Note: in this step there is functional genetic redundancy as the genomes compete in the formation of gene product.

Step F- SELECTION) The donor copy of the genetic element becomes non-functional (due to a regulatory or functional lesion); the recipient assumes full control of the genetic element and its products. Note: the donor copy is now fully redundant: by definition a pseudogene. The symbiosis is by now fully obligate.

Step G- LOSS) The moribund (pseudogene) copy of the genetic element is altered beyond recognition by mutation or otherwise lost from the donor genome.

These steps may be viewed as a formalization of the ideas on "levels of partner integration" (Margulis, 1976) and "rules for endosymbiont evolution" (Thornley and Harington, 1981), which until our analysis here had not been used to explain specific cases of molecular evolution.

While each of the steps in this scenario is essential for what we have termed the "net transfer" of a genetic element, the order of the steps is not. Only the sequence: DUPLICATION (of a genetic element); TRANSMIGRATION (of duplicate copies); COMPETITION (between copies integrated within different genomes); LOSS (of redundant copy), is crucial. Clearly, each of these processes may have occurred in the absence of (and independent of) the others, but only in the order given will a net transfer result.

ASPECTS OF DONOR AND RECIPIENT IN GENE TRANSFER

A genetic element may be incorporated into a genome in two ways. If a product (RNA, peptide, or protein) of the genetic element has been previously available, the ability of the recipient to process and utilize this product may precede the gene's integration. An entirely new genetic element, formed by the integration (recombination) act itself, must be incorporated into a preexisting recipient transcription unit before its potential products have been available to the recipient or subjected to natural selection. The success of such a genetic element will depend upon its expressibility, and the ease with which its products are processed and used.

For the net transfer of any protein-coding gene to occur, the recipient genome must "speak the donor's language;" in molecular terms this means the recipient must be capable of transcribing the entire gene, processing the messenger, translating it into the correct amino acid sequence, and performing any further processing and transport necessary to the function of the protein. As a result a genome with its transcription and translation closely coupled, without an RNA-splicing apparatus, could rarely acquire protein-coding genes from a genome with mosaic genes, transcription and translation uncoupled, and an elaborate splicing apparatus. Thus, with respect to a given donor and a given recipient, the potential for net transfer of protein-coding genes is an all or none matter.

The selection pressures on genes for stable RNAs are different, because they produce no protein products for the scrutiny of natural selection. Genes for stable RNAs can only transfer from a donor to a recipient genome if the stable RNAs for which they code are homologous enough to be recognized by the recipient protein synthetic system. Apparently, extreme barriers prevent stable RNAs themselves from transferring from their own protein synthetic system to any other (Gray and Doolittle, 1982).

Selection pressures governing the transfer of intervening sequence or "spacer" elements are less severe. The introduction of this type of genetic element will be selectively neutral or only mildly deleterious more often than in the case of "coding elements." Unless the newly acquired noncoding sequence alters the expression of other genes in the recipient genome, any deleterious effects of the transfer may not become apparent for many generations.

[*Editors' Note:* Material has been omitted at this point.]

GLOSSARY

Donor: A discrete genetic system which undergoes the net loss of a genetic element or elements.

Genetic element: An independent nucleic acid sequence; the unit of net transfer between genomes. probability of survival under changed conditions.

Intervening sequence (Intron): A discrete, continuous nucleic acid sequence which interrupts the coding regions of a cistron, with respect to a specific gene product.

Recipient: A discrete genetic system which undergoes the net gain of a genetic element or elements.

Stable RNA: Transfer RNA (tRNA) or ribosomal RNA (rRNA) or their unprocessed precursors. Note: for simplicity, no other types of RNA (such as stable messenger RNAs) are included in this definition.

REFERENCES

Gray, M.W., and Doolittle, W.F. (1982). Microbiol. Rev. 46:1-42.

Margulis, L. (1976). Exp. Parasitol. 39:277-349.

Thornley, A.L., and Harington, A. (1981). J. Theor. Biol. 91:515-523.

Part IV
ORIGIN OF PLASTIDS

Editors' Comments on Papers 19 through 24

19 LEWIN and WITHERS
Extraordinary Pigment Composition of a Prokaryotic Alga

20 BOYNTON et al.
Transmission, Segregation and Recombination of Chloroplast Genes in Chlamydomonas

21 GIBBS
The Chloroplasts of Euglena *May Have Evolved from Symbiotic Green Algae*

22 BONEN and DOOLITTLE
Ribosomal RNA Homologies and the Evolution of the Filamentous Blue-Green Bacteria

23 STERN and PALMER
Extensive and Widespread Homologies Between Mitochondrial DNA and Chloroplast DNA in Plants

24 SCHIFF
Excerpts from *Origin and Evolution of the Plastid and Its Function*

The generic word *plastid* is recommended rather than *chloroplast* when referring to the photosynthetic organelles of eukaryotes. In the macroscopic algal protoctists (rhodophytes, phaeophytes, and chlorophytes) there are three distinct types of plastids: rhodoplasts, phaeoplasts, and chloroplasts, respectively. When one considers the rest of the protoctistan microalgae, the number is probably far greater (Papers 19, 21, 22, and 24). The plastids are distinguished by pigments as well as enzymatic systems and are very likely polyphyletic. Lewin and Withers in Paper 19, Gibbs in Paper 21, and Bonen and Doolittle in Paper 22 suggest free-living counterparts for chloroplasts, chrysoplasts, and euglenoplasts. The general problem is reviewed by Raven in Paper 5 and by Schiff in Paper 24.

Plastids as independent entities with their own genetic system are discussed in Paper 20. Paper 23 is an example of gene transfer between mitochondria and plastids (chloroplasts) in corn. The frequency of such transfers from organelle to organelle and between organelle and nucleus is not yet known as the research in this area is still very new. For other examples of putative or demonstrated gene transfer see Papers 7, 8, 10, 11, 12, 17, 19, and 23, and references in these papers.

EXTRAORDINARY PIGMENT COMPOSITION OF A PROKARYOTIC ALGA

R. A. Lewin and N. W. Withers

WE report an anomalous unicellular alga with a prokaryotic cellular organisation like that of a blue-green alga, but with a pigment composition characteristic of green algae and higher plants.

Synechocystis didemni Lewin is a unicellular marine alga so far found only associated with colonial ascidians (*Didemnum* spp.) on the coasts of Baja California, Mexico[1]. Under the light microscope it looks like a blue-green alga, and by electron microscopy of thin sections we have confirmed that the cellular organisation is prokaryotic, without membrane-limited nucleus, plastids or mitochondria (ref. 2 and R. A. L. and M. Schulz-Baldes, unpublished). Typically, blue-green algae—photosynthetic prokaryotes which can evolve oxygen when illuminated—contain at least one red or blue bilin pigment (phycoerythrin or phycocyanin) and only one chlorophyll, *a*, *S. didemni*, however, is leaf-green in colour, apparently

736 R. A. Lewin and N. W. Withers

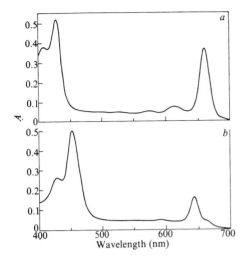

Fig. 1 Spectral absorption curves of solutions of *S. didemni* chlorophylls in ether, provisionally identified as *a* and *b*. For details of extraction and separation methods, see Table 2.

lacks bilin pigments, and contains two chlorophylls, *a* and *b*—features characteristic of green algae and land plants. Preliminary evidence for this unique combination of a prokaryotic organisation with a pigment complement so far only known in eukaryotic plants has been published elsewhere[2]. We offer here confirmation of the presence of chlorophylls *a* and *b*, together with other lipid-soluble (but no water-soluble) pigments.

So far it has not proved possible to culture cells of *S. didemni*; consequently we had to analyse living material collected from the wild. In March 1975, didemnid colonies bearing *S. didemni* were gathered in a tropical lagoon on Isla S. José, Baja California from dimly illuminated mangrove roots (maximum light intensity around noon, 300–500 lx; water temperature about 20–25 °C). The surfaces were brushed as described[2], and the algal cells removed were concentrated by centrifugation. We collected three lots of algal cells, each of a few hundred milligrams (wet weight). The pigments from small subsamples were extracted and examined chromatographically on the day of collection; the rest of each sample was stored at −20 °C for careful examination in the laboratories of the Scripps Institution of Oceanography several days later. (There was no reason to believe that storage in this way effected any appreciable change in the pigment composition of the samples.)

After thawing, the cells were washed in water, and the supernatant, essentially colourless, was examined for possible traces of bilin pigments. An absorption spectrum revealed no evidence for any pigmented material of this sort. The lipid-soluble pigments were then extracted and examined by thin-layer chromatography (TLC) and spectrophotometry (Fig. 1 and Tables 1 and 2).

The evidence for the existence of chlorophyll *b*, as well as *a*, in *S. didemni* seems unequivocal. This second green pigment cochromatographed with chlorophyll *b* from the chlorophyte *Codium fragile* and exhibited absorption maxima (Table 2) and band ratios typical for *b* (ref. 3). In the two samples examined the ratio *a*:*b* was found to be comparable with that of other algae which contain *b* (Table 1). The presence of *b* in our preparations could hardly be attributable to contamination of the samples by other algae, as microscopic examination revealed that the cells in our suspensions were almost all *S. didemni*, with no evident chlorophytes or euglenophytes. There was a small proportion (less than 1%) of diatoms or diatom debris, but these were mainly empty frustules, and only a few retained pigmented contents. The second chlorophyll of diatoms, *c*, has an R_f and other properties quite dissimilar from those of *b*. No *c* was detected in our material.

The predominant carotenoid in *S. didemni* is β carotene, representing some 65% of the total absorption attributable to yellow pigments. This proportion is higher than that normally encountered in chlorophytes (for example, 15% for *Chlorella pyrenoidosa*, which also has α carotene), but is close to that found in certain cyanophytes (for example, 46% for *Anacystis nidulans*[5], 63% for *Microcoleus vaginatus*[6]). The other carotenoids of *S. didemni* were presumably xanthophylls, perhaps including myxoxanthophyll. The amounts were insufficient for critical analysis, which must evidently await the availability of much larger quantities of cell material, if and when this can be obtained in culture.

Although it has been generally accepted that all photosynthetic cyanophytes (blue-green algae) possess bilin pigments, in an unidentified blue-green alga associated with ascidians of the Great Barrier Reef little or no phycocyanin or phycoerythrin was detected[7]. It has also been generally accepted that blue-green algae lack *b* (ref. 8), although spectral data from blue-green algal mats collected in Yellowstone Park indicated that about 15% of the chlorophyll was present as *b* (ref. 9). Since those algae had grown at temperatures above 48 °C, they could hardly have been contaminated by chlorophytes or euglenophytes, none of which has been reported to grow in such hot water. Among the predominantly filamentous algae Inman[9] noted the presence of *Synechococcus lividus* (Chroococcales) and of "a unicellular green alga, probably *Chlorella* sp.". From what we now know of such habitats we can be almost certain that the latter alga was *Cyanidium caldarium*, which does not contain *b* (ref. 10).

In the original description of the type species of the genus, *Synechocystis aeruginosa*[11], the cells were described as being of a blue-green colour. Drouet and Daily[12], who re-examined Sauvageau's material, reported that it contained cells which seemed to belong to the Chlorococcales (Chlorophyta); but it is hard to believe that Sauvageau could have mistaken a unicellular cyanophyte in this way. Further conclusions may only be drawn after examination of the pigments of an authentic

Table 1 Comparative pigment data for *S. didemni* and other algae

	Phycobilins (g per 10^{14} cells)	Chlorophylls		*a*:*b*	β carotene/ total carotenoids	Calculated from data in ref.
		a (g per 10^{14} cells)	*b* (g per 10^{14} cells)			
S. didemni (from white *Didemnum**)		95	22	4.36	—	
S. didemni (from grey *Didemnum**)		47	6.9	6.92	0.65	
Anacystis nidulans	1,300	60	0	—	0.46	5, 14, 15
Chlorella pyrenoidosa		40	7.6	5.25	0.15	4, 16, 17
Euglena gracilis		200	33.3	6.0	0.11	18, 19, 20

*The species could not be determined.
Cell counts were made with a Levy haemocytometer. Carotenoids and chlorophylls were estimated spectrophotometrically using standard extinction coefficients[3,13].

Table 2 R_f values and absorption maxima (nm) of *S. didemni* pigments

R_f on sucrose (TLC)	Colour	Acetone	Solvent Ethyl ether	n-Hexane	Identity
0.92	Yellow	425, 452, 480	—	420, 448, 475	β carotene*
0.72	Blue-green	430, 662	431, 662	—	Chlorophyll *a*
0.61	Orange	425, 451, 480	—	—	Unidentified xanthophyll
0.48	Yellow-green	457, 645	452, 642	—	Chlorophyll *b*
0.19	Yellow	440, 470	—	—	Unidentified xanthophyll
0.02	Orange	445	—	—	Unidentified xanthophyll

*Failed to separate from authentic β carotene (from *Daucus carota*) in a mixed chromatogram on Al_2O_3 + MgO developed with 6% ethyl acetate in *n*-hexane[22].

Pigments were extracted by sonicating the cells in a small volume of cold acetone. Cell residues were removed by centrifugation, and the pigments from the supernatant were transferred to ethyl ether by addition of cold, saturated saline. The ether extract was concentrated under N_2 and subjected to TLC on sucrose, using 1.3% *n*-propanol in ligroine (boiling point 63–75 °C)[21].

strain of *S. aeruginosa* (we are attempting to obtain a culture of this species from the Department of Genetics, University of Moscow), and of the pigments of other cyanophytes grown in natural conditions, notably those encountered by *S. didemni* in mangrove swamps and by other species which survive in hot springs.

In conclusion, we have confirmed that natural collections of *S. didemni* contain chlorophylls *a* and *b*, β carotene, at least three xanthophylls, and no demonstrable water-soluble phycobilin pigment. In these respects it seems to be unique among cyanophytes which have been examined to date.

We thank Dr F. T. Haxo for advice and guidance with the chromatography, and the Foundation for Ocean Research for making the RV Dolphin available for collecting material.

[1] Lewin, R. A., and Cheng, L., *Phycologia*, **14** (in the press).
[2] Lewin, R. A., *Phycologia*, **14** (in the press).
[3] Strain, H. H., Thomas, M. M., and Katz, J. J., *Biochim. biophys. Acta*, **75**, 306 (1963).
[4] Allen, M. B., Goodwin, T. W., and Phagpolngarm, S., *J. gen. Microbiol.*, **23**, 93 (1960).
[5] Halfen, L. N., and Francis, G. W., *Archs Mikrobiol.*, **81**, 25 (1972).
[6] Goodwin, T. W., *J. gen. Microbiol.*, **17**, 467 (1957).
[7] Newcomb, E. H., and Pugh, T. D., *Nature*, **253**, 533 (1975).
[8] Allen, M. B., French, C. S., and Brown, J. S., in *Comparative Biochemistry of Photoreactive Systems* (edit. by Allen, M. B.), 33 (Academic, London 1960).
[9] Inman, O. L., *J. gen. Physiol.*, **23**, 661 (1940).
[10] Allen, M. B., *Archs Mikrobiol.*, **32**, 270 (1959).
[11] Sauvageau, C., *Bull. Soc. bot. France*. **39**, 104 (1892).
[12] Drouet, F., and Daily, W. A., *Butler Univ. Bot. Stud.*, **12**, 218 (1956).
[13] Davies, B. H., in *Chemistry and Biochemistry of Plant Pigments* (edit. by Goodwin, T. W.), 489 (Academic, London, 1965).
[14] Evans, E. L., and Allen, M. M., *J. Bact.*, **113**, 403 (1973).
[15] Allen, M. M., *J. Bact.*, **96**, 831 (1968).
[16] Myers, J., and Graham, J. R., *Pl. Physiol., Lancaster*, **48**, 282 (1971).
[17] Jørgensen, E. G., *Physiologia Pl.*, **22**, 1307 (1969).
[18] Brown, J. S., Alberte, R. S., Thornber, J. P., and French, C. S., *Carnegie Instn Yb.*, 694 (1974).
[19] Holowinsky, A. W., and Schiff, J. A., *Pl. Physiol., Lancaster*, **45**, 339 (1970).
[20] Krinsky, N. I., and Goldsmith, T. H., *Archs biochem. Biophys.*, **91**, 271 (1960).
[21] Jeffrey, S. W., *Biochim. biophys. Acta*, **162**, 271 (1968).
[22] Chapman, D. J., and Haxo, F. T, *J. Phycol.*, **2**, 89 (1966).

TRANSMISSION, SEGREGATION AND RECOMBINATION OF CHLOROPLAST GENES IN CHLAMYDOMONAS
John E. Boynton, Nicholas W. Gillham, Elizabeth H. Harris, Constance L. Tingle, Karen Van Winkle-Swift, and G.M.W. Adams, Duke University, Durham, NC 27706

Chlamydomonas reinhardtii, a eukaryotic unicellular green alga with a single chloroplast and several mitochondria, is an exemplary organism for investigation of organelle genetics[1-4]. Both chloroplast and mitochondrial DNAs have been characterized physically and appear to be present in many copies per cell. The chloroplast genome contains sufficient unique sequence DNA to code for several hundred proteins of average molecular weight, whereas the mitochondrial DNA has less than 10% of this coding capacity. Phenotypes of chloroplast mutants include resistance to or dependence on antibacterial antibiotics, in many cases attributable to direct effects on chloroplast ribosomes[5], as well as temperature sensitivity and requirement for acetate[1,3,4]. With very few exceptions[3,6,14], the allelic relationships of chloroplast gene mutants with similar phenotypes are not known.

Mutations in C. reinhardtii show one of three distinct patterns of inheritance. Nuclear gene mutations, which fall into 16 linkage groups[7], regularly segregate 2:2 at meiosis in reciprocal crosses and do not segregate somatically. Putative mitochondrial gene mutations[8,9] are transmitted by either parent to the meiotic progeny in a non-Mendelian fashion. Chloroplast gene mutations are usually transmitted uniparentally by the maternal (mt^+) parent[1-4,10]; the physical basis for this uniparental transmission is not completely understood. A few zygotes transmit chloroplast genes from both parents (biparental zygotes) or, rarely, only from the mt^- parent (paternal zygotes). The frequency of these exceptional zygotes can be greatly increased by UV irradiation of mt^+ gametes prior to mating[11,12]. Segregation and recombination of chloroplast genes have been studied among the progeny of biparental zygotes by pedigree analysis,[3,13,14] by analysis of successive generations of zoospore progeny growing in liquid culture[3,15] and by zygote clone analysis[1,12,14,16]. The relative merits of different methods for mapping chloroplast genes are discussed by Adams et al[1]. In the zygote clone analysis method used in our laboratory, 64 progeny clones are selected randomly from 50-100 biparental zygotes. Recombination frequencies are calculated as (recombinant progeny)/(total progeny from all biparental zygotes), and in this respect our method resembles the random diploid method used for mitochondrial genes in yeast[17]. The few remaining heteroplasmic progeny are excluded from calculations.

Using pedigree analysis, Sager and Ramanis[3,13,14] have reported a 1:1 ratio of parental chloroplast alleles among the progeny of biparental zygotes, and have interpreted this as evidence for the presence of only two copies of the chloroplast genes per vegetative cell. Recently they have also found the same 1:1 ratio in zygote clones irrespective of the UV dose used to increase the frequency of biparental zygotes[14]. In contrast, we have found a biased output in zygote clones favoring chloroplast genes from the maternal parent[12] (Fig. 1). With increasing UV dose transmission of paternal chloroplast genes also increases, and the average ratio of maternal to paternal genes in a population of biparental zygotes approaches 1:1. However, individual biparental zygotes vary widely in output of paternal and maternal chloroplast genomes, as defined by allelic ratios, and are not normally distributed around the mean of 0.5 which presumably represents equal allelic input. The extreme classes are as frequent as those with a ratio of 0.5.

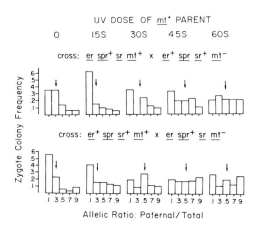

FIGURE 1. Allelic ratios among biparental zygotes in reciprocal crosses. The mt^+ parent was treated with various doses (seconds) of UV prior to mating. Zygote clones were grouped into classes based on the fraction of progeny in each clone carrying the alleles from the paternal parent. From Gillham et al. 1974[12]

er = er-u-37
spr = spr-u-1-6-2
sr = sr-u-2-60

We have also examined the meiotic progeny of complete tetrads produced from biparental zygotes (Table 1). Among 1012 meiotic products, 50% were homoplasmic at the end of the second meiotic division for the three chloroplast markers studied. The ratio of products homoplasmic for maternal and paternal chloroplast markers approximates 3:1, and is clearly not consistent with Sager's two-copy model. Assuming that the entire chloroplast genome is transmitted en bloc, including the region delimited by these three markers, then tetrads in which the meiotic products show a predominance of maternal chloroplast genomes comprise 70% of the total. Thus our results indicate that these biparental zygotes must have contained more than two copies of the chloroplast genome prior to meiosis, with the bias favoring the maternal parent. If each meiotic product receives at least two copies of the chloroplast genome, then in the most extreme cases one has to assume a minimum of 7 maternal:1 paternal genome (21% of observed tetrads) and 1 maternal:7 paternal (2% of observed tetrads). We do not yet know whether our stocks continue to show a biased segregation of chloroplast genes during the first two mitotic divisions following meiosis. Neither the consistent maternal bias seen in the progeny of biparental zygotes (Fig. 1) nor that seen among the meiotic products (Table 1) can be accounted for by differential growth of specific genotypes.

TABLE 1. Inheritance of chloroplast genomes in meiotic products of biparental zygotes. Analysis is based on 1012 meiotic products from 253 complete tetrads formed by biparental zygotes in the crosses shown in Tables 2 and 3. A biparental zygote is here defined as one in which identifiable chloroplast markers (er-u-37, spr-u-1-6-2, sr-u-2-60) from both parents can still be scored at the end of meiosis. Each meiotic product is presumed to have a minimum of two copies of the chloroplast genome.

Cross: M/M mt^+ X P/P mt^-, where M/M indicates maternal genotype, P/P indicates paternal genotype, and M/P indicates biparental genotype (carrying chloroplast markers from both parents).

CHLOROPLAST GENOTYPES OF THE FOUR MEIOTIC PRODUCTS			MINIMUM NUMBER OF CHLOROPLAST GENOMES IN ZYGOTE AFTER REPLICATION		OBSERVED FREQUENCY OF TETRAD TYPE	
					NUMBER	%
Zygotes with maternal bias						
3 M/M	1 M/P	----	7 M	1 P	52	21
2 M/M	2 M/P	----	6 M	2 P	60	24
3 M/M	----	1 P/P	6 M	2 P	0	0
1 M/M	3 M/P	----	5 M	3 P	51	20
2 M/M	1 M/P	1 P/P	5 M	3 P	13	5
Zygotes with no bias						
1 M/M	2 M/P	1 P/P	4 M	4 P	27	11
----	4 M/P	----	4 M	4 P	15	6
2 M/M	----	2 P/P	4 M	4 P	3	1
Zygotes with paternal bias						
----	3 M/P	1 P/P	3 M	5 P	12	5
1 M/M	1 M/P	2 P/P	3 M	5 P	3	1
----	2 M/P	2 P/P	2 M	6 P	11	4
1 M/M	----	3 P/P	2 M	6 P	1	0.5
----	1 M/P	3 P/P	1 M	7 P	5	2

By application of the Visconti-Delbrück[18] theory of phage genetics, the relative input ratio of maternal and paternal chloroplast genomes in individual biparental zygotes should influence the frequency of recombination of chloroplast genes[1,12]. The same analysis may be applied to mitochondrial genes in yeast[19]. Data from reciprocal crosses involving three chloroplast markers with different phenotypes support this hypothesis (Table 2): Progeny of the most paternally and most maternally skewed biparental zygotes show substantially less recombination than progeny from biparental zygotes with more nearly equal allelic ratios. Since biparental zygotes with nearly equal maternal and paternal ratios of chloroplast genes are relatively infrequent, one cannot realistically use only progeny from these zygotes for mapping chloroplast genes. Instead our approach is to keep the extent of maternal bias constant by using standardized conditions for gametogenesis, UV irradiation and mating.

TABLE 2. Recombination analysis of the progeny of biparental zygote clones arising from reciprocal crosses. Data are pooled into 5 allelic ratio classes from a series of crosses in which the mt^+ parent received from 0 to 60 s UV prior to mating.

(Cross: er-u-37 spr-u-1-6-2$^+$ sr-u-2-60 X er-u-37$^+$ spr-u-1-6-2 sr-u-2-60$^+$)

PATERNAL ALLELIC RATIO	TOTAL SAMPLED		% RECOMBINATION		
OF BIPARENTAL ZYGOTE	BIPARENTAL ZYGOTES	PROGENY	er--sr	er--spr	spr--sr
0.01 to 0.20	249	15,791	2.9	2.2	1.4
0.21 to 0.40	79	4,961	9.8	7.5	4.4
0.41 to 0.60	60	3,820	7.1	5.6	3.5
0.61 to 0.80	41	2,586	6.6	6.3	2.0
0.81 to 0.99	47	2,975	4.1	3.8	1.6
overall	476	30,133	5.0	4.0	2.2

To demonstrate that the parental combination of chloroplast markers has no intrinsic effect on recombination of chloroplast genes, we have made all possible reciprocal crosses of stocks carrying markers for resistance to streptomycin, erythromycin, and spectinomycin (Table 3). From among the progeny of the initial pair of reciprocal crosses, we selected recombinant genotypes which served as parents for the five subsequent crosses. The seven crosses not only produced the same map order for the three chloroplast markers involved, but also yielded remarkably similar recombination frequencies for all three intervals.

TABLE 3. Effect of marker combination on recombination of the chloroplast genes er-u-37, spr-u-1-6-2, and sr-u-2-60 analyzed by the zygote clone method. In each cross, mt^+ gametes were subjected to 15 s UV irradiation prior to mating. Approximately 3200 progeny from 50 biparental zygotes were analyzed in each cross.

	CROSS			% RECOMBINATION		
	mt^+	X	mt^-	er--sr	er--spr	spr--sr
1.	er spr$^+$ sr	X	er$^+$ spr sr$^+$	4.3	4.0	0.6
2.	er$^+$ spr sr$^+$	X	er spr$^+$ sr	5.0	4.8	1.5
3.	er$^+$ spr$^+$ sr	X	er spr sr$^+$	5.2	4.2	1.5
4.	er spr sr$^+$	X	er$^+$ spr$^+$ sr	6.0	5.7	1.3
5.	er$^+$ spr sr	X	er spr$^+$ sr$^+$	4.1	3.8	2.2
6.	er spr$^+$ sr$^+$	X	er$^+$ spr sr	4.8	3.9	2.7
7.	er$^+$ spr$^+$ sr$^+$	X	er spr sr	3.7	2.8	2.3
	overall			4.7	4.2	1.7

Thus although the zygote clone analysis protocol is potentially subject to problems with differential growth of certain genotypes[1], such problems do not seriously affect our ability to map chloroplast genes. We feel that since a large number of progeny can be quickly and accurately scored by the zygote clone method, and since segregation is almost entirely complete at the time of analysis (less than 1% of total progeny heteroplasmic for any marker), this method is valid for chloroplast gene mapping in Chlamydomonas. In mapping by pedigree analysis[3,13,14] one faces the problem that many chloroplast genes are still heteroplasmic at the time the analysis is done.

To establish a map of the chloroplast genes for antibiotic resistance that affect chloroplast ribosomes, we first tested the known chloroplast markers with similar antibiotic resistance phenotypes for allelism. Somatic segregation of chloroplast genes in diploids as well as haploids prevents the use of the classical dominance/complementation test used in vegetative diploids to determine whether one or more gene functions are involved. Instead allelism must be judged entirely on the ability of two antibiotic resistant mutants with similar phenotypes to produce recombinant progeny which are antibiotic sensitive[6]. Table 4 illustrates two streptomycin resistant mutants which recombine at an appreciable frequency (1.6%) and are not alleles (cross 1), in contrast to two which show less than 0.03% recombination and appear to be alleles (cross 2).

TABLE 4. Results from typical crosses made to test chloroplast mutants with similar phenotypes for allelism. The mt$^+$ gametes received 15 s UV prior to mating.

CROSS	NUMBER OF PROGENY								% RECOMBINATION	
	er + sr	+ nr sr	er nr sr	+ + +	+ + sr	er + +	+ nr +	er nr +	sr-sr*	er-nr
1	3353	863	16	0	194	0	25	10	1.6	4.9
2	1576	4853	96	0	272	0	0	0	<0.029	5.4

Cross 1: sr-u-2-60 er-u-37 mt$^+$ X sr-u-sm3 nr-u-2-1 (spr-u-1-27-3) mt$^-$
Cross 2: sr-u-sm5 nr-u-2-1 mt$^+$ X sr-u-sm-3a er-u-1a mt$^-$

*(number of streptomycin sensitive progeny X 2 / total progeny) X 100

Among mutants from our laboratory and from Sager and Bogorad, we have now unequivocally demonstrated 4 discrete loci for streptomycin resistance and 2 for erythromycin resistance in the chloroplast genome (Fig. 2)[6]. At least 4 independently isolated mutant alleles have been identified at one streptomycin locus (Table 5). Similarly, all three chloroplast mutants resistant to spectinomycin and the mutant resistant to neamine appear to be alleles at a single locus[6,20].

Eight different mapping crosses were analyzed by the zygote clone method to establish the order and relative position of these seven loci on the chloroplast genome. The frequencies of parental and recombinant progeny appear in Table 6. In 7 of the 8 crosses, a bias exists favoring the maternal genotype, and in some instances reciprocal recombinant types occur in vastly unequal frequencies. The genotypes er spr nr sr and spr nr sr appear to be at a selective disadvantage whether they are recombinant (Table 6) or parental (Table 7). Nonetheless these

FIGURE 2. Tentative map of chloroplast ribosome genes in <u>Chlamydomonas reinhardtii</u> based on zygote clone analysis of allele test and mapping crosses. Parentheses indicate preliminary data from a small number of progeny.

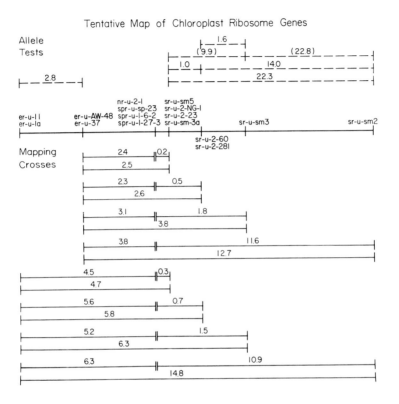

TABLE 5. Recombination analysis of chloroplast gene mutants conferring antibiotic resistance on chloroplast ribosomes: Allelism at one streptomycin locus. Recombination of each mutant with <u>sr-u-2-60</u> is shown for comparison.

NUMBER OF STREPTOMYCIN-SENSITIVE RECOMBINANTS/ TOTAL PROGENY					
	sr-u-sm-3a	sr-u-2-23	sr-u-2-NG-1	sr-u-sm5	sr-u-2-60
sr-u-sm-3a	0/2884	0/3175	0/3163	0/6158	16/4252
sr-u-2-23		1*/2928	0/3127	0/1280	12/3129
sr-u-2-NG-1			0/3175	0/640	30/2805
sr-u-sm5				--	18/6400
sr-u-2-60					0/3182

*possibly either a wild type contaminant or a back-mutation

TABLE 6. Results of mapping crosses between mutants at 7 different chloroplast gene loci, using the zygote clone method. Possible recombinant classes where no progeny were observed are not listed. spr = spr-u-1-27-3 and nr = nr-u-2-1.

CROSS		NUMBER OF PROGENY								
		PARENTAL		RECOMBINANT						
mt$^+$ (15 s UV) X mt$^-$		+	er	er	+	er	+	+	er	+
		+	nr	+	nr	+	nr	+	nr	+
		+	spr	+	spr	+	spr	+	spr	spr
		sr	+	sr	+	+	sr	+	sr	+
sr-u-sm-3a	X er-u-37 nr spr	3921	2878	116	155	1	2	0	0	0
sr-u-2-60	X er-u-37 nr spr	5439	755	64	76	9	0	21	2	0
sr-u-sm3	X er-u-37 nr spr	2891	2904	40	115	24	9	62	12	0
sr-u-sm2	X er-u-37 nr spr	3257	2208	72	84	85	1	649	1	0
sr-u-sm-3a	X er-u-1a nr spr	4063	2017	88	197	2	0	17	0	0
sr-u-2-60	X er-u-1a nr spr	4056	1505	157	158	14	3	26	1	1
sr-u-sm3	X er-u-1a nr spr	3707	2128	36	275	7	3	74	7	0
sr-u-sm2	X er-u-1a nr spr	2089	1058	82	108	46	0	360	0	1

crosses have yielded a highly consistent map order and remarkably similar map distances for the 7 loci involved (Fig. 2). Comparison of our map with the linear map of Sager and Ramanis[13], which is based on recombination frequencies observed in pedigrees, gives the identical gene order for four markers which these maps have in common, i.e. er-u-11/er-u-1a (ery), nr-u-2-1 (nea), sr-u-sm3 (sm3) and sr-u-sm2 (sm2). However, this order does not agree well with the circular map subsequently proposed by Sager on the basis of data derived primarily from co-segregation analysis and marker segregation in liquid culture[3,14,15,21], unless the circle is broken between ery and sm2. We are presently attempting to resolve these discrepancies by additional allele test and mapping crosses with the mutants which are at variance. The linkage and arrangement of the chloroplast genes which are known to affect chloroplast ribosomes in Chlamydomonas is also of interest in view of the clustering of ribosomal genes observed in bacteria[22].

In the case of yeast mitochondrial genetics most segregation and recombination analysis has been carried out on vegetative diploid progeny from mitotic zygotes[23]. To allow more direct comparison between chloroplast genetics of Chlamydomonas and mitochondrial genetics of yeast, we have also studied the transmission, segregation, and recombination of chloroplast genes in vegetative diploids (mitotic zygotes) of Chlamydomonas[24]. Gillham[16,25] found that vegetative diploids isolated using the method of Ebersold[26] transmit chloroplast genes from both parents at a higher frequency than is normally observed for meiotic zygotes. Van Winkle-Swift has shown that this high biparental transmission of chloroplast genes depends upon the newly formed mitotic zygote dividing somatically soon after gamete fusion. Isolation of mitotic zygotes under conditions which delay the first somatic division following mating, e.g. incubation in darkness (Fig. 3), leads to a pronounced drop in the frequency of biparental diploids, and in the most extreme instances results in diploid populations exhibiting predominantly maternal inheritance[24,27]. Delayed division apparently allows the chloroplast genomes of the mt^- parent to be lost systematically, or at least not transmitted to the diploid progeny. While routine procedures for selection of mitotic zygotes promote the rapid onset of somatic division, meiotic zygotes are not usually induced to divide until 6-7 days after mating. We are presently investigating whether premature germination of meiotic zygotes increases the transmission of chloroplast genes from the paternal parent.

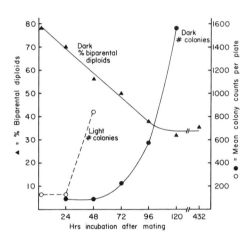

FIGURE 3. Decline in frequency of biparental diploids (mitotic zygotes carrying chloroplast gene markers from both maternal and paternal parents) when first mitotic division is delayed by incubation in darkness. Respreading experiments with plates incubated in light compared to plates incubated in dark indicate time of first division.

We have also compared recombination of four chloroplast gene markers in two crosses involving vegetative diploids with a cross involving meiotic zygotes (Table 7). The gene order observed in all three crosses is the same as found previously (Table 4, Fig. 2). The somewhat higher recombination frequencies observed for all intervals in these crosses compared to those in Table 2 may be due to the additional presence of the nr-u-2-1 marker in the quadruple antibiotic resistant stocks used as one parent in each of these crosses. As discussed above, stocks carrying nr in combination with sr and spr grow slowly, possibly because all three mutations alter the same chloroplast ribosomal subunit and in combination impair its function in the absence of antibiotics. This preferentially lowers the frequency of one parental class (Table 7), so that the overall frequency of recombinants appears higher.

TABLE 7. Comparison of recombination of the chloroplast genes sr-u-2-60, nr-u-2-1, spr-u-1-27-3 and er-u-37 in zygote clones and vegetative diploid clones. Cross 3 was not scored for spectinomycin resistance.

CROSS	PARENTAL		RECOMBINANT								% RECOMBINATION		
	+ + + +	er nr spr sr	er + + +	+ nr + +	+ + spr sr	er nr + sr	er + spr +	+ nr spr +	er + + sr	+ nr spr sr	er-sr	er-nr	nr-sr
1	5,406	241	208	21	252	103	89	77	1	0	9.1	6.2	8.2
2	5,930	6	97	5	249	33	46	33	0	1	6.0	2.8	5.6
3	10,467	191	277	35	131	85	55	50	--	--	4.7	3.7	2.8

Cross 1: wild type mt^+ (15 s UV) X er nr spr sr mt^- (zygote clones)
Cross 2: wild type arg-7 mt^+ X er nr spr sr arg-2 mt^- (vegetative diploid clones)
Cross 3: er nr spr sr arg-2 mt^+ X wild type arg-7 mt^- (vegetative diploid clones)

In conclusion, Chlamydomonas is at present the only photosynthetic organism in which one can map chloroplast genes by formal genetic methods. Although certain discrepancies in map order exist between Sager's laboratory and our own, we concur in the demonstration of a single chloroplast linkage group. The majority of the chloroplast mutations studied to date confer antibiotic resistance on chloroplast ribosomes and are thought to code for specific ribosomal proteins[5]. Even after the known chloroplast mutants and the cistrons coding for chloroplast ribosomal RNA are accounted for, there remains in the chloroplast genome sufficient information to code for a significant number of additional proteins. Isolation and characterization of a broad spectrum of acetate-requiring (non-photosynthetic) mutants in the chloroplast genome should enable us to determine which other photosynthetic components are coded by chloroplast genes and how these genes map with respect to the chloroplast ribosomal genes described here.

REFERENCES

1. Adams, G.M.W., K.P. Van Winkle-Swift, N.W. Gillham & J.E. Boynton, 1976, in The Genetics of Algae, R.A. Lewin, ed. Botanical Monographs Vol. 12, Blackwell Scientific, Oxford, pp. 69-118.
2. Gillham, N.W., 1974, Ann. Rev. Genet. 8: 347-391.
3. Sager, R., 1972, Cytoplasmic Genes and Organelles, Academic Press, New York, 405 pp.
4. Sager, R., 1974, in Algal Physiology and Biochemistry, W.D.P. Stewart, ed., Botanical Monographs Vol. 10, U. California Press, Berkeley, pp. 314-345.
5. Gillham, N.W., J.E. Boynton, E.H. Harris, S.B. Fox, & P.L. Bolen, this volume.
6. Conde, M.F., J.E. Boynton, N.W. Gillham, E.H. Harris, C.L. Tingle & W.L. Wang, 1975, Molec. gen. Genet. 140: 183-220.
7. Hastings, P.J., E.E. Levine, E. Cosbey, M.O. Hudock, N.W. Gillham, S.J. Surzycki, R. Loppes & R.P. Levine, 1965, Microb. Genet. Bull. 23: 17-19.
8. Alexander, N.J., N.W. Gillham & J.E. Boynton, 1974, Molec. gen. Genet. 112: 225-228.

9. Wiseman, A., N.W. Gillham & J.E. Boynton, 1975, Genetics 80: s84-s85.
10. Sager, R., 1954, Proc. Nat. Acad. Sci. USA 40:356-362.
11. Sager, R. & Z. Ramanis, 1967, Proc. Nat. Acad. Sci. USA 58: 931-937.
12. Gillham, N.W., J.E. Boynton & R.W. Lee, 1974, Genetics 78: 439-457.
13. Sager, R. & Z. Ramanis, 1970, Proc. Nat. Acad. Sci. USA 65: 593-600.
14. Sager, R. & Z. Ramanis, 1976, Genetics 83: 303-321.
15. Sager, R. & Z. Ramanis, 1976, Genetics 83: 323-340.
16. Gillham, N.W., 1969, Amer. Naturalist 103: 355-388.
17. Coen, D., J. Deutsch, P. Netter, E. Petrochilo & P.P. Slonimski, 1970, Symp. Soc. Exp. Biol. 24: 449-496.
18. Visconti, N. & M. Delbrück, 1953, Genetics 38: 5-33.
19. Dujon, B., P.P. Slonimski & L. Weill, 1974, Genetics 78: 415-437.
20. Boynton, J.E., E.H. Harris, C.L. Tingle, S.B. Fox & N.W. Gillham, 1975, J. Cell Biol. 67: 41a.
21. Singer, B., R. Sager & Z. Ramanis, 1976, Genetics 83:341-354.
22. Jaskunas, S.R., M. Nomura & J. Davies, 1974, in Ribosomes, ed. M. Nomura, A. Tissieres & P. Lengyel, Cold Spring Harbor Laboratory, New York, pp. 333-368.
23. Birky, C.W., Jr., 1976, in Genetics and Biogenesis of Mitochondria and Chloroplasts, ed. C.W. Birky, Jr., P.S. Perlman & T.J. Byers, Ohio State University Press, Columbus, pp. 182-224.
24. Van Winkle-Swift, K.P., 1976, The transmission, segregation and recombination of chloroplast genes in diploid strains of Chlamydomonas reinhardtii, Ph.D. thesis, Duke University.
25. Gillham, N.W., 1963, Nature 200: 294.
26. Ebersold, W.T., 1963, Genetics 48: 888.
27. Van Winkle-Swift, K.P., 1976, Genetics, abstract in press.

ACKNOWLEDGEMENTS

We thank Sue Fox for her able assistance. This work was supported by grants NSF-GB22769 and NIH-GM19427, and by NIH RCDA awards GM70453 to J.E.B. and GM70437 to N.W.G.

The chloroplasts of *Euglena* may have evolved from symbiotic green algae

SARAH P. GIBBS

Department of Biology, McGill University, Montreal, P.Q., Canada H3A 1B1

Received April 5, 1978

GIBBS, S. P. 1978. The chloroplasts of *Euglena* may have evolved from symbiotic green algae. Can. J. Bot. **56**: 2883–2889.

It is proposed that the chloroplasts of *Euglena* have arisen from the progressive reduction of endosymbiotic green algae. The theory is supported by the presence of a third membrane around the chloroplasts of *Euglena* which is not endoplasmic reticulum (ER) in origin and may be derived from the plasmalemma of the original symbiont. In addition, *Euglena* is the only organism in which chloroplast loss can be induced experimentally. Dinoflagellate chloroplasts are also surrounded by three membranes, and it is proposed that they too evolved from symbiotic eucaryotic algae.

GIBBS, S. P. 1978. The chloroplasts of *Euglena* may have evolved from symbiotic green algae. Can. J. Bot. **56**: 2883–2889.

Les chloroplastes d'*Euglena* pourraient devoir leur origine à la réduction progressive d'algues vertes endosymbiotiques. Cette théorie s'appuie sur l'existence d'une troisième membrane autour des chloroplastes d'*Euglena*, qui ne provient pas du réticulum endoplasmique, mais dont l'ancêtre pourrait être le plasmalemme du symbiote originel. De plus, l'*Euglena* est le seul organisme chez lequel on peut provoquer expérimentalement la disparition du chloroplaste. Les chloroplastes des Dinoflagellés sont également entourés de trois membranes et pourraient eux aussi résulter de l'évolution de la symbiose avec des algues eucaryotiques.

The classification of *Euglena* has long been an enigma. Since green algae and *Euglena* were the only algae known which had both chlorophyll *a* and chlorophyll *b* and as it was considered unlikely that chlorophyll *b* could have arisen twice in evolution, systematists classified *Euglena* either in the Chlorophyta with the green algae (12) or in a division of its own, the Euglenophyta, which they placed adjacent to the Chlorophyta (40, 55, 59). Yet *Euglena* in almost all its ultrastructural characteristics other than those of its chloroplasts is strikingly different from green algae. It is bounded by a proteinaceous pellicle, rather than a polysaccharide-containing cell wall; its storage product is paramylon, a β-1:3-linked glucose polymer, not starch which is α-1:4 linked; it has such animal-like characteristics as an anterior gullet and in some euglenoids, trichocysts are present; its eyespot lies free in the cytoplasm rather than enclosed in a chloroplast, and this eyespot is associated with a photoreceptor contained in a swelling on the main flagellum; and finally *Euglena* in both its nuclear structure and mitosis is distinctly different from any green alga. The chromosomes of *Euglena* remain condensed during interphase, and in addition the nucleolus does not disperse during mitosis but instead elongates and divides in two.

These many differences between *Euglena* and green algae indicate that *Euglena*, despite the similarities of its chloroplast pigments and chloroplast ultrastructure (29) to those of green algae, is in fact not closely related to green algae at all (40, 64). I would like to suggest a solution to this paradox. I believe that the chloroplasts of euglenoids have arisen during evolution from endosymbiotic green algae, i.e., a second symbiosis has occurred. Although still disputed by some authors (11, 66), it is now generally accepted that chloroplasts evolved from symbiotic procaryotic algae (8, 27, 47, 67). Thus early in evolution green algae could have acquired their chloroplasts from endosymbiotic procaryotic algae. Much later in evolution unicellular green algae could well have invaded a number of euglenoid flagellates. For some time these symbioses remained very much like that which occurs today between *Chlorella* and *Paramecium bursaria* (36). Ultimately, however, since the useful part of the symbiotic alga to the *Euglena* was its chloroplast, there was a progressive loss of cell wall, cell nucleus, mitochondria, Golgi apparatus, ribosomes, until all that persisted was the algal chloroplast and the surrounding cell membrane. This is why *Euglena* chloroplasts today are surrounded by three membranes.

The Chloroplasts of Euglena are Surrounded by Three Membranes, None of Which is ER Derived

It has been known for a number of years that the chloroplasts of all euglenoid species studied are surrounded by three membranes (16, 29, 37, 40, 41, 43). However, I and my colleagues have, for a long time, misinterpreted what these three membranes signified.

In red and green algae, as well as in all green plants, chloroplasts are always surrounded by an envelope consisting of two membranes. In all the other classes of algae, except the euglenoids and dinoflagellates which have three membranes around their plastids, chloroplasts are surrounded by four membranes, two of the chloroplast envelope and two of a layer of endoplasmic reticulum which completely encloses the chloroplast (9, 28). In species where the single chloroplast lies against the nucleus, the nuclear envelope forms one part of this ER sac. This is illustrated in Fig. 1 which is a section of a greening cell of the chrysophyte alga, *Ochromonas danica*. The point of continuity between chloroplast ER and the outer membrane of the nuclear envelope is indicated by the double arrow. Ribosomes which may be arranged in polysomes (29) are present on the outer membrane of chloroplast ER, but are never found on the inner membrane. The cell in Fig. 1 is fixed by a simultaneous glutaraldehyde – osmium tetroxide method (25) which does not preserve cytoplasmic ribosomes well so almost none of the ribosomes on the outer membrane of the chloroplast ER have been preserved. In Fig. 3b, which is a cell of *Ochromonas* fixed by conventional methods, the ribosomes on the chloroplast ER are well preserved.

In the eight classes of algae which have a layer of ER around their chloroplasts, there is a row of vesicles and tubules, or in one class scattered vesicles, lying between the chloroplast ER and the chloroplast envelope, usually at a specific location, for example, over the pyrenoid or where the chloroplast lies closely appressed to the nucleus (Fig. 1). Recently I have shown that the most likely function of these vesicles is to transport chloroplast-bound proteins from the lumen of the chloroplast ER to the lumen of the chloroplast envelope (30). Plastid-bound proteins appear to be synthesized on the polysomes of the chloroplast ER, passing during synthesis into the lumen of the ER cisterna. Vesicles pinch off from the chloroplast ER (Fig. 1, arrow) and presumably deliver their contents to the lumen of the chloroplast envelope by fusing with the outer membrane of the chloroplast envelope. From there, plastid-bound proteins enter the chloroplast by crossing the inner membrane of the chloroplast envelope.

Because of the presence of chloroplast ER in eight of the classes of algae, I hypothesized in 1970 (29) that the three membranes found around the chloroplasts of *Euglena* and dinoflagellates originated from a fusion of two of the four membranes of the chloroplast envelope – chloroplast ER complex. It is now accepted that this is not the case in the dinoflagellates (6, 15). However, in the case of *Euglena*, several inconclusive pieces of evidence (6, 41) have been advanced in support of this hypothesis, so that in the most recent review of algal chloroplast structure (6), *Euglena* chloroplasts are diagrammed as having their outermost membrane covered with ribosomes and connected to the nuclear envelope. That this is almost certainly not the case I will demonstrate below.

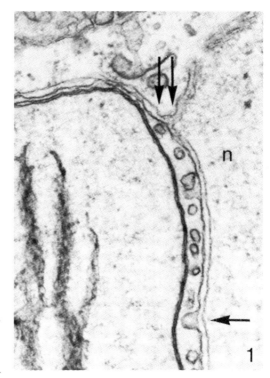

FIG. 1. Section of the chloroplast and adjacent nucleus (n) of a greening cell of *Ochromonas danica*. The chloroplast is surrounded by four membranes, two of the chloroplast envelope and two of chloroplast ER. The chloroplast ER is continuous with the nuclear envelope at the double arrow. A vesicle believed to be carrying plastid-bound proteins can be seen pinching off the nuclear envelope at the single arrow. Simultaneous glutaraldehyde – osmium tetroxide fixation. × 76 000.

First, however, I wish to show that the chloroplasts of *Euglena* are indeed surrounded by three membranes. Figure 2a is a light-grown cell of *Euglena gracilis* var. *bacillaris*. Here the chloroplast envelope unquestionably consists of three membranes which are equally spaced from each other (seen most clearly at the arrows). Often, however, the outer membrane has a more undulating course than the inner two membranes (Fig. 2b).

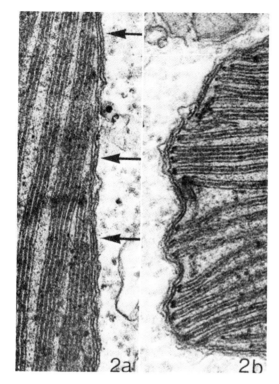

FIG. 2. Sections of chloroplasts of *Euglena gracilis* var. *bacillaris*. The chloroplasts are enclosed by three membranes, which can be seen particularly clearly at the arrows. Standard glutaraldehyde – osmium tetroxide fixation. × 100 000.

Occasionally, the inner two membranes but not the outer one can be seen to infold into the chloroplast matrix.

There are three reasons why it is unlikely that any of the membranes surrounding *Euglena* chloroplasts are derived from chloroplast ER. First the outermost membrane does not have ribosomes attached to it. In Fig. 2a, cytoplasmic ribosomes are clearly visible, but none are seen to attach to the outer membrane of the chloroplast. Sometimes a row of ribosomes paralleling the outer membrane of the chloroplast can be seen in the cytoplasm, but there is always a distinct space between the ribosomes and the outer chloroplast membrane (Fig. 3a). Bisalputra (6) has reported that there are ribosomes attached to the outer membrane of euglenoid chloroplasts. However, examination of his micrograph of *Phacus* (Fig. 4-9 in Ref. 6) shows that there is a distinct space between most of the cytoplasmic ribosomes and the outer membrane of the plastid with only an occasional ribosome actually appearing to make contact with the membrane. For all eight classes of algae which have chloroplast ER, there are pictures in the literature which show a row of ribosomes on the outer membrane of the chloroplast ER in which each ribosome can be seen to touch the membrane. The reader is referred to the following micrographs which should be compared with that of *Phacus*: for the Cryptophyceae, Fig. 3 in Ref. 53; for the Raphidophyceae, Fig. 7 in Ref. 33; for the Haptophyceae, Fig. 8 in Ref. 7; for the Chrysophyceae, Fig. 2 in Ref. 10; for the Bacillariophyceae, Fig. 10 in Ref. 32; for the Xanthophyceae, Fig. 2 in Ref. 26; for the Eustigmatophyceae, Fig. 5 in Ref. 2; and for the Phaeophyceae, Fig. 8 in Ref. 3.

Second, there has never been observed in *Euglena* any attachment like that illustrated in Fig. 1 between the outer two membranes surrounding a chloroplast and the outer membrane of the nuclear envelope. Nor are there any connections between the outer membrane of the chloroplast and the rough ER of the cytoplasm as has been shown especially well for *Chroomonas* (53), *Hymenomonas* (54), *Ochromonas* (60), *Nitzschia* (23), *Biddulphia* (32), *Botrydium* (24), and *Laminaria* (13). There is one picture in the literature, Fig. 10 in Ref. 41, in which some ER membranes appear to be in contact with the outer membrane of a *Euglena* chloroplast. However, the ER membranes in this figure are cut obliquely and a clear connection between them and the outer membrane of the chloroplast is not visible.

Third, there are never any vesicles present between any of the three membranes surrounding a *Euglena* chloroplast. Since these vesicles are believed to play a vital role in the transport of plastid-bound proteins from their site of synthesis on the polysomes of the chloroplast ER to the chloroplast matrix, their absence in *Euglena* argues against the outer membrane of *Euglena* chloroplasts being an ER membrane. Thus it is unlikely that any of the membranes surrounding *Euglena* chloroplasts is an ER membrane. I propose instead that the inner two membranes are the chloroplast envelope and that the outer membrane is derived from the plasmalemma of a green algal symbiont. Stewart and Mattox (61) have also suggested that euglenoid algae might have obtained their chloroplasts by the evolutionary reduction of an eucaryotic green alga, but they did not consider that the outermost chloroplast membrane might represent the plasmalemma of the symbiont. Originally each green algal symbiont would presumably have been separated from the host cytoplasm by the phagocytic vacuole membrane of the *Euglena*, its cell wall, and

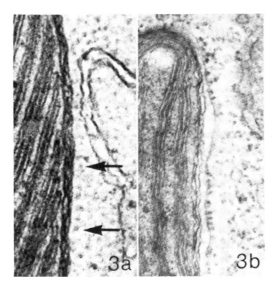

FIG. 3. (a) Section of a chloroplast of *Euglena gracilis* var. *bacillaris*. Between the arrows a row of cytoplasmic ribosomes can be seen running parallel to the chloroplast. Note that the ribosomes are not attached to the outer membrane of the chloroplast. Compare this figure with Fig. 3b where ribosomes can be seen to be attached to the outer membrane of the chloroplast ER. Standard glutaraldehyde – osmium tetroxide fixation. × 129 000. (b) Section of a chloroplast of a light-grown cell of *Ochromonas danica*. Ribosomes are present on the outer membrane of the chloroplast ER. Standard glutaraldehyde – osmium tetroxide fixation. × 52 400.

its plasmalemma. With time the algal cell wall and the phagocytic vacuole membrane would have been lost. I believe it is the phagocytic vacuole membrane, not the cell membrane, which has been lost, for in present day symbioses, the phagocytic vacuole membrane is very labile and is often lost. For example, in *Paramecium aurelia*, lambda and sigma symbiotic bacteria are enclosed in a vacuole membrane, whereas kappa and mu particles lie free in the *Paramecium* cytoplasm (4). Anderson et al. (1) studied four species of *Rickettsia* grown in the same cell line and observed that one was located in host vacuoles, whereas three had no vacuole membrane between their cell walls and the host cell's cytoplasm. In the dinoflagellates *Peridinium balticum* (65) and *P. foliaceum* (35), which have a symbiotic chrysophyte alga in their cytoplasm, the symbiote is separated from its host by one membrane only, and this membrane is almost certainly the plasmalemma of the symbiote. There are, as far as I know, no present-day symbioses where the phagocytic vacuole membrane has persisted and the plasmalemma of the symbiote has disappeared.

Other Evidence that the Chloroplasts of Euglena have Evolved from Symbiotic Green Algae

Besides the presence of a third membrane around their chloroplasts, there are a number of facts about the euglenoid flagellates which support the hypothesis that euglenoid chloroplasts evolved from symbiotic green algae. For example, *Euglena* is the only plant in which chloroplast loss can be induced. This can be done with heat, ultraviolet (UV), streptomycin, and several other inhibitors. Kivic and Vesk (37) have looked at a number of strains of *Euglena gracilis* which were bleached by a variety of methods. In 15 strains, proplastid-like structures were observed, but in three strains, including a streptomycin-bleached one, the plastids were completely eliminated. I have also studied six bleached strains of *Euglena gracilis* by electron microscopy and found proplastids in five strains, but none at all in the sixth, a UV-bleached mutant. It seems likely that chloroplasts acquired relatively recently from symbiotic eucaryotes would be more susceptible to loss than chloroplasts acquired early in evolution.

In addition, the euglenoid flagellates as well as the dinoflagellates are groups of algae which possess large numbers of colorless species. These colorless species seldom have any trace of plastids (40, 50), whereas frequently in other groups, for example, in crytomonads (57), greens (38, 48, 52), and diatoms (56), the colorless species have definite plastids. This suggests that euglenoids were originally colorless organisms which could have obtained their chloroplasts by the capture and progressive reduction of plastid-containing eucaryotic algae. The present-day colorless species which are closely related to the plastid-containing species could have readily lost chloroplasts which had been so obtained.

I have already presented the argument that the origin of euglenoid chloroplasts from symbiotic green algae nicely explains how organisms as different as *Euglena* and green algae could have similar chloroplast pigments and similar chloroplast ultrastructure. Lewin's (44) recent discovery of a procaryotic alga which contains both chlorophyll *a* and *b* (45) negates the pigment part of this argument for a procaryotic alga possessing chlorophylls *a* and *b* could have independently invaded two unrelated eucaryotic groups. However, as I have reviewed earlier (29), the chloroplasts of the Euglenophyceae are like those of the Chlorophyceae and Prasinophyceae and unlike those of all other classes of algae in three of their ultrastructural characteristics. The lamellae characteristically consist of from two to six appressed thylakoids, the thylakoids are always tightly appressed, and stacks of thylakoids are formed by the invagination and (or) evagination of existing thylakoids. It is unlikely that this 'green' type of chloroplast ultrastructure has evolved twice; thus I feel that the symbiotic green alga hypothesis best explains why *Euglena* chloroplasts are so similar to those of green algae. Note that in this article I have not split the Prasinophyceae from the green algae since they are identical in chloroplast structure.

The Chloroplasts of Dinoflagellates may also have Arisen from the Progressive Reduction of a Symbiotic Eucaryotic Alga

The envelope of the chloroplasts of a wide variety of dinoflagellate species has been shown to consist of three membranes (5, 14–20, 22, 49, 62, 63). It is accepted (6, 15) that the third or outer membrane is not an ER membrane, for it never has ribosomes attached to it, it is never connected to the nuclear envelope or rough ER of the cytoplasm and there are never any vesicles between it and the inner two membranes. I would like to suggest that here, too, this third membrane has evolved from the plasmalemma of an eucaryotic endosymbiont. I am not the first to suggest that the chloroplasts of dinoflagellates evolved from a progressive reduction of a symbiotic eucaryotic alga. Tomas and Cox (65) discovered that all the chloroplasts of the dinoflagellate *Peridinium balticum* belonged to a symbiotic chrysophyte alga which ramifies throughout the host cell. Their discovery led them to suggest (65) that the chloroplasts of other photosynthetic dinoflagellates could have arisen from symbiotic chrysophyte algae. Loeblich (46) has also proposed that the chloroplasts of the few fucoxanthin-containing species of dinoflagellates arose from captured fucoxanthin-containing eucaryotic algae. Since all dinoflagellate chloroplasts are surrounded by three membranes, I believe they all have arisen by the reduction of endosymbiotic eucaryotic algae, but it is difficult to decide which group of algae were the endosymbionts. I do not believe it was the Chrysophyceae because dinoflagellate chloroplasts never have girdle lamellae as the chloroplasts of chrysophytes invariably do. Possibly the invading symbiont was a haptophycean alga since their chloroplast structure is similar to that of present-day dinoflagellates. However, most dinoflagellate chloroplasts contain a unique carotenoid pigment, peridinin. Thus it is necessary to postulate either that peridinin evolved after the acquisition of chloroplasts from a symbiotic eucaryotic alga or else that the peridinin-containing eucaryotic invaders no longer exist today.

The dinoflagellates are a particularly interesting group with respect to symbioses. In both the species which today harbour a chloroplast-containing chrysophyte symbiont, *Peridinium balticum* and *P. foliaceum*, the dinoflagellate cytoplasm contains a large eyespot which is surrounded by three membranes (21, 65). Thus these two species long ago may have had a chloroplast-containing eucaryotic endosymbiont from which they have kept only the single chloroplast containing the eyespot granules. More recently a second invasion by an eucaryotic alga has occurred and at present the entire endosymbiont persists. Although not generally known, there is another species of dinoflagellate which appears to contain an endosymbiotic cell. This dinoflagellate, which is either a *Gymnodinium* or a *Peridinium* (51), contains a typical dinoflagellate nucleus (Fig. 4 in Ref. 51), numerous dinoflagellate chloroplasts, the remnants of a former symbiosis, and what I interpret to be a small eucaryotic symbiont. This symbiont is delimited from the rest of the cell by one or more membranes and consists of a large eucaryotic nucleus surrounded by ribosomes and double membrane limited bodies which could be rudimentary mitochondria or rudimentary chloroplasts.

Have the Chloroplasts of any other Class of Algae Arisen from an Eucaryote Symbiont?

Lee (39) has recently proposed that all algae with chloroplast endoplasmic reticulum evolved from the endosymbiosis by a ciliate of a *Cyanophora*-like eucaryote. In Lee's theory, the outer membrane of chloroplast ER is the phagocytic vacuole membrane of the host ciliate, the inner membrane of chloroplast ER is the plasmalemma of the eucaryotic symbiont, and the row of vesicles and tubules the remnants of the eucaryote's cytoplasm. At present I do not feel there is enough evidence to accept Lee's theory. However, in the Cryptophyceae, there is a nucleus-like body, called a nucleomorph, between the chloroplast ER and chloroplast envelope (31, 34, 53), and Greenwood et al. (31) have suggested that cryptophycean chloroplasts may have arisen from the endosymbiosis of a eucaryotic cell. Very little has been published yet on nucleomorphs, but if they do indeed turn out to be remnants of nucleii then it would be very likely that cryptophycean chloroplasts arose from endosymbiotic eucaryotes and one would have to consider the possibility by extension that this could have also happened in all the algae with chloroplast ER.

Biochemical Implications

The possibility that *Euglena* chloroplasts are derived from symbiotic green algae has wide-spread biochemical implications. It means that most studies on chloroplast biology have been carried out exclusively or almost exclusively on 'green' organisms, the green algae, green algal symbionts in *Euglena*, and green plants. An example which comes readily to mind is that all studies on the molecular size, conformation, and kinetic complexity of chloroplast DNA have been done on *Euglena*, green algae, or higher plants. These investigations have shown that chloroplast DNA is a circular molecule approximately 40 μm long. It is possible that if other plants beside 'green' ones had been studied, more variable results might well have been found. It is time scientists interested in chloroplast biology paid more attention to the other groups of algae.

1. ANDERSON, D. R., H. E. HOPPS, M. F. BARILE, and B. C. BERNHEIM. 1965. Comparison of the ultrastructure of several rickettsiae, ornithosis virus, and *Mycoplasma* in tissue culture. J. Bacteriol. **90**: 1387–1404.
2. ANTIA, N. J., T. BISALPUTRA, J. Y. CHENG, and J. P. KALLEY. 1975. Pigment and cytological evidence for reclassification of *Nannochloris oculata* and *Monallantus salina* in the Eustigmatophyceae. J. Phycol. **11**: 339–343.
3. BAKER, J. R. J., and L. V. EVANS. 1973. The ship-fouling alga *Ectocarpus* II. Ultrastructure of the unilocular reproductive stages. Protoplasma, **77**: 181–189.
4. BEALE, G. H., A. JURAND, and J. R. PREER. 1969. The classes of endosymbiont of *Paramecium aurelia*. J. Cell Sci. **5**: 65–91.
5. BIBBY, B. T., and J. D. DODGE. 1974. The fine structure of the chloroplast nucleoid in *Scrippsiella sweeneyae* (Dinophyceae). J. Ultrastruct. Res. **48**: 153–161.
6. BISALPUTRA, T. 1974. Plastids. *In* Algal physiology and biochemistry. *Edited by* W. D. P. Stewart. University of California Press, Berkeley. pp. 124–160.
7. BLANKENSHIP, M. L., and K. M. WILBUR. 1975. Cobalt effects on cell division and calcium uptake in the coccolithophorid *Cricosphaera carterae* (Haptophyceae). J. Phycol. **11**: 211–219.
8. BONEN, L., and W. F. DOOLITTLE. 1975. On the prokaryotic nature of red algal chloroplasts. Proc. Natl. Acad. Sci. U.S.A. **72**: 2310–2314.
9. BOUCK, G. B. 1965. Fine structure and organelle associations in brown algae. J. Cell Biol. **26**: 523–537.
10. BROWN, R. M., W. E. FRANKE, H. KLEINIG, H. FALK, and P. SITTE. 1970. Scale formation in chrysophycean algae I. Cellulosic and noncellulosic wall components made by the Golgi apparatus. J. Cell Biol. **45**: 246–271.
11. CAVALIER, T. The origin of nuclei and of eukaryotic cells. Nature (London), **256**: 463–468.
12. CHRISTENSEN, T. 1964. The gross classification of algae. *In* Algae and man. *Edited by* D. F. Jackson. Plenum Press, New York. pp. 59–64.
13. DAVIES, J. M., N. C. FERRIER, and C. S. JOHNSTON. 1973. The ultrastructure of the meristoderm cells of the hapteron of *Laminaria*. J. Mar. Biol. Assoc. U.K. **53**: 237–246.
14. DODGE, J. D. 1968. The fine structure of chloroplasts and pyrenoids in some marine dinoflagellates. J. Cell Sci. **3**: 41–48.
15. DODGE, J. D. 1971. Fine structure of the Pyrrophyta. Bot. Rev. **37**: 481–508.
16. DODGE, J. D. 1973. The fine structure of algal cells. Academic Press, New York.
17. DODGE, J. D. 1974. A redescription of the dinoflagellate *Gymnodinium simplex* with the aid of electron microscopy. J. Mar. Biol. Assoc. U.K. **54**: 171–177.
18. DODGE, J. D. 1975. A survey of chloroplast ultrastructure in the Dinophyceae. Phycologia, **14**: 252–263.
19. DODGE, J. D., and B. T. BIBBY. 1973. The Prorocentrales (Dinophyceae) I. A comparative account of fine structure in the genera *Prorocentrum* and *Exuviaella*. Bot. J. Linn. Soc. **67**: 175–187.
20. DODGE, J. D., and R. M. CRAWFORD. 1968. Fine structure of the dinoflagellate *Amphidinium carteri* Hulbert. Protistologica, **4**: 231–242.
21. DODGE, J. D., and R. M. CRAWFORD. 1969. Observations on the fine structure of the eyespot and associated organelles in the dinoflagellate *Glenodinium foliaceum*. J. Cell Sci. **5**: 479–493.
22. DODGE, J. D., and R. M. CRAWFORD. 1970. The morphology and fine structure of *Ceratium hirundinella* (Dinophyceae). J. Phycol. **6**: 137–149.
23. DRUM, R. W. 1963. The cytoplasmic fine structure of the diatom *Nitzschia palea*. J. Cell Biol. **18**: 429–440.
24. FALK, H. 1967. Zum Feinbau von *Botrydium granulatum* Grev. (Xanthophyceae). Arch. Mikrobiol. **58**: 212–227.
25. FALK, H. 1969. Rough thylakoids: polysomes attached to chloroplast membranes. J. Cell Biol. **42**: 582–587.
26. FALK, H., and H. KLEINIG. 1968. Feinbau und Carotinoide von *Tribonema* (Xanthophyceae). Arch. Mikrobiol. **61**: 347–362.
27. FLAVELL, R. 1972. Mitochondria and chloroplasts as descendants of prokaryotes. Biochem. Gen. **6**: 275–291.
28. GIBBS, S. P. 1962. Nuclear envelope – chloroplast relationships in algae. J. Cell. Biol. **14**: 433–444.
29. GIBBS, S. P. 1970. The comparative ultrastructure of the algal chloroplast. Ann. N.Y. Acad. Sci. **175**: 454–473.
30. GIBBS, S. P. 1979. The route of entry of cytoplasmically synthesized proteins into chloroplasts of algae possessing chloroplast ER. J. Cell Sci. In press.
31. GREENWOOD, A. D., H. B. GRIFFITHS, and U. J. SANTORE. 1977. Chloroplasts and cell compartments in Cryptophyceae. Br. Phycol. J. **12**: 119.
32. HEATH, I. B., and W. M. DARLEY. 1972. Observations on the ultrastructure of the male gametes of *Biddulphia levis* Ehr. J. Phycol. **8**: 51–59.
33. HEYWOOD, P. 1972. Structure and origin of flagellar hairs in *Vacuolaria virescens*. J. Ultrastruct. Res. **39**: 608–623.
34. HIBBERD, D. J. 1977. Observations on the ultrastructure of the cryptomonad endosymbiont of the red-water ciliate *Mesodinium rubrum*. J. Mar. Biol. Assoc. U.K. **57**: 45–61.
35. JEFFREY, S. W., and M. VESK. 1976. Further evidence for a membrane-bound endosymbiont within the dinoflagellate *Peridinium foliaceum*. J. Phycol. **12**: 450–455.
36. KARAKASHIAN, S. J., M. W. KARAKASHIAN, and M. A. RUDZINSKA. 1968. Electron microscopic observations on the symbiosis of *Paramecium bursaria* and its intracellular algae. J. Protozool. **15**: 113–128.
37. KIVIC, P. A., and M. VESK. 1974. An electron microscope search for plastids in bleached *Euglena gracilis* and in *Astasia longa*. Can. J. Bot. **52**: 695–699.
38. LANG, N. J. 1963. Electron-microscopic demonstration of plastids in *Polytoma*. J. Protozool. **10**: 333–339.
39. LEE, R. E. 1977. Evolution of algal flagellates with

chloroplast endoplasmic reticulum from the ciliates. S. Afr. J. Sci. **73**: 179–182.
40. LEEDALE, G. F. 1967. Euglenoid flagellates. Prentice-Hall, Englewood Cliffs, New Jersey.
41. LEEDALE, G. F. 1968. The nucleus in *Euglena*. *In* The biology of *Euglena*. Vol. 1. *Edited by* D. E. Buetow. Academic Press, New York. pp. 185–242.
42. LEEDALE, G. F. 1970. Phylogenetic aspects of nuclear cytology in the algae. Ann. N.Y. Acad. Sci. **175**: 429–453.
43. LEEDALE, G. F. 1975. Envelope formation and structure in the euglenoid genus *Trachelomonas*. Br. Phycol. J. **10**: 17–41.
44. LEWIN, R. A. 1977. *Prochloron*, type genus of the Prochlorophyta. Phycologia, **16**: 217.
45. LEWIN, R. A., and N. W. WITHERS. 1975. Extraordinary pigment composition of a prokaryotic alga. Nature (London), **256**: 735–737.
46. LOEBLICH, A. R. 1976. Dinoflagellate evolution: speculation and evidence. J. Protozool. **23**: 13–28.
47. MARGULIS, L. 1970. Origin of eukaryotic cells. Yale University Press, New Haven, CT.
48. MENKE, W., and B. FRICKE. 1962. Einige Beobachtungen an *Prototheca ciferrii*. Port. Acta Biol. **6**: 243–252.
49. MESSER, G., and Y. BEN-SHAUL. 1969. Fine structure of *Peridinium westii* Lemm., a freshwater dinoflagellate. J. Protozool. **16**: 272–280.
50. MIGNOT, J. P. 1966. Structure et ultrastructure de quelques euglénomonadines. Protistologica, **2**: 51–117.
51. MIGNOT, J. P. 1970. Remarques sur le développement du reticulum endoplasmique et du système vacuolaire chez les Gymnodiniens. Protistologica, **6**: 267–281.
52. MOORE, J., M. C. CANTOR, P. SHEELER, and W. KAHN. 1970. The ultrastructure of *Polytomella agilis*. J. Protozool. **17**: 671–676.
53. OAKLEY, B. R., and J. D. DODGE. 1976. The ultrastructure of mitosis in *Chroomonas salina* (Cryptophyceae). Protoplasma, **88**: 241–254.
54. OUTKA, D. E., and D. E. WILLIAMS. 1971. Sequential coccolith morphogenesis in *Hymenomonas carterae*. J. Protozool. **18**: 285–297.
55. ROUND, F. E. 1965. The biology of the algae. Edward Arnold, London.
56. SCHNEPF, E. 1969. Leukoplasten bei *Nitzschia alba*. Oesterr. Bot. Z. **116**: 65–69.
57. SEPSENWOL, S. 1973. Leucoplast of the cryptomonad *Chilomonas paramecium*. Exp. Cell Res. **76**: 395–409.
58. SIEGESMUND, K. A., W. G. ROSEN, and S. R. GAWLIK. 1962. Effects of darkness and of streptomycin on the fine structure of *Euglena gracilis*. Am. J. Bot. **49**: 137–145.
59. SILVA, P. C. 1962. Classification of algae. *In* Physiology and biochemistry of algae. *Edited by* R. A. Lewin. Academic Press, New York.
60. SLANKIS, T., and S. P. GIBBS. 1972. The fine structure of mitosis and cell division in the chrysophycean alga *Ochromonas danica*. J. Phycol. **8**: 243–256.
61. STEWART, K. D., and K. R. MATTOX. 1975. Comparative cytology, evolution, and classification of the green algae with some consideration of the origin of other organisms with chlorophylls *a* and *b*. Bot. Rev. **41**: 104–135.
62. TAYLOR, D. L. 1969. Identity of zooxanthellae isolated from some Pacific Tridacnidae. J. Phycol. **5**: 336–340.
63. TAYLOR, D. L. 1971. On the symbiosis between *Amphidinium klebsii* (Dinophyceae) and *Amphiscolops langerhansi* (Turbellaria: Acoela). J. Mar. Biol. Assoc. U.K. **51**: 301–313.
64. TAYLOR, F. J. R. 1974. Implications and extensions of the serial endosymbiosis theory of the origin of eukaryotes. Taxon, **23**: 229–258.
65. TOMAS, R. N., and E. R. COX. 1973. Observations on the symbiosis of *Peridinium balticum* and its intracellular alga. I. Ultrastructure. J. Phycol. **9**: 304–323.
66. UZZELL, T., and C. SPOLSKY. 1974. Mitochondria and plastids as endosymbionts: a revival of special creation? Am. Sci. **62**: 334–343.
67. ZABLEN, L. B., M. S. KISSIL, C. R. WOESE, and D. E. BUETOW. 1975. Phylogenetic origin of the chloroplast and prokaryotic nature of its ribosomal RNA. Proc. Natl. Acad. Sci. U.S.A. **72**: 2418–2422.

22

Copyright © 1978 by Springer-Verlag, New York
Reprinted from *J. Mol. Evol.* **10**:283-291 (1978)

Ribosomal RNA Homologies and the Evolution of the Filamentous Blue-Green Bacteria

Linda Bonen and W. Ford Doolittle

Department of Biochemistry, Dalhousie University, Halifax, Nova Scotia, Canada

Summary. Ribosomal RNA (rRNA) sequence homology (as determined by comparisons of T1 oligonucleotide catalogs of ^{32}P-labeled 16S rRNAs) has been used to assess phylogenetic relationships within the filamentous and unicellular blue-green bacteria, and to identify regions of evolutionary conservatism within blue-green bacterial 16S rRNAs. *Nostoc* and *Fischerella*, representatives of two morphologically distinct and highly differentiated orders, are shown to be as closely related (on the basis of RNA sequence homology) as typical members of the non-blue-green bacterial genus *Bacillus*. They are further shown to be (on the same basis) indistinguishable from typical unicellular members of a subgroup of the unicellular blue-green bacterial order Chroococcales. These results have general implications for studies of the origin of differentiated prokaryotes and of evolutionary change in prokaryotic macromolecules. In particular, they provide indirect evidence that the divergences of contemporary major prokaryotic groups are truly ancient ones.

Key words: Blue-green bacteria — 16S rRNA homologies — Ribosome structure and function.

Introduction

Filamentous blue-green bacteria (blue-green "algae") are uniquely suited to the study of the evolution of prokaryotic structural differentiation. They appear in fossil deposits more than two billion years old, and diverse filamentous and unicellular blue-green species dominate the Precambrian era, which has been termed on this basis the "Age of Blue-green Algae" (Schopf, 1974). Many fossil forms closely resemble, in morphology, their modern descendants, and the origin of these different lineages can thus be placed within the geological time scale.

Contemporary filamentous blue-greens exhibit a great diversity of structurally and functionally differentiated cell types. Blue-green bacterial taxonomy rests on morphology, and filamentous species have been divided on this basis into at least two (Nostocales and Stigonematales) and frequently more major orders. There are, however, reasons to believe that taxonomies based on morphology do not accurately reflect phylogenetic diversity. Unicellular blue-greens, classically grouped because of their simple structure in a single order (Chroococcales), show a range of DNA GC contents (35—71 moles percent; Stanier et al., 1971) nearly as great as that shown by all bacteria, while the GC contents of the DNA of characterized filamentous forms range only between 39 and 51 moles percent (Edelman et al., 1967). Furthermore, filamentous variants of normally unicellular forms (Ingram and Van Baalen, 1970; Kunisawa and Cohen-Bazire, 1970) and branched variants of normally unbranched filamentous strains (Singh and Tiwari, 1969) can arise from simple mutational events, and drastic changes in morphology of both filamentous and unicellular species can be provoked by environmental manipulation (Evans et al., 1976; Lazaroff, 1973).

It is of interest to obtain quantitative (molecular) measurements of genetic divergence between filamentous species of known (fossil) antiquity, and between such species and unicellular blue-greens, with which they presumably share a common ancestor (Desikichary, 1973). Doing so should not only provide an indication of the reliability of morphology as an index of prokaryotic evolutionary diversity, but might also permit correlation of quantitative measures of genetic homology with times of phylogenetic divergence deduced from the fossil record. Such correlation is not possible with bacteria other than blue-greens, since these lack an extensive interpretable fossil record. Yet it is necessary for any phylogeny which assumes (as most do) that the surviving major groups of prokaryotes are the direct descendants of the Precambrian prokaryotic community whose metabolic activities not only transformed the oceans and atmosphere of the primitive earth, but made possible the evolution of eukaryotic cellular systems.

Materials and Methods

Growth, Labeling and Purification of RNA, and Oligonucleotide Cataloging. Nostoc (strain MAC) (from L.O. Ingram) and *Fischerella ambigua* (from S.E. Stevens, Jr.) were grown autotrophically either as liquid cultures (*Nostoc*) or on sterile strips of dialysis tubing on top of medium (Bonen et al., 1976) solidified with 1.25% agar (*Fischerella*). Labeling was effected (24 h) in a phosphate-free glycyl-glycine buffered medium (Bonen et al., 1976) containing 1—2 mCi carrier-free ^{32}P-orthophosphate (New England Nuclear) per ml. Cells were lysed (after homogenization in a glass-and-Teflon homogenizer) in an Aminco French pressure cell at $1-2 \times 10^7$ kg/m^2. 16S rRNA was purified either directly from 2.8% SDS polyacrylamide gels of total phenol-extracted RNA (Bonen et al., 1976; Bonen and Doolittle, 1975); or from 30S ribosomal subunits resolved on 15—30% linear sucrose gradients (*Fischerella*). Procedures for T1 nuclease digestion, two-dimensiona electrophoresis and sequencing of the resolved T1 oligonucleotides have been described previously (Bonen and Doolittle, 1975; Fox et al., 1977; Woese et al., 1976).

Calculation of Similarity Coefficients and Construction of Dendrogram. Our definition of similarity coefficient (S value) has been discussed in detail elsewhere (Bonen and Doolittle, 1976) and is more simply $S = 200 N_{ij} \div (N_i + N_j)$, where N_i = total number of residues represented by oligonucleotides of five or more residues in the catalog of organism i, N_j = total number of residues represented by oligonucleotides of five or more residues in the catalog of organism j, and N_{ij} = total number of residues represented by all oligonucleotides of five or more residues coincident between the two catalogs. It differs from the definition of Fox et al. (1977) only in the inclusion of pentanucleotide sequences and in being normalized so that maximum similarity (i.e. identity) yields an S value of 100. Dendrograms were constructed from a matrix of S values (Table 2) using the unweighted pair-group method (with arithmetic averages) of Sneath and Sokal (1973).

Results and Discussion

Oligonucleotide Catalogs for Nostoc and Fischerella 16S rRNAs. Woese and his collaborators (Fox et al., 1977; Woese et al., 1975, 1976) and we (Bonen and Doolittle, 1975, 1976) have demonstrated the utility of assembling catalogs of the sequences of all T1 ribonuclease-generated 16S rRNA oligonucleotides in analyses of phylogenetic relationships between organisms. For such analyses, purified ^{32}P-labeled 16S rRNAs are digested with T1 ribonuclease, and the resulting oligonucleotides (ca. 500, comprising 100–150 unique sequences) are resolved by two-dimensional electrophoresis (Sanger et al., 1965; Woese et al., 1976). Each oligonucleotide is sequenced by "secondary" and "tertiary" digestion using ribonucleases A, U_2 and "T3" (Woese et al., 1976). Pairwise comparisons are then made between catalogs of the 16S rRNAs derived from different organisms, and similarity coefficients ("S values") which reflect the number of oligonucleotide sequences common to the two catalogs are derived (Bonen and Doolittle, 1976). From matrices of such pairwise measures of rRNA sequence homology, dendrograms ("trees") which reflect the relatedness of all the organisms under consideration can be constructed using common numerical taxonomic procedures (Sneath and Sokal, 1973).

We have previously applied these techniques to the 16S rRNAs of three unicellular blue-green bacteria designated by Stanier et al. (1971) to be of "typological group IIA" (strains 6308, 6701 and 6714), the 16S rRNA of one unicell designated by them to be of "typological group IA" (strain 6301), and the chloroplast 16S rRNA of a primitive eukaryotic red alga. Our results (Bonen and Doolittle, 1975, 1976) confirmed the typological assignments made by Stanier and allowed us to conclude that red algal chloroplasts were of blue-green bacterial (and likely type IIA blue-green bacterial) origin. We here extend the analysis to two filamentous species chosen as representative of different orders. The first, *Nostoc* (strain "MAC") (Hoare et al., 1971) is a member of the Nostocales, an order which usually produces simple, unbranched, heterocyst-bearing filaments. The second, *Fischerella ambigua,* is a member of the Stigonematales, the most "advanced" blue-green bacterial order (Desikichary, 1973) showing truly-branched filaments and structurally diverse cell types. The numerous and strong morphological differences between these two filamentous species and between each and unicellular forms of the order Chroococcales should be apparent in Figure 1.

Catalogs of the 16S rRNAs of *Nostoc* and *Fischerella* are presented in Table 1, which also indicates the presence or absence of oligonucleotides of identical sequence in 16S rRNAs from the unicellular species previously characterized. (Sequences of oligonucleotides shorter than five residues, which contribute little to phylogenetic analyses, have been omitted).

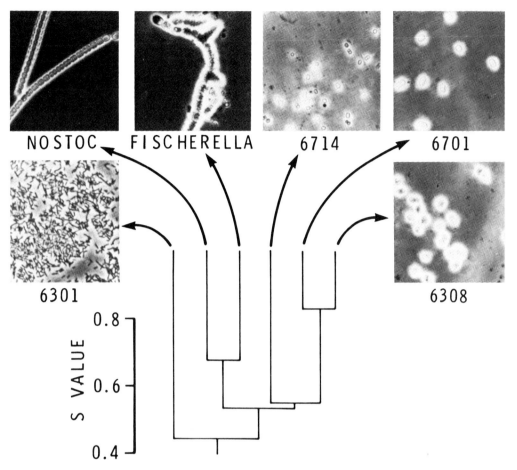

Fig. 1. Phylogenetic relationships of filamentous and unicellular blue-green bacteria. A dendrogram relating *Nostoc*, *Fischerella*, and four unicellular blue-green bacteria was constructed from similarity coefficients (S values) shown in Table 2. Phase contrast photomicrographs are of equivalent (ca. 600X) magnification. That for *Nostoc* indicates three separate filaments. That for *Fischerella* shows a portion of a filament with at least two "true" branches

Phylogenetic Relationships Among Blue-Greens. Similarity coefficients for the 15 possible pairwise comparisons between the six blue-green bacterial 16S rRNAs are shown in Table 2, and a dendrogram constructed from them (Sneath and Sokal, 1973) is presented in Figure 1. From this table and figure we draw the following conclusion:

(1) *Nostoc* and *Fischerella*, although morphologically very different and considered on this basis as members of separate orders, are in fact extremely closely related. The similarity coefficient relating them (S = 67) is higher than values relating any two unicells (all of which classically comprise a single order) except for that relating 6701 and 6308, which we have tentatively designated as members of a single genus (Bonen and Doolittle, 1973). The evolutionary affinity between *Nostoc* and *Fischerella* is in fact much stronger than that between 6714 and either 6701 or 6308, although these three strains are by our criteria and those of Stanier et al. (1971) members of a restricted subgroup

Table 1. Oligonucleotide catalogs of Nostoc and Fischerella 16S rRNAs[a]

Oligonucleotide	Presence in: N	F	U	Oligonucleotide	Presence in: N	F	U
Pentamers				*Hexamers* (cont'd)			
CCC*CG[b]	1	1	4	UACCUG	1	1	2
CCCCG	0	0–1	0	CUCUAG	0	1	3
CCCAG	1	(1)	4	CCUUAG	(1)	(1)	3
ACCCG	1	0	4	AUCCUG	1	1	4
CAACG	1–2	1	4	AAUUCG	1	1	1
C*AACG[b]	1	1	4	AAUCUG	2	1	3
ACACG	1	1	4	UUAAAG	1	1	1
AAACG	1	1	4	CUCUUG	1	1	3
UCCCG	1	1	4	UCUUCG	(1)	0	0
CUACG	1	1	0	UAAUUG	1	1	0
UACCG	1	1	4				
UCACG	0	1	2	*Heptamers*			
CUCAG	1–2	1–2	4				
CAUCG	2–1	1	3	AACACCG	(1)	1	0
UCCAG	0	1	0	AACACAG	1	1	3
UAACG	1	1	1	CAACUCG	1	1	2
UCAAG	1	1	4	UAAACCG	1	0	0
CAAUG	1	1	4	U*AACAAG[b]	1	1	4
AAUCG	1–2	1	4	CACUUCG	0	1	0
AUCAG	1–2	1	3–4	UAUCCCG	1	1	3
AACUG	0	1	0	CAUACUG	1	1	3
ACAUG	1	1	0	UAAUACG	1	1	4
UAAAG	2–1	1	4	CAUUAAG	1	0	1
AAAUG	1	1	4	AUACUAG	1	1	3
CUCUG	1	1	0	AU,AU,C,AG	1	0	0
CCUUG	1	1	4	UUAUCCG	1	1	4
UCCUG	0	1	0	UUAACUG	0	1	0
CUUAG	1	0	0	UACUUAG	0	1	0
UUCAG	1	2–1	4	AAUCUUG	0	1	0
UAUCG	1	1–2	3	UCUAUUG	0	1	0
UCUAG	1	0	0	UUUUUAG	1	0	2
ACUUG	2–1	0	0	unsequenced	–	1	–
UUAAG	2–3	3–4	4				
AUUAG	2	2–1	4	*Octamers*			
AAUUG	2	1–2	4	ACAAACCG	(1)	1	0
UCUUG	1	0	2	CCACACUG	1	1	4
				CAAUACCG	1	1	3
Hexamers				CACUCUAG	1	0	2
				AAUUCCUG	1	1	0
CACAAG	1	1	4	CUCUUUCG[c]	0	1	0
AAACCG	0	1	0				
ACAAAG	0	1	2				
C$_{0-1}$ACACAG	1	0	0	*Nonamers*			
CUAACG	1	1	1				
CCUAAG	1	0	0	CAACCCUCG	1	1	3
CAACUG	1	1	0	CAAAUCCCG	1	0	0
UAAACG	1	1	4	UACACACCG	1	(1)	4
ACACUG	1	1	3	CUACACACG	(1)	(1)	4
AAUCAG	1	1	0	AAACUCAAG	0	1	0
UAAAAG	1	0	2	CUAACUCCG	1	1	3
AAUAAG	0	1	2	CCUACCUAG	1	0	0
UUCCCG	1	1	4	A$_{0-1}$CACUCUAAG[c]	0	1	0
CCUUCG	1	1	0				
CUUCAG	1	0	0				

Table 1. (Fortsetzung)

Oligonucleotide	Presence in: N	F	U	Oligonucleotide	Presence in: N	F	U
ACUCCUACG	1	(1)	4	*Undecamers and*			
CUAAUACCG	1	1	2	*larger*			
CUCAACUAG	1	0	0				
AAUUUUCCG	1	1	3	A(AA,CUA,CA)CAG[c]	0	1	0
UUUAAUUCG	1	1	4	CA,CAA,CCACUG[c]	1	1	0
Decamers				CUUAACACAUG	(1)	1	3
				AACCUUACCAG	0	1	1
CAUACCCCAG	1	(1)	0	AACCUUACCAAG	1	0	3
CUA,CA,CCAAG[c]	0	1	0	UCAC(AAAC,AC)AG[c]	1	0	0
AAACUCAAAG	1	1	4	AAAU,AAC,$C_{\sim 2}$,U_{3-4},–			
CCCCCUUACG	(1)	1	3	AAAG	0	1	0
UCAC*ACCAUG[b]	1	1	4	UAAAC(C_{2-4},U_{4-5},–			
CAAAUCUCAG	0	1	0	CUCA)G	1	0	0
UACUACAAUG	1	1	4	UCA,UUA,CUCCAACCA––			
				UUCG	1	0	0
				Unsequenced large oligonucleotides, 5′- and 3′-termini	~5	~4	–

[a] Sequences of oligonucleotides of five or more residues in T1 ribonuclease digests of 16S rRNA from *Nostoc* or *Fischerella* are shown. Numerals in column "N" and "F" indicate number of molar equivalents of the oligonucleotide in 16S rRNAs from *Nostoc* and *Fischerella*, respectively. Numerals in column "U" indicate the number of unicellular blue-green algal species known to contain at least one copy of the oligonucleotide (four catalogs available; Bonen and Doolittle, 1976). Parentheses indicate that oligonucleotide is very probably, but not certainly, present.
[b] Asterisk indicates modified nucleoside. CCC*CG was formerly and incorrectly designated CC*CCG (C.R. Woese, personal communication).
[c] Sequence tentative.

Table 2. S values relating blue-green bacterial 16S rRNAs[a]

6301	Nostoc	Fischerella	6714	6701		
100	45	42	49	44	41	6301
	100	67	57	55	55	Nostoc
		100	52	50	49	Fischerella
			100	58	52	6714
				100	83	6701
					100	6308

[a] Similarity coefficients (S values) relating 16S rRNAs in each of 15 possible pairwise comparisons were calculated as described in Materials and Methods, and by Bonen and Doolittle (1976)

of the order Chroococcales. Most surprisingly, the *Nostoc-Fischerella* relationship is as strong as that found by Fox et al. (1977) (using the same methods) between *Bacillus subtilis* and *B. stearothermophilus*, and considerably stronger than relationships between several other members of the bacterial genus *Bacillus*. Thus, classical reliance on morphology in blue-green bacterial taxonomy results in a gross overestimate of the phylogenetic diversity of filamentous forms, as compared to unicellular species.

(2) *Nostoc* and *Fischerella* are, within experimental error, as closely related to the unicell 6714 (average S value 54.8) as this unicell is to the other cataloged members of its subgroup, 6701 and 6308 (average S value 55.1). On the basis of RNA sequence homology alone, therefore, there is no reason not to consider the two filamentous strains (and presumably other members of their respective orders) as morphologically differentiated species of the otherwise unicellular typological group IIA.

(3) The greatest phylogenetic diversity in blue-greens as a group is likely to be found among unicellular forms. It is of obvious importance to characterize a greater variety of unicells by the methods described here.

Sequence Conservation in Blue-Green Bacterial 16S rRNAs. Correlation of blue-green bacterial ribosomal RNA sequence homologies with phylogenetic divergences apparent in the fossil record can provide crude indications of rates of ribosomal RNA sequence change. These rates can be used to estimate the times of divergence of other (non-blue-green) bacterial groups which lack fossil records provided that (a) blue-green and other bacterial ribosomes have not been subjected to different selection pressures affecting the relationship of ribosome structure to function and (b) sequence changes which do not significantly affect function occur with equal frequency in blue-green and other bacteria.

Although it may not be possible to prove that the second condition holds, there is evidence which bears directly on the first. Oligonucleotides of special importance to prokaryotic ribosome function can be in principal identified as those which are retained in the 16S rRNAs of a great variety of prokaryotic species. It is possible to locate many such oligonucleotides within the 16S molecule if it is assumed that they occupy (in most prokaryotes) the same positions as they occupy in *Escherichia coli* 16S rRNA, whose primary sequence is almost completely known (Ehresmann et al., 1975). In an analysis of oligonucleotides conserved in 27 diverse bacterial 16S rRNAs, Woese et al. (1975) identified nine regions of high conservatism, six of which occur in the 3'-terminal half of the prokaryotic 16S molecule. The blue-greens (as a group) and the bacilli (as a group) are about equally remote from *E. coli* (Balch et al., 1977). If the ribosomes of the two groups had been subjected to different selection pressures altering relationships of structure to function, then it would be reasonable to expect that *E. coli* oligonucleotides conserved within one would differ from those conserved in the other, and that when these oligonucleotides are aligned against the primary sequence of *E. coli* 16S rRNA, different patterns should be apparent. The results of such alignment are shown in Figure 2. (Data on 16S oligonucleotide sequences from a diverse collection of bacilli are taken from Fox et al. [1977]). The patterns are substantially the same, and most *E. coli* oligonucleotides retained by a majority of blue-greens (above) or bacilli (below) also fall within the nine conservative regions identified in comparisons of 16S rRNAs from a much broader range of prokaryotes (hatched areas; Woese et al., 1975). Thus

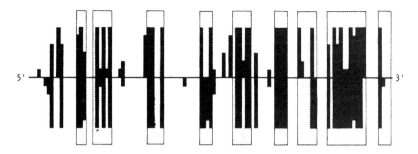

Fig. 2. Sequences conserved in the 16S rRNAs of blue-green bacteria and bacilli. Oligonucleotides simultaneously present in the 16S rRNAs of one or more blue-green bacteria and that of *E. coli* are indicated by black bars drawn above positions which these oligonucleotides are known to occupy in the *E. coli* 16S molecule (horizontal line). *E. coli* oligonucleotides similarly present in bacilli are indicated by black bars drawn below the horizontal line. Bar height indicates number of blue-green bacterial (maximum six) or *Bacillus* (maximum six) species in which the oligonucleotides are found. Hatched areas are those identified by Woese et al. (1975) as conserved in a diverse collection of 27 prokaryotic 16S molecules

structure-function relationships in blue-green bacterial 16S rRNAs are likely the same as those in their non-blue-green bacterial homologs.

Rates of Prokaryotic 16S rRNA Sequence Change and Bacterial Antiquity. The origin of the Stigonematales (presumably from the Nostocales) has been estimated from the fossil record to have occurred between 0.4 billion and one billion years ago, while the divergence of Nostocales from unicells is likely much more ancient (more than two billion years; Schopf, 1974). These two divergences are marked by S values of 67 and (at least) 53 respectively. The former is as great as S values relating members of the bacterial genus *Bacillus*. The latter is considerably greater than S values relating the bacilli and Enterobacteriaceae, for instance (Fox et al., 1977), or than S values relating members of either of these two groups and any blue-green (S = 30–40; Bonen and Doolittle, 1976; C.R. Woese, personal communication). It seems reasonable to conclude on this basis that the phylogenetic divergences of enterics and bacilli (or of either and blue-greens) are truly ancient ones, and predate the origin of filamentous blue-green bacteria. This conclusion is not a trivial one. Without an estimate of rates of prokaryotic macromolecular sequence change, there is simply no guarantee (Sneath, 1974) that most modern bacterial groups are not the products of much more recent divergence from a very limited number of survivors of the primitive, Precambrian prokaryotic community.

Acknowledgements. We thank the Medical Research Council and National Research Council of Canada for financial support, L.O. Ingram and S.E. Stevens, Jr. for strains, C.R. Woese for helpful discussion and the provision of unpublished data, R.A. Singer and D.O. Phillips for advice on the preparation of this manuscript, and C. Ehrhardt for help with the photomicrographs.

References

Balch, W.E., Magrum, L.J., Fox, G.E., Wolfe, R.S., Woese, C.R. (1977). J. Mol. Evol. (in press)

Bonen, L., Allen, G.V., Dobson, P.R., Doolittle, W.F. (1976). J. Bacteriol. **126**, 1020–1023

Bonen, L., Doolittle, W.F. (1975). Proc. Nat. Acad. Sci. USA **72**, 2310–2314

Bonen, L., Doolittle, W.F. (1976). Nature **261**, 669–673

Broda, E. (1975). The Evolution of the Bioenergetic Process. Oxford: Pergamon Press

Desikichary, T.V. (1973). Status of classical taxonomy. In: The Biology of the Blue-green Algae, N.G. Carr, B.A. Whitton, eds. pp. 473–481. Oxford: Blackwell Scientific Publications

Edelman, M., Swinton, D., Schiff, J.A., Epstein, H.T., Zeldin, B. (1967). Bacteriol. Revs. **31**, 315–331

Ehresmann, C., Stiegler, P., Mackie, G.A., Zimmermann, R.A., Ebel, J.P., Fellner, P. (1975). Nucleic Acids Res. **2**, 265–301

Evans, E.H., Foulds, I., Carr, N.G. (1976). J. Gen. Microbiol. **92**, 147–155

Fox, G.E., Pechman, K.R., Woese, C.R. (1977). Int. J. Systematic Bacteriol. **27**, 44–57

Hoare, D.S., Ingram, L.O., Thurston, E.H., Walkup, R. (1971). Arch. Mikrobiol. **78**, 310–321

Ingram, L.O., Van Baalen, C. (1970). J. Bacteriol. **102**, 784–789

Kunisawa, R., Cohen-Bazire, G. (1970). Arch. Mikrobiol **71**, 49–59

Lazaroff, N. (1973). Photomorphogenesis and Nostocacean development. In: The Biology of the Blue-green Algae, N.G. Carr, B.A. Whitton, eds., pp. 279–319. Oxford: Blackwell Scientific Publications

Sanger, F., Brownlee, G.G., Barrell, P.G. (1965). J. Mol. Biol. **13**, 373–398

Schopf, J.W. (1974). Paleobiology of the Precambrian: the age of blue-green algae. In: Evolutionary Biology, vol. 7, T. Dobzhansky, M.L. Hecht, W.C. Steere, eds., pp. 1–43. New York: Plenum Press

Singh, R.N., Tiwari, D.N. (1969). Nature **221**, 62–64

Sneath, P.H.A. (1974). Phylogeny of micro-organisms. In: Evolution in the Microbial World. Symposia of the Society for General Microbiology, vol. 24. (M.J. Carlile, J.J. Skehel, eds.), pp. 1–20. Cambridge: University Press

Sneath, P.H.A., Sokal, R.R. (1973). Numerical Taxonomy: The Principles and Practice of Numerical Classification. San Francisco: Freeman and Company

Stanier, R.Y., Kunisawa, R., Mandel, M., Cohen-Bazire, G. (1971). Bacteriol. Revs. **35**, 171–205

Woese, C.R., Fox, G.E., Zablen, L., Uchida, T., Bonen, L., Pechman, K., Lewis, B.J., Stahl, D. (1975). Nature **254**, 83–86

Woese, C., Sogin, M., Stahl, D., Lewis, B.J., Bonen, L. (1976). J. Mol. Evol. **7**, 197–213

Reprinted from Natl. Acad. Sci. (USA) Proc. 81:1946-1950 (1984)

Extensive and widespread homologies between mitochondrial DNA and chloroplast DNA in plants

(organelles/angiosperms/DNA transposition)

DAVID B. STERN AND JEFFREY D. PALMER*

Carnegie Institution of Washington, Department of Plant Biology, 290 Panama Street, Stanford, CA 94305

Communicated by Winslow R. Briggs, November 10, 1983

ABSTRACT We used hybridization techniques to demonstrate that numerous sequence homologies exist between cloned mung bean and spinach chloroplast DNA (ctDNA) restriction fragments and mtDNAs from corn, mung bean, spinach, and pea. The strongest cross-homologies are between clones derived from the ctDNA inverted repeat and mtDNA from corn and pea, although all the ctDNA clones tested hybridized to at least one mtDNA restriction fragment. Known chloroplast genes showing strong mtDNA homologies include those for the large subunit of ribulosebisphosphate carboxylase, which hybridizes to corn mtDNA, and the β subunit of the chloroplast ATPase, which hybridizes to mung bean mtDNA. Certain of these homologies were confirmed by using cloned spinach mtDNA restriction fragments as probes in reciprocal hybridizations to ctDNA. Several of these ctDNA-homologous mtDNA sequences were shown to be much more closely related to ctDNA from the same species than to that of a distantly related species. We interpret these differential homologies as evidence for relatively recent DNA sequence transfer events, suggesting that transposition between the two genomes is an ongoing evolutionary process.

The mitochondrial genomes of higher plants are large in comparison to their fungal and mammalian counterparts, varying from 218 kilobase pairs (kb) in the genus *Brassica* (1) to an estimated 2400 kb in muskmelon (2). In contrast, chloroplast genomes are highly conserved in size (120–180 kb) and in sequence arrangement (3). To date, no correlation has been demonstrated between mitochondrial genome size and the number of polypeptide products made by isolated mitochondria. It seems likely, therefore, that higher plant mtDNA consists largely of noncoding sequences (2, 4, 5).

Following the observation that corn mtDNA contains a 12-kb segment of the corn chloroplast DNA (ctDNA) inverted repeat (6), we wished to ascertain whether this phenomenon is restricted to corn or if it is a feature of other plant taxa. Here we present hybridization studies that extend these earlier results and demonstrate the widespread presence of ctDNA sequences in the mitochondrial genomes of four diverse species of angiosperms.

MATERIALS AND METHODS

Mitochondria were prepared from 1-wk-old dark-grown pea (*Pisum sativum* cv. Alaska), mung bean (*Vigna radiata* cv. berken), and corn (*Zea mays* B37-N) seedlings and from green spinach (*Spinacia oleracea*) leaves by treating mitochondria with DNase I by the method of Kolodner and Tewari (7). Chloroplasts were prepared either by the DNase I procedure (8) or by sucrose gradient centrifugation (9). DNAs were prepared from the purified, lysed organelles by two rounds of CsCl-ethidium bromide equilibrium centrifugation (9). Restriction endonuclease digestions, agarose gel electrophoresis, preparation of nitrocellulose filters, nick-translations, and hybridizations were carried out as described (9). All filters were washed in 0.3 M NaCl/30 mM trisodium citrate/0.1% sodium dodecyl sulfate at 65°C. Recombinant clones of spinach mtDNA were constructed by ligating (10) *Sal* I-digested mtDNA and *Sal* I-digested pUC8 (11) with T4 DNA ligase (Bethesda Research Ltd.), followed by transformation (12) into *Escherichia coli* JM83. The nick-translated 1670- and 3600-base-pair (bp) spinach ctDNA *Eco*RI fragments, isolated from agarose gels (13), were used to screen nitrocellulose replicates of the spinach mtDNA clone bank by colony hybridization (14).

RESULTS

To gain a general idea as to the extent of cross-homology between organellar DNAs, we first used an entire chloroplast genome as a hybridization probe against mtDNAs. Because cross-contamination between organellar DNAs is inevitable, ctDNA prepared from the same plant and digested with the same restriction enzyme (Sal I) was included in a lane next to each mtDNA track. In this way ctDNA *Sal* I fragments, the sizes of which are known for corn, mung bean, spinach, and pea (3), that contaminate the mtDNA could easily be visualized and disregarded. Total spinach ctDNA hybridized to a large number of *bona fide* mtDNA fragments from corn, mung bean, spinach, and pea (Fig. 1). The strongest hybridizations were to 12-kb and 14-kb corn mtDNA *Sal* I fragments and represented primarily the 12-kb ctDNA inverted repeat sequence described previously (6). There was also strong hybridization to a pea mtDNA *Sal* I fragment of about 18 kb and to two mung bean mtDNA *Sal* I fragments of 6.2 kb and 4.5 kb.

To determine how many regions of the chloroplast genome are represented in these mtDNAs, clones spanning most of mung bean ctDNA (10) were used as probes against gels similar to that shown in Fig. 1. The arrangement of these clones on the mung bean chloroplast genome and the results of the hybridizations are shown schematically in Fig. 2. Some of the most striking results were for MB 16.2 and MB 18.8 (Fig. 3), two mung bean clones that together contain the entire ctDNA inverted repeat. As expected (6), both of these clones hybridized strongly to corn mtDNA, but strong hybridization also was seen between MB 18.8 and pea mtDNA, with significant hybridization between MB 18.8 and mung bean mtDNA and between MB 16.2 and spinach mtDNA. Other examples of strong mitochondrial/chloroplast homol-

Abbreviations: kb, kilobase pairs; bp, base pairs; *atpB*, *atpE*, genes for the β and ε subunits, respectively, of the chloroplast ATPase; *rbcL*, gene for the large subunit of ribulosebisphosphate carboxylase; ctDNA, chloroplast DNA; mtDNA, mitochondrial DNA.
*Present address: Duke University, Department of Zoology, Durham, NC 27706

The publication costs of this article were defrayed in part by page charge payment. This article must therefore be hereby marked "*advertisement*" in accordance with 18 U.S.C. §1734 solely to indicate this fact.

Biochemistry: Stern and Palmer

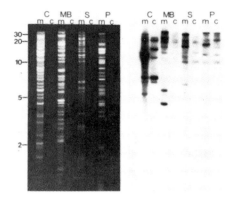

FIG. 1. Corn (C), mung bean (MB), spinach (S), and pea (P) mtDNA (m) and ctDNA (c) were digested with Sal I and electrophoresed in a 0.7% agarose gel. The gel was transferred to nitrocellulose and probed with nick-translated spinach ctDNA. The purified ctDNAs were loaded in an amount equal to the level of ctDNA contamination in each mtDNA preparation—i.e., the levels of ctDNA and mtDNA in adjacent lanes were empirically adjusted so that ctDNA·ctDNA hybridizations are of equal intensities for each pair of DNAs from a given plant. Size scale at left is in kb and was determined by using size markers consisting of phage λ DNA digested with Sal I, EcoRI, and HindIII.

ogies were between MB 13.3 (from the large single copy region) and corn mtDNA and between MB 16.5 (large single copy region) and spinach mtDNA and, more weakly, to corn mtDNA (Fig. 3).

An interesting case is the hybridization of MB 11.1 to the mtDNAs (Fig. 3). MB 11.1 hybridized to all four mitochondrial genomes and contains two identified chloroplast genes: atpB, which encodes the β subunit of the chloroplast ATPase; and atpE, which encodes the ε subunit of the same polypeptide complex (unpublished data). To identify the specific region of MB 11.1 responsible for each of the mtDNA homologies seen in Fig. 3 (MB 11.1 hybridization), a cloned spinach ctDNA BamHI fragment of 11.5 kb called "Bam 11.5," which contains most of the sequences in MB 11.1 and in addition contains rbcL [the gene for the large subunit of ribulosebisphosphate carboxylase (3, 18)], and also the EcoRI subclones of Bam 11.5 were used as probes. The hybridization of the entire cloned spinach BamHI fragment to a gel such as that shown in Fig. 1 is seen in Fig. 4, together with a map of the EcoRI sites and the positions of atpB, atpE, and rbcL. When the EcoRI subclones were used as probes, each of the mtDNA hybridizations could be traced to a specific region of Bam 11.5. For example, the 1750-bp EcoRI fragment, which consists almost entirely of rbcL gene sequence (18), hybridized strongly to a corn mtDNA Sal I fragment of 12.5 kb (Fig. 4). Note that MB 11.1 did not hybridize to this particular corn mtDNA Sal I fragment (Fig. 3) because in mung bean rbcL is located in a neighboring Pst I fragment of 7.5 kb (16).

The 1980-bp EcoRI fragment, which contains all of atpB and 45 bp of atpE, hybridized only to mung bean mtDNA (Fig. 4). In contrast, the 1670-bp EcoRI fragment, which includes the rest of atpE (19), hybridized to mtDNA fragments from all four genomes (Fig. 4). We have determined since (data not shown) that all of the hybridization between Eco 1670 and the mtDNAs can be attributed to the 1250-bp Xba I–EcoRI subfragment of Eco 1670, which consists almost entirely of non-atpE coding sequence (19). Finally, the 3600-bp EcoRI fragment hybridized to corn and spinach mtDNAs (Fig. 4). Thus, each of the hybridizations to the 11.5-kb spinach ctDNA BamHI clone (Fig. 4) can be accounted for by using smaller, more specific clones.

More rigorous verification of these interorganellar homologies was accomplished by cloning specific ctDNA-homologous mtDNA restriction fragments and showing that they hybridize to ctDNA fragments in a reciprocal fashion. We chose to investigate the hybridization of the spinach ctDNA clone Bam 11.5, specifically its EcoRI subfragments of 3600 and 1670 bp, to spinach mtDNA fragments of 10.5, 6.7, and 5.5 kb (Fig. 4). Hybridizations involving these three cloned spinach mtDNA fragments led to three major results. (i) Spinach Bam 11.5 hybridized to each of the three cloned inserts with the same relative intensity seen in its hybridization to total spinach mtDNA (compare Fig. 4 to Fig. 5). (ii) Each of the mtDNA clones hybridized to a fragment of a size identical to its own insert in a Sal I digest of total mtDNA, to the contaminating ctDNA Sal I fragment of 22 kb (3), to the spinach Bam 11.5 insert, and to the 11.5-kb fragment in a BamHI digest of total spinach ctDNA (Fig. 5). (iii) The 6.7- and 5.5-kb spinach mtDNA fragments did not cross-hybridize. Preliminary mapping studies indicated that these two fragments are adjacent in the mitochondrial genome. Therefore, the simplest explanation for the hybridization of Eco 3600 to both fragments is that there is a single region of mitochondrial homology to Eco 3600 which includes the Sal I site separating the two mtDNA fragments.

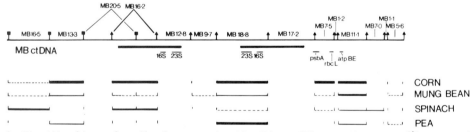

FIG. 2. (Upper) Map of the mung bean chloroplast genome adapted from Palmer and Thompson (15). Pst I sites (↑) and Sal I sites (▼) are indicated. The clones used in this study are shown on the map—for example, MB 18.8 is a clone of an 18.8-kb Pst I fragment. The positions of the genes are according to Palmer and Thompson (15), Palmer et al. (16), and unpublished data. The heavy lines just beneath the ctDNA map indicate the inverted repeats. (Lower) Schematic representation of the hybridization strength of the ctDNA clones to mtDNA. Hybridizations were classified as very weak (-----), weak (—), strong (▬) or very strong (■). Clones not tested were MB 12.8, MB 17.2, and MB 20.5 (whose sequences are largely contained within the inverted repeat and were mostly represented by MB 18.8 and MB 16.2) and also MB 1.2 [whose sequence is present within the spinach ctDNA clone Bam 11.5 used in Figs. 4 and 5 (16)]. Hybridizations are shown directly underneath the mung bean ctDNA fragment used as a probe. The fact that MB 18.8 hybridizes to all the mtDNAs, at least weakly, may be indicative of homology between the mitochondrial and chloroplast ribosomal RNA genes, which has been found to be 62% in a 664-bp region of corn mtDNA (17).

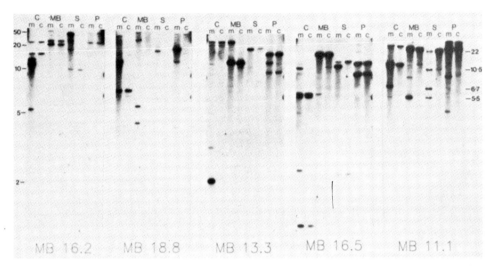

FIG. 3. Gels similar to that shown in Fig. 1 were transferred to nitrocellulose and probed with the nick-translated ctDNA clones given below each panel. Due to degradation of the mung bean ctDNA, the largest ctDNA fragments are not visible in the ctDNA tracks for the MB 16.2, MB 18.8, and MB 13.3 probes, although these mung bean ctDNA restriction fragments are visible in the accompanying mtDNA lane. Spinach ctDNA has been overloaded relative to the level of contamination in the mtDNA preparation for the MB 16.5 and MB 11.1 probes and has been underloaded for the MB 13.3 and MB 18.8 probes. There is a contaminating DNA fragment of 12 kb visible in the three right-hand mung bean ctDNA lanes that has homology to vector (pBR322) sequences.

DISCUSSION

The major finding that emerges from this study is that interorganellar DNA transfer is a general phenomenon in plants. Previously, observation of cross-homology between mtDNA and ctDNA had been restricted to a single, specific sequence in corn (6). Here we have demonstrated the pervasive nature of ctDNA sequences in the mitochondrion, to the extent that every ctDNA sequence tested reacted with one or more mtDNA restriction fragments (Fig. 2).

That the chloroplast and mitochondrial genomes share extensive sequence homology does not in itself lead to any conclusions regarding the direction of sequence movement between the two organelles. We note, however, that these shared sequences are present in all higher plant chloroplast genomes so far examined, are highly conserved in nucleotide sequence and arrangement within the chloroplast (3), and almost certainly are transcribed and functional within the chloroplast (ref. 20; see last paragraph of *Discussion*). In contrast, these sequences have a more or less random representation within the mitochondria of the four species examined (Fig. 2). Based on these observations, we feel that many, if not most, of these shared sequences have been transferred from the highly constrained chloroplast genome (3, 20) or its progenitor into the mitochondrial genome, which is more variable in size (2).

Timing of Interorganellar Sequence Transfer. Do these ctDNA–mtDNA homologies represent recent transfer events between the organelles, or were these sequences present in the mtDNAs of all four species prior to their divergence, possibly even reflecting a partial common origin for chloroplast and mitochondrion? If a ctDNA sequence were transferred into the mitochondrion relatively recently compared to the divergence of two species, then we would expect that

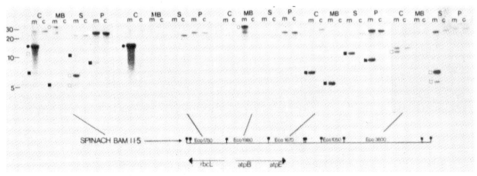

FIG. 4. (*Upper*) Gels similar to that shown in Fig. 1 were transferred to nitrocellulose and probed with the nick-translated spinach ctDNA clones indicated on the physical map (*Lower*), which shows the positions of the *Eco*RI subclones of spinach Bam 11.5 that were used as hybridization probes and the positions of identified genes. ●, mtDNA homologies to Eco 1750; ○, mtDNA homologies to Eco 1980; ■, mtDNA homologies to Eco 1670; and □, mtDNA homologies to Eco 3600. *Eco*RI fragments not tested were the two small *Eco*RI fragments between Eco 1670 and Eco 1050 and the two terminal *Eco*RI–*Bam*HI fragments at either end of Bam 11.5. Eco 1050 did not hybridize to any mtDNA fragments (data not shown).

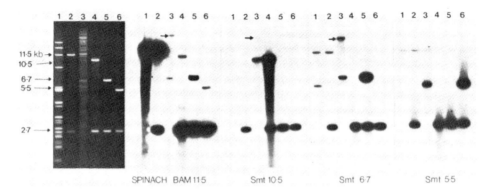

FIG. 5. Hybridizations of nick-translated spinach mtDNA and ctDNA clones to BamHI-digested spinach ctDNA (lanes 1), BamHI-digested spinach Bam 11.5 (lanes 2), Sal I-digested spinach mtDNA (lanes 3), Sal I-digested Smt 10.5 (lanes 4), Sal I-digested Smt 6.7 (lanes 5), and Sal I-digested Smt 5.5 (lanes 6); Smt, spinach mtDNA clone, followed by the size of the insert in kb. Hybridization probes are given below each panel. In each case, with the exception of Smt 5.5, a fragment of about 20 kb (→) is identified in the mtDNA track. This fragment is a Sal I fragment from contaminating ctDNA (see Fig. 4 and text). Smt 6.7 appears to hybridize to a ctDNA fragment other than Bam 11.5. This is consistent with the hybridization of MB 18.8 to a 6.7-kb spinach mtDNA fragment (Fig. 3), which we have confirmed by reciprocal hybridizations between Smt 6.7 and MB 18.8 (data not shown). In addition, the relatively strong hybridization of Smt 6.7 to the 20-kb region of spinach mtDNA reflects not only ctDNA contamination but also sequence homology between Smt 6.7 and a *bona fide* spinach mtDNA fragment of 19.5 kb (unpublished data). The 2.7-kb restriction fragment seen in lanes 2, 4, 5, and 6 is the cloning vector pUC 8 (11).

these mitochondrially located transferred sequences should be significantly more closely related to ctDNA from one species than from the other. Experimental results compatible with this hypothesis were obtained when mung bean and spinach ctDNA clones (MB 11.1 and Bam 11.5, respectively), which contain similar sequences (3), were hybridized to mung bean and spinach mitochondrial DNAs. MB 11.1 hybridized very strongly to a 5.5-kb mung bean mtDNA restriction fragment and less intensely to 5.5-, 6.7-, and 10.5-kb spinach mtDNA fragments (Fig. 3). Reciprocal hybridization intensities were obtained with Bam 11.5 (Fig. 4). Relative to their hybridization with MB 11.1, all three spinach mtDNA fragments hybridized more strongly to Bam 11.5 than did the mung bean 5.5-kb mtDNA fragment. Additionally, whereas the three spinach mtDNA fragments hybridized with approximately equal intensity to MB 11.1 (Fig. 3), the enhancement of hybridization to spinach Bam 11.5 appeared to be greater for the 6.7-kb spinach mtDNA fragment than for the 5.5- and 10.5-kb spinach mtDNA fragments.

Thus, it is possible to envision a hierarchical timing of DNA sequence transfers, where the ctDNA-homologous portion of the 6.7-kb spinach mtDNA fragment was transferred from the chloroplast more recently than the corresponding parts of the 10.5- and 5.5-kb spinach fragments and where all three spinach homologies represent events that occurred in a spinach-specific lineage subsequent to the divergence of mung bean and spinach. Alternatively, variable rates and patterns of sequence evolution within the mung bean and spinach chloroplast genomes could account for the differential homologies observed. In particular, the portions of MB 11.1 and Bam 11.5 that have homology to the 6.7-kb spinach mtDNA fragment may be more closely related than are the parts of those ctDNA clones that hybridize to the 10.5- and 5.5-kb spinach mtDNA Sal I fragments.

Among the mtDNAs examined, that of corn has the greatest overall amount of homology to ctDNA (Figs. 1–4). Much of this strong cross-homology can be explained by invoking relatively recent interorganellar DNA transfer. In the case of corn mtDNA hybridization to the ctDNA inverted repeat (Figs. 2 and 3), absolute identity of restriction sites has been reported over a 12-kb region of interorganellar homology (6). In addition, mapping studies indicate that the strong hybridization of corn mtDNA to ctDNA *rbcL* sequences (Fig. 4) reflects a corn mtDNA *rbcL* sequence that is virtually identical to its corn ctDNA homolog (21).

Mechanism of Interorganellar Sequence Transfer. There are two types of mechanism that could account for random and widespread sequence transfer between cytoplasmic organelles. One type would require direct physical contact. Membrane continuities and other associations between chloroplast and mitochondrion have been reported for several species, including barley (22), corn (23), *Hyptis suaveolens* (23), *Pteris vittata* (24), tobacco (25), *Panicum schenckii* (26), and *Euglena* (27). Moreover, Wildman et al. (28) observed various physical interactions between the two organelles in cinematic studies of living cells. Membrane continuities might facilitate intermolecular recombination; evidence for both intermolecular and intramolecular recombination has been accumulated for both mtDNA (1, 29–32) and ctDNA (33, 34). Enclosure of the mitochondrion by the chloroplast (29) suggests transformation of the mitochondrion as a likely mechanism for ctDNA uptake.

On the other hand, exchange of DNA sequences between organelles may not require direct physical contact. DNA released into the cytoplasm from broken or lysed chloroplasts may be taken up randomly by the mitochondrion by transformation. Alternatively, there may exist in the cytoplasm specific vector molecules capable of transferring sequences between organelles. If a vector is facilitating interorganellar DNA transfer, it might resemble a transposable element (35) or perhaps a transducing phage, in analogy to tobacco mosaic virus, which can produce upon infection a small proportion of pseudovirions, which have been shown to contain RNA homologous to both ctDNA and nuclear DNA (36). If ctDNA–mtDNA exchange is governed by a vector molecule, this might impose a selectivity on which sequences were transferred. For instance, a ctDNA sequence flanked by strong recombination sites might be transposed more frequently, as might a sequence with close homology to a viral recognition sequence, such as the encapsidation initiation sequence of tobacco mosaic virus (37).

We feel it unlikely that these integrated ctDNA sequences play any biological role in the mitochondrion. Significantly, the strongest cross-homologies we have observed are to ctDNA sequences quite unlikely to have a function in the mitochondrion: *rbcL* (corn) and the ctDNA ribosomal RNA genes (corn and pea). It has been suggested that sequences from within the ctDNA inverted repeat might play a functional role within the corn mitochondrial genome because alterations in this ctDNA-homologous sequence are observed in the mtDNA of cytoplasmic male sterile corn (6). However, these alterations now appear to represent only a small fraction of the numerous rearrangements that distinguish fertile and male-sterile mtDNAs in both corn (38) and petunia (32) and, thus, are probably unrelated causally to the expression of male sterility. We feel that the widespread presence of ctDNA sequences in plant mtDNA is best regarded as a dramatic demonstration of the dynamic nature of interactions between the chloroplast and the mitochondrion, similar to the ongoing process of interorganellar DNA transfer already documented between mitochondrion and nucleus (39–43) and between chloroplast and nucleus (44).

We thank Dr. W. F. Thompson in whose laboratory this work was performed, and H. Edwards for providing the spinach ctDNA *Eco*RI clones. This work was supported in part by National Institutes of Health Training Grant GM 07276-08 to D.B.S. and National Science Foundation Grant PCM-81-09795 to Dr. W. F. Thompson. This is Carnegie Institution of Washington Department of Plant Biology publication no. 822.

1. Palmer, J. D. & Shields, C. R. *Nature (London),* (1984) **307,** 437–440.
2. Ward, B. L., Anderson, R. S. & Bendich, A. J. (1981) *Cell* **25,** 793–803.
3. Palmer, J. D. & Thompson, W. F. (1982) *Cell* **29,** 537–550.
4. Gray, M. W. (1982) *Can. J. Biochem.* **60,** 157–171.
5. Leaver, C. J. & Gray, M. W. (1982) *Annu. Rev. Plant Physiol.* **33,** 373–402.
6. Stern, D. B. & Lonsdale, D. M. (1982) *Nature (London)* **299,** 698–702.
7. Kolodner, R. & Tewari, K. K. (1972) *Proc. Natl. Acad. Sci. USA* **69,** 1830–1834.
8. Kolodner, R. & Tewari, K. K. (1975) *Biochim. Biophys. Acta* **402,** 372–390.
9. Palmer, J. D. (1982) *Nucleic Acids Res.* **10,** 1593–1605.
10. Palmer, J. D. & Thompson, W. F. (1981) *Gene* **15,** 21–26.
11. Vieira, J. & Messing, J. (1982) *Gene* **19,** 259–268.
12. Dagert, M. & Erlich, S. D. (1979) *Gene* **6,** 23–28.
13. Maxam, A. M. & Gilbert, W. (1980) *Methods Enzymol.* **65,** 499–560.
14. Grunstein, M. & Hogness, D. S. (1975) *Proc. Natl. Acad. Sci. USA* **72,** 3961–3965.
15. Palmer, J. D. & Thompson, W. F. (1981) *Proc. Natl. Acad. Sci. USA* **78,** 5533–5537.
16. Palmer, J. D., Edwards, H., Jorgensen, R. A. & Thompson, W. F. (1982) *Nucleic Acids Res.* **10,** 6819–6832.
17. Chao, S., Sederoff, R. R. & Levings, C. S., III (1983) *Plant Physiol.* **71,** 190–193.
18. Zurawski, G., Perrot, B., Bottomley, W. & Whitfield, P. (1981) *Nucleic Acids Res.* **9,** 3251–3270.
19. Zurawski, G., Bottomley, W. & Whitfield, P. R. (1982) *Proc. Natl. Acad. Sci. USA* **79,** 6260–6264.
20. Poulsen, C. (1983) *Carlsberg Res. Commun.* **48,** 57–80.
21. Lonsdale, D. M., Hodge, T. P., Howe, C. J. & Stern, D. B. (1983) *Cell* **34,** 1007–1114.
22. Wellburn, F. A. M. & Wellburn, A. R. (1979) *Planta* **147,** 178–179.
23. Montes, G. & Bradbeer, J. W. (1976) *Plant Sci. Lett.* **6,** 35–41.
24. Crotty, W. J. & Ledbetter, M. C. (1973) *Science* **182,** 839–841.
25. Wildman, S. G., Jope, C. & Atchison, B. A. (1974) *Plant Physiol.* **54,** 231–237.
26. Brown, R. H., Rigsby, L. L. & Akin, D. E. (1983) *Plant Physiol.* **71,** 437–441.
27. Calvayrac, R., Laval-Martin, D., Briand, J. & Farineau, J. (1981) *Planta* **153,** 6–13.
28. Wildman, S. G., Hongladarom, T. & Honda, S. I. (1962) *Science* **138,** 434–436.
29. Lazarus, C. M., Earl, A. J., Turner, G. & Kuntzel, H. (1980) *Eur. J. Biochem.* **106,** 633–641.
30. Nagy, F., Török, I. & Maliga, P. (1981) *Mol. Gen. Genet.* **183,** 437–439.
31. Vierny, C., Keller, A., Begel, O. & Belcour, L. (1982) *Nature (London)* **297,** 157–159.
32. Boeshore, M. L., Lifshitz, I., Hanson, M. R. & Izhar, S. (1983) *Mol. Gen. Genet.* **190,** 459–467.
33. Lemieux, C., Turmel, M. & Lee, R. W. (1981) *Curr. Genet.* **3,** 97–103.
34. Palmer, J. D. (1983) *Nature (London)* **301,** 92–93.
35. Starlinger, P. (1980) *Plasmid* **3,** 241–259.
36. Siegel, A. (1971) *Virology* **46,** 50–59.
37. Jonard, G., Richards, K. E., Guilley, H. & Hirth, L. (1977) *Cell* **11,** 483–493.
38. Lonsdale, D. M., Fauron, C. M.-R., Hodge, T. P., Pring, D. R. & Stern, D. B. (1983) in *Genetic Rearrangement, Proceedings of the 5th John Innes Symposium,* eds. Chater, K. F., Cullis, C. A., Hopwood, A., Johnston, A. W. B. & Woolhouse, H. W. (Croom Helm, London), pp. 183–205.
39. Van den Boogaart, P., Samallo, J. & Agsteribbe, E. (1982) *Nature (London)* **298,** 187–189.
40. Farrelly, F. & Butow, R. A. (1983) *Nature (London)* **301,** 296–301.
41. Gellissen, G., Bradfield, J. Y., White, B. N. & Wyatt, G. R. (1983) *Nature (London)* **301,** 631–634.
42. Jacobs, H. T., Posakony, J. W., Grula, J. W., Roberts, J. W., Xin, J., Britten, R. J. & Davidson, E. H. (1983) *J. Mol. Biol.* **165,** 609–632.
43. Kemble, R. J., Mans, R. J., Gabay-Laughnan, S. & Laughnan, J. R. (1983) *Nature (London)* **304,** 744–747.
44. Scott, N. & Timmis, J. N. (1983) *Nature (London)* **305,** 65–67.

Copyright © 1981 by The New York Academy of Sciences
Reprinted from pages 166, 167, 172-174, 186, and 187-188 of N.Y. Acad. Sci. Ann. **361**:166-192 (1981)

ORIGIN AND EVOLUTION OF THE PLASTID AND ITS FUNCTION

Jerome A. Schiff

Institute for Photobiology of Cells and Organelles
Brandeis University
Waltham, Massachusetts 02154

INTRODUCTION: PLASTIDS AND MITOCHONDRIA AS PROKARYOTIC RESIDENTS IN EUKARYOTIC CELLS

Any modern discussion of the evolution of cellular organelles such as the mitochondrion and chloroplast must begin with a consideration of their prokaryotic properties.[1] The more primitive free-living cells or prokaryotes, lack microscopically visible internal organelles. Some of their properties are summarized in TABLE 1. Eukaryotic cells contain conspicuous organelles when viewed in the light microscope, and electron microscopy shows that these organelles (among them the mitochondrion and the chloroplast) are surrounded by limiting membranes. As information has accumulated, biologists have realized that the mitochondria and chloroplasts resemble prokaryotic cells within eukaryotic cells (TABLE 1).

These organelles are semiautonomous. They perpetuate themselves by division but are subject to regulation by the eukaryotic cell and, in turn, regulate the cell in which they are residents. Genetic studies with higher plants, algae and fungi, some of them quite early in this century, have shown that the genes determining organelle phenotypes are situated in both the nuclear genome and an organelle's genome.[1,2]

The fact that plastids and mitochondria resemble prokaryotic cells within eukaryotic cells suggests that they may have originated from the endosymbiotic invasion of primitive nonphotosynthetic, nonrespiratory eukaryotic cells by free-living prokaryotes.[3,4] In this way, cyanobacteria, *Prochloron*,[5] and as yet undiscovered pigmented prokaryotes may have established the evolutionary lines leading, respectively, to the chloroplasts of: the red algae; green algae, higher plants and *Euglenas;* and other modern algae such as the brown and goldenbrown organisms. Similarly, mitochondria may have begun as the endosymbiotic establishment of primitive aerobic bacteria in such primitive eukaryotic cells.[3,4]

TABLE 1

SOME PROPERTIES OF PROKARYOTES COMPARED WITH MITOCHONDRIA, PLASTIDS, AND THE EUKARYOTIC CELLS CONTAINING THEM [1]

Property	Prokaryotes (Bacteria, Blue-Green Algae or Cyanobacteria)	Mitochondria	Plastids	Rest of Eukaryotic Cell
Size	~ 1–10 μm	~ 1–2 μm	~ 1–10 μm	~ 1–100 μm
Membrane-bounded organelles within	None	None	None	Yes
Cell Wall	Usually present	Absent	Absent	Present or Absent
Endoplasmic Reticulum and Golgi	Absent	Absent	Absent	Present
Ribosomes	70S type	70S type (but variable in size)	70S type	80S type
Ribosomal Proteins	~ 55	18–107	~ 55	~ 75
tRNAs	Complete set	Unique complete set	Unique complete set	Unique complete set
Inhibition of Protein Synthesis by Chloramphenicol, Streptomycin, etc.	Yes	Yes	Yes	No
Inhibition of Protein Synthesis by Cycloheximide, etc.	No	No	No	Yes
DNA Genome	Single molecule, circular	Single molecule, often circular	Single molecule, often circular	Many molecules in chromosomes
Spindle or Comparable Mitotic Mechanism	No	No	No	Present
Reproduction	Fission, fragmentation, directional parasexual recombination	Fission, fragmentation	Fission	Mitosis, meiosis, sexual reproduction
Phosphorylative Cell Repiration (when present)	In cell membrane	In inner membrane	—	No, in mitochondria
Photosynthesis (when present)	In cell membrane, extensions, or free thylakoids	—	In thylakoids	No, in chloroplasts

[*Editors' Note:* Material has been omitted at this point.]

Several evolutionary patterns of photocontrol can be distinguished in contemporary organisms (TABLE 3). Among the prokaryotes, two major patterns in phototactic and phototropic responses can be discerned. One begins with the use of the photosynthetic pigments of the anoxygenic photosynthetic bacteria and the oxygenic cyanobacteria as a means of finding the light for photosynthesis; the action spectrum for phototaxis in these organisms is generally the same as the action spectrum for photosynthesis.[24,25] The ATP produced from photophosphorylation is undoubtedly made available for cell motility. It would have been very adaptive during evolution for cells to use the same pigments for phototaxis and photosynthesis since this would insure that the light, once found, would be of a quality suitable for photosynthesis. Later in the evolution of algal eukaryotes, a separation of perception and photosynthesis took place leading to the stigma/photoreceptor system for control of flagellar movement, as in *Euglena*.[26-29] With this anatomic specialization came a specialization in pigments as well since blue light is effective in orienting these organisms, but red light is not. The unknown blue-light absorbing pigments variously suggested to be flavoproteins carotenoids, and other molecules are collectively called "cryptochrome" as a short-hand notation for our ignorance.[30] This system was apparently preserved in the evolution of multicellular plants as a cryptochrome system in the ferns and the tip of the stem, or coleoptile, of Angiosperms where it controls the transport of auxin, leading to differential growth below the tip and consequent growth towards or away from the light.[31,32]

Also found among the prokaryotes is a bacteriorhodopsin system in *Halobacterium* that mediates phototaxis, ion transport, and ATP formation (TABLE 3). This may well have become the perceptual system of invertebrates and vertebrates culminating in the rhodopsin of the rods and cones of vertebrates and the squid and octopus.[34] Although a nervous system serves as a mediator between the absorption of light in the thylakoids of the rods and cones and the muscles of these organisms, the system is analogous to the more primitive association of perceptor and flagellum or cilium in the more primitive cells. It's worth remembering that in the rods and cones, there are immobile cilia that may be vestigial remnants of the motile structures in more primitive organisms.

The developmental cycle of *Nostoc* and chromatic adaptation in other cyanobacteria (TABLE 3) are controlled by primitive phytochromes or "prophytochromes," which are based on red-green reversible chromoproteins.[35-39] Red light induces the formation of seriate filaments in *Nostoc;* green light reverses the effect of red.[37] Similarly, phycocyanin synthesis is induced by orange-red light and phycoerythrin synthesis by green light.[38,39] It seems likely

TABLE 3
PHOTOCONTROL AND PHYLOGENY

	Phototaxis, Phototropism, Vision	Chloroplast Development Non Chloroplast	Chloroplast	Photomorphogenesis	Chromatic Adaptation	Plastid Rotation
Eukaryotes						
Higher Animals	Rhodopsin	—	—	—	—	—
Higher Plants Angiosperms	Cryptochrome	Phytochrome	Protochlorophyll(ide)	Phytochrome (± Cryptochrome?)	—	—
Ferns	Phytochrome? Cryptochrome?	?	?	Phytochrome-Cryptochrome	—	—
Algae	Cryptochrome	Cryptochrome	Protochlorophyll(ide)	Phytochrome-Cryptochrome	—	Phytochrome-Cryptochrome
Fungi	Cryptochrome	—	—	Cryptochrome-Phytochrome	—	—
Prokaryotes						
Cyanobacteria (Blue-Green Algae)	Photosynthetic Pigments	—	—	Prophytochrome (red-green reversible)	Prophytochrome (red-green reversible)	—
Photosynthetic Bacteria	Photosynthetic Pigments	—	—	—	—	—
Halobacteria	Bacteriorhodopsin	—	—	—	—	—

[Cryptochrome: Near UV-Blue; Phytochrome: Red Far-Red Reversible; Protochlorophyll(ide): Blue-Red]. ("Cryptochrome" designates pigments unidentified or tentatively identified as flavoproteins or carotenoids).

that these prophytochromes evolved after open-chain tetrapyrrole pigments were available from the evolution of phycoerythrin and phycocyanin as photosynthetic accessory pigments. Linkage of modified open-chain tetrapyrroles to other proteins would have led to the formation of the prophytochromes for cellular control functions. Later, these prophytochromes could have become modified during evolution to become the familiar red/far-red reversible phytochrome chromoprotein of *Mougeotia* and multicellular plants.[40-42] Phytochrome and cryptochrome frequently act in concert to control cellular morphogenesis as in prothalial development in ferns, spore germination in fungi, and various responses in higher plants.[31, 40, 41]

This brings us, finally, to the control of chloroplast development by light (TABLE 3). In algae such as *Euglena* and *Scenedesmus*, many early functions connected with chloroplast development occur external to the chloroplast during the lag period and are under control of blue-light absorbing cryptochrome systems.[30, 43] In higher plants this function appears to have been largely assumed by the red/far-red phytochrome system.[44] The later, plastid-localized functions of the linear period of plastid development in both *Euglena* and higher plants appear to be under the control of blue and red light absorbed by protochlorophyll(ide) or some pigment with a similar absorption spectrum. Since the photoconversion of protochlorophyll(ide) to chlorophyll(ide) is often an obligatory step in chlorophyll formation, this control point is very strategically located.

[*Editors' Note:* Material has been omitted at this point. Only the references cited in the preceding excerpts are reproduced below.]

REFERENCES

1. SCHIFF, J. A. 1980. Development, inheritance and evolution of plastids and mitochondria. *In* The Biochemistry of Plants. N. Edward Tolbert, Ed. Vol. **1:** 209–272. Academic Press, Inc. New York, N.Y.
2. RHOADES, M. M. 1946. Plastid mutations. Cold Spring Harbor Symp. Quant. Biol. **11:**202–207.
3. MARGULIES, L. 1970. Origin of eucaryotic cells. Yale University Press. New Haven, Conn.
4. SCHIFF, J. A. 1973. The development, inheritance and origin of the plastid in *Euglena. In* Advances in Morphogenesis. M. Abercrombie & J. Brachet, Eds. Vol. **10:**265–312. Academic Press, Inc. New York, N.Y.
5. LEWIN, R. 1977. *Prochloron*, type genus of the prochlorophyta. Phycologia **16:**217.

24. CLAYTON, R. 1964. Phototaxis in microorganisms. *In* Photophysiology. A. Geise, Ed. Vol. **2**:51–77.
25. NULTSCH, W. 1970. Photomotion of microorganisms and interaction with photosynthesis. *In* Photobiology of Microorganisms. P. Halldal, Ed.: 213–251. John Wiley & Sons, Inc. New York, N.Y.
26. DIEHN, B. 1969. Action spectra of phototactic responses in *Euglena*. Biochim. Biophys. Acta **177**:136–143.
27. LEEDALE, G. F. 1967. Euglenoid Flagellates. Prentice-Hall, Inc. Englewood Cliffs, N.J.
28. BUETOW, D. E., ED. 1968. The Biology of Euglena V. Academic Press, Inc. New York, N.Y.
29. FEINLIEB, M. 1978. Photomovement in microorganisms. Photochem. Photobiol. **27**:849–854.
30. SENGER, H., Ed. 1980. The Blue Light Syndrome. Springer-Verlag. Heidelberg, West Germany.
31. HOWLAND, G. P. & M. E. EDWARDS. 1979. Photomorphogenesis of fern gametophytes. *In* The Experimental Biology of Ferns. A. F. Dyer, Ed. :394–427. Academic Press, Inc. New York, N.Y.
32. BRIGGS, W. 1964. Phototropism in higher plants. *In* Photophysiology. A. Giese, Ed. Vol. **1**:223–271. Academic Press, Inc. New York, N.Y.
33. STOCKENIUS, W., R. H. LOZIER & R. A. BOGOMOLNI. 1979. Bacteriorhodopsin and the purple membrane of halobacteria. Biochim. Biophys. Acta **505**:215–278.
34. DARTNALL, H. J. A., Ed. 1972. Handbook of Sensory Physiology. Vol. VII/I. Springer-Verlag. Heidelberg, West Germany.
35. VOGELMAN, T. C. & J. SCHEIBE. 1978. Action spectra for chromatic adaptation in the blue-green alga *Fremyella diplosiphon*. Planta **143**:233–240.
36. BJORN, G. S. 1979. Action spectra for *in vivo* and *in vitro* conversions of phytochrome b, a reversibly photochromic pigment in a blue-green alga and its separation from other pigments. Physiol. Plantarum **46**:281–286.
37. LAZAROFF, N. & J. A. SCHIFF. 1962. Action spectrum for developmental photoinduction of the blue-green alga *Nostoc muscorum*. Science **137**:603–604.
38. FUJITA, Y. & A. HATTORI. 1962. Photochemical interconversion between precursors of phycobilin chromoproteids in *Tolypothrix tenuis*. Plant Cell Physiol. **3**:209–220.
39. HAURY, J. & L. BOGORAD. 1977. Action spectra for phycobiliprotein synthesis in a chromatically adapting cyanophyte *Fremyella diplosiphon*. Plant Physiol. **60**:835–839.
40. SHROPSHIRE, W. 1977. Photomorphogenesis. *In* The Science of Photobiology. K. Smith, Ed.: 281–312. Plenum Publishing Corp. New York, N.Y.
41. SHROPSHIRE, W., JR. 1972. Phytochrome, a photochromic sensor. *In* Photophysiology. A. Giese, Ed. Vol. **7**:33–72.
42. HAUPT, W. & SCHONBOHM, E. 1970. Light-oriented chloroplast movements. *In* Photobiology of Microorganisms. P. Halldal, Ed.: 283–307. John Wiley & Sons. New York, N.Y.
43. SCHIFF, J. A. 1978. Photocontrol of chloroplast development in *Euglena*. *In* Chloroplast Development. G. Akoyunoglou J. H. Argyroudi-Akoyunoglou, Eds.: 747–767. Elsevier North-Holland, Inc. Amsterdam.
44. VIRGIN, H. 1972. Chlorophyll biosynthesis and phytochrome action. *In* Phytochrome. K. Mitrakos & W. Shropshire, Jr., Eds.: 371–406. Academic Press, Inc. New York, N.Y.

Part V
ORIGIN OF NUCLEOCYTOPLASM

Editors' Comments
on Papers 25 and 26

25 STARR
Bdellovibrio *as Symbiont: The Associations of Bdellovibrios with other Bacteria Interpreted in Terms of a General Scheme for Classifying Organismic Associations*

26 MARGULIS
Mitochondria: Acquisition by Whom?

What were the original prokaryotes (pre-eukaryotes) initially associated with the symbionts, which became plastids, mitochondria, and motility organelles (undulipodia)? The topic is reviewed by Margulis in Paper 26.

Most modern bacteria are walled, and phagocytosis is unknown in bacteria. This leads to a difficulty in proposing a mechanism by which an intimate endosymbiosis between bacteria might have occurred. Much experimental work remains to be done. The following discussion explains three hypotheses that have been suggested.

    1. Phagocytotic bacteria do not exist now but may have existed 2 billion years ago. Although mycoplasmas are not phagocytotic, they are wall-less bacteria, which may have been more accessible hosts to potential symbionts. Mycoplasmas have several striking similarities to the nucleo-cytoplasm of eukaryotes discussed by Searcy, Stein, and Green (1978).

    2. *Bdellovibrio* bacteria have mechanisms by which they bore into a host bacterium's periplasmic space, where they reproduce and then finally burst out as described by Starr in Paper 25. This *bdellovibrio* habit may be fairly common in nutrient-poor waters. It is not known whether *bdellovibrio*-type bacteria exist that do not destroy their hosts but establish more long-term symbioses. The problem is that the currently used assay for *bdellovibrio* is the presence of damage to host bacteria. *Bdellovibrio* are similar to mitochondria metabolically. The *bdellovibrio*-habit may have been a preadaptation for the endosymbiotic origin of an organelle.

    3. There are many examples of bacterial consortia (associations between two or more bacteria in wall-to-wall contact), although due to the difficulty of culturing such associations, we may be aware only of a fraction of the number of cases present in nature. Also, due to the difficulty of cultivation, little is known about the mechanisms of nutrient exchange and reproduction in such relationships. The idea that the nucleus, the plastid, and the mitochondrion originated as a three-way bacterial consortium with the cytoplasm and outer membrane secreted secondarily is worth considering.

REFERENCE

Searcy, D. G., D. B. Stein, and G. R. Green, 1978, Phylogenetic Affinities between Eukaryotic Cells and a Thermoplasmic Mycoplasma, *Biosystems* **10:**19-28.

25

Copyright © 1975 by the Society for Experimental Biology

Reprinted from pages 95–104 and 121–124 of *Symbiosis,* Symposia of the Society for Experimental Biology, No. 29, Society for Experimental Biology, London, England, 1975, pp. 93–124

BDELLOVIBRIO AS SYMBIONT: THE ASSOCIATIONS OF BDELLOVIBRIOS WITH OTHER BACTERIA INTERPRETED IN TERMS OF A GENERALIZED SCHEME FOR CLASSIFYING ORGANISMIC ASSOCIATIONS

By M. P. STARR

Department of Bacteriology, University of California, Davis, California 95616, USA

[*Editors' Note:* In the original, material precedes this excerpt.]

THE ASSOCIATIONS OF SYMBIOSIS-COMPETENT BDELLOVIBRIOS WITH OTHER BACTERIA

The discovery of bdellovibrios and other prefactory matters

The bacteria which now comprise the genus *Bdellovibrio* (Stolp & Starr, 1963) were discovered 'accidentally' in 1962 by Heinz Stolp (Stolp & Petzold, 1962; Stolp, 1973) when he was attempting to isolate bacterial viruses (bacteriophages; phages). Stolp had inoculated soil filtrates onto the confluent cellular growths (lawns) of the bacterium *Pseudomonas phaseolicola* being cultivated, in the conventional double-layer method, on nutrient agar in Petri plates. Following the usual 24 hour incubation period, the lawns showed no cleared zones (bacteriophage plaques) where the bacteria would have been lysed by phages in the otherwise opaque bacterial lawns. For some undetermined reason, Stolp did not discard the plates – as one would expect a phage worker to do because phages develop only in young, growing bacteria – but he re-examined them after another 24 hours had passed. The later examination was crucial for, at this time, plaques were evident in the *Pseudomonas* lawns; these plaques continued to increase in size for about a week. Because of the delay in starting plaque formation and the continuing increase in plaque size, Stolp concluded that conventional bacteriophages were probably not the cause of the plaques. He then – quite exceptionally for a phage researcher – examined material from the plaques in a phase-contrast microscope.

A large number of rapidly motile tiny microbes, in addition to a few cells of the larger pseudomonad, were demonstrated by these microscopic observations of material removed from the developing plaques. The small organisms collided with the much larger pseudomonad cells, adhered to their surfaces, and seemed to cause them to lyse. The lytic agent was clearly not a phage: the agent did not pass through a filter with very small pore size (200 nm) through which phages can pass, nor did plaques form on a streptomycin-containing medium with a lawn of streptomycin-resistant pseudomonad cells, whereas phages do yield plaques under these conditions.

A comprehensive study (Stolp & Starr, 1963) of these unusual microbes led to the conclusion that they are indeed unique bacteria. Based on the ability to enter into this peculiar interbacterial association, and the bacteriolytic and other properties including the relatively small size of these bacteria, a new genus, *Bdellovibrio*, was established (Stolp & Starr, 1963), with the specific epithet, *bacteriovorus*, for the single species then recognized. The generic name reflects the organism's mode of initial attachment to other bacteria and its shape ('*Bdello-*' is derived from the Greek word for a leech and '-*vibrio*' refers to its comma shape), while '*bacteriovorus*' refers to the fact that *Bdellovibrio* seemed to devour the bacteria with which it was associated.

It might be useful to examine the epistemological implications of this early work on *Bdellovibrio*. In view of the presently known ubiquity of *Bdellovibrio* in soils and waters the world over, it is quite likely that many bacteriologists had 'seen-without-seeing' bdellovibrios, because they had no prior knowledge or concept of such creatures. Then came the 'chance' observation in 1962 by Stolp of the late-forming plaques in bacterial lawns. His microscopic examination, as already related, was crucial because it provided the first 'seeing' glimpse of these highly motile, tiny, vibrioid bacteria which attached to other bacteria and seemed to cause them to lyse. Taking into account that another unusual and (at the time) little-known bacterium, *Caulobacter*, was known to attach sometimes to other bacteria, the early notion (Stolp & Petzold, 1962) that *Bdellovibrio* might be some sort of *Caulobacter*-like creature is understandable. However, this notion was soon corrected. Comparative study (Stolp & Starr, 1963) of several strains of *Bdellovibrio* in 1962-1963 extended the factual base; the conceptual base, unfortunately, hardened into an erroneous notion – namely, 'ectoparasitism' – which conceived of *Bdellovibrio* as being able to act antagonistically upon its symbiont only from outside the symbiont. Despite numerous clues – now patently clear from hindsight – it took another two years before the error was corrected and the ability of *Bdellovibrio* to enter other bacterial cells and to develop in this intramural (intraperiplasmic) locus was established experimentally (Starr & Baigent, 1966; Scherff, DeVay & Carroll, 1966).

The account which follows presents very briefly the salient facts about *Bdellovibrio*, emphasizing those which bear on its association with other bacteria. Further details and other aspects are treated in recent review articles (Shilo, 1969, 1973*a*; Starr & Seidler, 1971; Starr & Huang, 1972; Stolp, 1973; Varon, 1974). This first part of the presentation deals only with the significant associations of *Bdellovibrio* cells with the cells of other bacteria. That is to say, we shall consider here only those aspects of the life of the bdellovibrios which are symbiotic (in the sense given by the statement in the Introduction). Bdellovibrios which are capable of entering into symbiotic associations with other bacteria will be termed herein 'symbiosis-competent' ('S-C'; conventionally: 'host-dependent' or 'H-D' or 'parasitic' or 'predatory'). Non-symbiotic aspects of *Bdellovibrio* life, and terminological points pertaining thereto, are handled separately in a subsequent section on p. 104. The associations of symbiosis-competent bdellovibrios with other bacteria will be dissected into a series of approximately chronological stages; however, the exact timing is in many cases unknown or varies with the experimental system. Hence, the form of presentation should be viewed primarily as a convenient didactic device, rather than as an accurate chronology.

'Recognition' and 'chemotaxis'

The initial association of *Bdellovibrio* swarmers (the tiny, highly motile, usually vibrioid cells) with other bacteria may involve some sort of 'recognition' of a suitable associant cell. This 'recognition' (if indeed it exists) might be mediated by a 'chemotaxis'. However, diligent efforts over several years in various laboratories including ours have not yielded experimental results in accordance with the reasonable expectations stemming from this hypothesis. These expectations are that there should be massive, selective locomotion of bdellovibrios toward congenial bacteria (potential symbionts) or extracts prepared therefrom used as 'bait' in the Adler (1966, 1973) capillary-tube procedure and various modifications thereof. Although we have sometimes found in our own work statistically valid increases in numbers of bdellovibrios moving toward congenial bacteria, as compared with uncongenial bacteria, used as bait, generally rather small numbers of bdellovibrios swam even to the congenial bacterial bait; moreover, the results in replicate experiments have sometimes been contradictory. Hence, we have been disinclined to publish our ambiguous and still unexplainable findings. A recent publication (Straley & Conti, 1974) reports similar 'weak chemotactic responses' of bdellovibrios with bacterial preparations, remarks that 'possible explanations are being explored' and then turns to the chemotaxis of bdellovibrios toward yeast extract – a subject which is not yet directly relevant to the present issue of trying to decide whether or not there is 'recognition' by *Bdellovibrio* of a congenial bacterial symbiont and a 'directed motility' of the *Bdellovibrio* toward it, controlled by some sort of 'chemotaxis'. So, although the hypothesis is tempting, there is little yet to support it – with the possible exception of some (also still perplexing) work on 'symbiont specificity' to which we will now turn.

'Symbiont specificity'

The solid facts about the specificity of the symbiotic associations of bdellovibrios with other bacteria ('symbiont specificity' in my sense;

conventionally: 'host specificity') are not numerous and what has been reported is still fairly confusing. A given strain of *Bdellovibrio* is generally reported to enter into associations only with a fairly limited array of rather closely related bacteria. Some *Bdellovibrio* strains are reportedly limited to a single symbiont strain; others will enter into associations with many different strains belonging to one or more bacterial genera. Symbiont specificities of bdellovibrios are reported or summarized in several works (Stolp & Petzold, 1962; Stolp & Starr, 1963; Uematsu & Wakimoto, 1970; Starr & Seidler, 1971; Stolp, 1973; Taylor, Baumann, Reichelt & Allen, 1974). In some cases, the reported differences in ability of a given *Bdellovibrio* strain to attack various bacteria extended to cases wherein the attacked and unattacked bacteria were practically indistinguishable on other grounds even by an experienced bacteriologist! The bases and indeed the full extent of this symbiont specificity remain largely unknown; moreover the observed 'specificity' will often vary with the assay condition. Up to now, only Gram-negative bacteria have been conclusively shown to serve as symbionts for the presently known bdellovibrios (but see Burger, Drews & Ladwig, 1968, and Gromov & Mamkaeva, 1972, for associations of purported bdellovibrios with Gram-positive bacteria and with green algae of the genus *Chlorella*).

An interesting start has been made in understanding one possible basis of symbiont specificity, in a study (Varon & Shilo, 1969a) wherein *Salmonella* and *Escherichia* mutants with various blocks in their syntheses of cell wall lipopolysaccharides were used as prospective symbionts for *Bdellovibrio*. The bdellovibrios attached better to rough strains having the complete lipopolysaccharide core but lacking the type-specific O antigens than they did either to the smooth wild type (having the type-specific O antigens) or to the extremely rough strains with defects in the lipopolysaccharide core.

Attachment

Regardless of the uncertainty about whether there is 'recognition' of a suitable associant cell and whether there is a 'chemotaxis', there is no question about the existence of a violent collision of a motile *Bdellovibrio* swarmer with a prospective associant cell, followed by attachment. The bdellovibrio swarmer, which is provided with an unusual sheathed flagellum (Seidler & Starr, 1968), has been estimated to swim at the astonishing speed of 100 cell-lengths per second (Stolp, 1967 a, b). The bdellovibrio swarmer may shove the larger bacterial cell, with some ten to twenty times the mass of the *Bdellovibrio* cell, over a distance corresponding to several cell-lengths. Loss of motility or flagellar sheath integrity, induced in the *Bdellovibrio* by any of several procedures (Stolp & Starr, 1963; Varon & Shilo, 1968; Abram & Davis, 1970; Dunn, Windom, Hansen & Seidler, 1974), results in total inability of the *Bdellovibrio* to attach to other bacteria and thus to enter into a detectable symbiotic association with them.

When an actively motile preparation of symbiosis-competent bdellovibrio is added to a suspension of susceptible ('congenial') bacterial cells, attachments by the bdellovibrio cells to the cells of the other bacterial organism (the 'associant' or the 'symbiont'; conventionally: the 'host' or the 'prey') begin immediately. The bdellovibrio attaches by its anterior, aflagellated end – which has some interesting and unique structures (Abram & Davis, 1970). Often, more than one bdellovibrio attaches to a single symbiont cell. The kinetics of attachment and various other factors pertaining to attachment are presented in the still unsurpassed pioneering study by Varon & Shilo (1968). The most recent work on the subject (Dunn *et al.*, 1974) reports the isolation and many interesting properties of temperature-sensitive 'attachment mutants' of *Bdellovibrio*. These mutants could be further exploited as a powerful experimental tool in the clarification of still unresolved questions about attachment.

The early stages of contact may be non-specific (Dunn *et al.*, 1974): for example, symbiosis-competent bdellovibrios sometimes collide with and attach to non-susceptible bacteria or even to the glass coverslips of microscopic slide preparations. The attachment to an uncongenial bacterial cell and even to a congenial symbiont cell may be reversible in the earliest stages; that is, the attachment may be aborted and the bdellovibrio may swim away to another bacterial cell. It is not known whether these sorts of attachments have any functional, anatomical, or behavioral relationship to the usual symbiotic connection.

Penetration by bdellovibrios into cells of other bacteria

After a successful *Bdellovibrio*-symbiont attachment has been established, the next noticeable morphological aspect of the association involves the penetration of the symbiont cell by the *Bdellovibrio* cell. This penetration requires the formation (by mechanisms which are still not known with certainty) in the symbiont cell wall of a rather undersized pore through which the bdellovibrio eventually either squeezes by its own locomotor activity (Starr & Baigent, 1966; Stolp, 1973) or possibly is pulled 'passively' into the other bacterial cell (Abram, Castro e Melo & Chou, 1974); perhaps both penetration procedures operate. The fate of this pore (hole) in the symbiont's cell wall seems not to be known: is it repaired or plugged and, if so, how?

Although not all of the following notions are based on equally solid experimental evidence (and, indeed, some of them may have been fairly effectively refuted), the pore formation and penetration have been claimed to result from one or more of the following events: ballistic damage to the cell wall in the initial violent collision of the *Bdellovibrio* with the symbiont cell (Stolp, 1973), enzymatic action by the *Bdellovibrio* on the symbiont's cell wall (Huang & Starr, 1973*a*; Fackrell & Robinson, 1973; Engelking & Seidler, 1974), autolysis mechanisms induced in the symbiont by the *Bdellovibrio* (Stolp & Starr, 1965), damage to the cell wall by the mechanical drilling or swivelling motion of the attached *Bdellovibrio* prior to penetration (Starr & Baigent, 1966; Stolp, 1973), and other less explicitly stated factors. What are thought to be rebuttals of some of these notions have appeared. When the motility of the *Bdellovibrio* is retarded by the use of viscous media (thus reducing the ballistic impact) both attachment and penetration are claimed (Abram *et al.*, 1974) still to occur. Loss of *Bdellovibrio* motility immediately after attachment (thus making a drilling motion impossible) has been reported (Horowitz, Kessel & Shilo, 1974) and also vehemently denied (Stolp, 1967*a, b*, 1973). Abram *et al.* (1974) state that 'the motion and the impact between parasite and host are neither sufficient nor absolute prerequisites for attachment and penetration'. Then these authors deny on other grounds that the rotational or swivelling or other active motility of the *Bdellovibrio* plays any significant role in its penetration – which latter process they claim is a 'passive act of the parasite', firmly bonded to the symbiont's protoplast. This – by its retraction or that of the cell wall – pulls the *Bdellovibrio* into the periplasmic space. The notion of autolysis induction in the symbiont as a factor in penetration has been said (Shilo, personal communication) to be ruled out by the experiments summarized in Table 3 of Varon & Shilo (1968) with streptomycin-resistant *E. coli* and streptomycin-sensitive *Bdellovibrio* in which 'the inhibitory effect of streptomycin on invasion was expressed fully'.

Although I expect some significant findings about penetration to follow the use of temperature-sensitive 'penetration mutants' (Dunn *et al.*, 1974), my own view at the moment is that making the pore in the symbiont's cell wall may possibly be aided by the purported ballistic damage and the swivelling or drilling motion of the *Bdellovibrio*, but that there is now both morphological and biochemical evidence to indicate that the penetration process is facilitated or mediated by enzymatic action of the bdellovibrio and by the purported retraction of the symbiont's cell wall or its cytoplasmic membrane to which latter structure the penetrating bdellovibrio is said to be firmly bonded. Clearly this entire situation demands further clarification.

Regardless of the mechanismic bases for pore formation and penetration, when these processes have been effected the *Bdellovibrio* cell becomes lodged in the periplasmic region between the symbiont's cytoplasmic membrane and cell wall. This is one of the rare points on which all *Bdellovibrio* workers seem to be in agreement: the cytoplasmic membrane of the symbiont cell is not breached (except to the extent of the localized lesions reported by Snellen & Starr, 1974) and the *Bdellovibrio* cell occupies an intramural (intraintegumental; intraperiplasmic) position within the symbiont cell. Re-stating this important point in other words, the *Bdellovibrio* cell is not inside the protoplast of the symbiont cell, although it is within a space (the periplasmic region) between the cell wall and the cytoplasmic membrane of the symbiont cell.

Early effects of the symbiosis

Symbiont cells that ordinarily are motile have been reported to stop swimming within moments after their initial contacts with *Bdellovibrio* swarmers; precisely how the bdellovibrio brings about this particular dysfunctioning of its symbiont is not yet known, but it may well be a concomitant of the elimination of the symbiont's respiratory potential (Rittenberg & Shilo, 1970). Much more often reported, with many *Bdellovibrio*-symbiont combinations, is the subsequent conversion of the attacked cell into a swollen, often spherical body at an early stage following attachment and before complete penetration is microscopically apparent. Although I have been chided rather strongly (Rittenberg, personal communication) about my use of the words 'before complete penetration' in the previous sentence (he says that 'the bdelloplast forms after complete penetration'), I can only reply that my views on the matter are supported unequivocally by many observations, including those depicted in Figure 1 of Starr & Baigent (1966) and Figure 9 of Seidler & Starr (1969a). These swollen or spherical bodies are probably not identical with spheroplasts – because, unlike some spheroplasts, they are not osmotically sensitive (Starr & Seidler, 1971; Starr & Huang, 1972) – hence, many *Bdellovibrio* workers prefer to use the terms 'spherical (or swollen) body' or 'bdelloplast' (rather than 'spheroplast') in referring to them. When the symbiont cell is a short rod, the entire cell is converted into such a bdelloplast; in a longer symbiont cell and at low multiplicities of infection, the swelling is more localized and a kind of multiple 'ballooning' is often seen. The formation of these bdelloplasts and balloons has been related circumstantially to enzymatic action of the *Bdellovibrio* on the rigid layer (murein) of the symbiont cell wall (Huang & Starr, 1973a).

Local damage has been reported to occur on the symbiont's cell surface (even though the *Bdellovibrio* cell may subsequently have become detached) in the form of holes and pitted areas in that surface of the symbiont's cell wall which had been in contact with the bdellovibrio (Shilo, 1969). Most recently, localized damage to the symbiont's cytoplasmic membrane in the vicinity of the entered *Bdellovibrio* has been shown in ultrastructural studies of the association (Snellen & Starr, 1974).

In addition to the ultrastructural membrane and wall damage, the cessation of motility, and the formation of bdelloplasts and other swollen bodies, there are signs of other serious dysfunctionings of the symbiont's cells in the earlier stages of their association with *Bdellovibrio*. Attachment of the *Bdellovibrio* cell to the cell wall of its symbiont has been shown to be sufficient to cause cessation of protein and nucleic acid synthesis in the symbiont, even before penetration has become microscopically apparent (Varon, Drucker & Shilo, 1969; Shilo, 1973*a*, *b*). The cytoplasmic membrane of the entered symbiont cell also functions abnormally (becomes less selectively permeable, but in what appears to be a controlled fashion) already in the earliest stages of its association with the *Bdellovibrio* (Rittenberg & Shilo, 1970; Crothers & Robinson, 1971). This controlled leakage has recently been related circumstantially to the aforementioned localized ultrastructural damage effected by the bdellovibrio on the cytoplasmic membrane of the entered symbiont (Snellen & Starr, 1974), but it must be emphasized that there is as yet no direct experimental evidence about actual leakage at these localized damaged sites. The respiratory potential of the symbiont is essentially eliminated within a few minutes after *Bdellovibrio* attachment (Rittenberg & Shilo, 1970). Many of these detrimental early effects on the symbiont may, quite unexpectedly, be reversible, as is suggested by the exciting finding of Mielke and Stolp (personal communication) that they could 'cure *Bdellovibrio* infections' by means of deoxycholate; further details on this point are given in a subsequent section (pp. 116–17).

Development of Bdellovibrio *inside its symbiont*

Once in the periplasmic locus, the small and vibrioid *Bdellovibrio* cell begins its intramural growth phase and, as a second stage in its characteristic dimorphism, elongates into a C- or helical-shaped cell. The helical (more usually, but incorrectly in a geometric sense, called 'spiral') cell may be ten or more times longer than the entering vibrioid swarmer cell; its size seems to be directly proportional to the size of the symbiont cell within which it is developing. During this intramural growth phase, there is a progressive – but meticulously regulated (Matin & Rittenberg, 1972; Hespell, Rosson, Thomashow & Rittenberg, 1973) – disorganization and dissolution and utilization, by the *Bdellovibrio*, of the symbiont's cytoplasm, nucleoplasm, and other components. The digested symbiont components are converted at very high efficiency into *Bdellovibrio* components (Hespell *et al.*, 1973; Hespell, Thomashow & Rittenberg, 1974). The rather large, helical bdellovibrio cell undergoes multiple constriction prior to its segmentation into several vibrioid daughter (progeny swarmer) cells. At about this time, the typical sheathed flagellum (Seidler & Starr, 1968) develops on each progeny swarmer (Burnham, Hashimoto & Conti, 1968) and the swarmers can then be seen, by phase-contrast microscopy, swimming within the swollen and usually heavily disorganized ('ghosted') bdelloplast.

The final stage: release of Bdellovibrio *progeny*

The means by which the *Bdellovibrio* progeny swarmers leave the ghosted symbiont cell are not entirely clear. The juvenile swarmers are rapidly motile within the ghosted symbiont cell; perhaps they are able to break down its remnants mechanically by virtue of this motility. In many instances, the destruction of the symbiont cell is so extensive that no significant structural barrier remains to retard departure of the *Bdellovibrio* progeny ('swarmers') and reinitiation of another symbiotic cycle when a congenial associant is found. In other cases, the remnants of the symbiont, after the *Bdellovibrio* swarmers have left, appear to be fairly substantial.

Exhabitational symbiosis by bdellovibrios

Rittenberg (personal communication) has recently informed me about the interesting associations of bdellovibrios with very small bacteria, *Bordetella pertussis*, approximately the same size as *Bdellovibrio*. Here something akin to an exhabitational symbiosis sometimes occurs, possibly because the relatively equal sizes of the associants precludes complete entrance of the *Bdellovibrio*. Another tendency toward an exhabitational symbiotic mode has recently been related to me by Abram (personal communication). By manipulation of the ionic and other environmental conditions, penetration (but not attachment) of the *Bdellovibrio* is inhibited and the attached bdellovibrio develops in this exhabitational locus. However, in both of these exhabitational cases, it was not clear whether or not the symbiont's cell wall is breached (I expect it is) and whether or not the bdellovibrio bonds tightly to the symbiont's cytoplasmic membrane (I expect it does, if the purported bonding in the inhabitational mode is confirmed and generalized). If my expectations are realized, then the spatial situations in these cases are neither purely exhabitational nor entirely inhabitational, and they must be classed in intermediate positions within the exhabitational-inhabitational continuum (see earlier article) in the same way as the haustoria of some fungi.

These expectations seem indeed to have been realized by the latest news from Abram (personal communication) wherein she describes a *Bdellovibrio–Acinetobacter* system in which there is simultaneously an array of relative spatial arrangements ranging from the usual intramural inhabitation, through partial inhabitation (only the anterior end of the *Bdellovibrio* enters), to full exhabitation. In many or all of these relative spatial arrangements, the *Bdellovibrio* has been reported to go through its characteristic dimorphic developmental cycle; fairly substantial *Bdellovibrio* progeny yields were observed in some. But her generosity in sharing this exciting news with me should be partially compensated by allowing her to disclose the full details of this fantastic story elsewhere!

Important, indeed essential, for the present enterprise is that these findings by Rittenberg and Abram (separate personal communications) add the occurrence in *Bdellovibrio*-symbiont systems of exhabitational symbiosis, and various intermediates between inhabitational and exhabitational symbiosis, to the better-known, intraperiplasmic, inhabitational kind. These findings support the utility of arranging criteria about organismic associations into continua. The same might be said about the continuum extending from symbiosis to non-symbiosis: some of the cases related here and some of those discussed elsewhere in this essay in connection with non-symbiotic development of bdellovibrios, may involve relationships intermediate between symbiosis and some sorts of non-symbiosis ('saprotrophy' or 'scavenging' – in my terminology).

[*Editors' Note:* Material has been omitted at this point.]

REFERENCES

ABRAM, D., CASTRO E MELO, J. & CHOU, D. (1974). Penetration of *Bdellovibrio bacteriovorus* into host cells. *J. Bact.*, **118**, 663–680.

ABRAM, D. & DAVIS, B. K. (1970). Structural properties and features of parasitic *Bdellovibrio bacteriovorus*. *J. Bact.*, **104**, 948–965.

ADLER, J. (1966). Chemotaxis in bacteria. *Science, Wash.*, **153**, 708–716.

— (1973). A method for measuring chemotaxis and use of the method to determine optimum conditions for chemotaxis by *Escherichia coli*. *J. gen. Microbiol.*, **74**, 77–91.

AHMADJIAN, V. & HALE, M. E. [eds.] (1973). *The Lichens*. New York: Academic Press.

ALTHAUSER, M., SAMSONOFF, W. A., ANDERSON, C. & CONTI, S. F. (1972). Isolation and preliminary characterization of bacteriophage for *Bdellovibrio bacteriovorus*. *J. Virol.*, **10**, 516–524.

DE BARY, A. (1879). *Die Erscheinung der Symbiose*. Strassburg: Verlag von Karl J. Trübner.

BUCHNER, P. (1965). *Endosymbiosis of Animals with Plant Organisms*. New York: John Wiley & Sons, Inc.

BURGER, A., DREWS, G. & LADWIG, R. (1968). Wirtskreis und Infektionscyclus eines neu isolierten *Bdellovibrio bacteriovorus*-Stammes. *Arch. Mikrobiol.*, **61**, 261–279.

BURNET, M. (1962). *Natural History of Infectious Disease*, third edition. London: Cambridge University Press.

BURNHAM, J. C., HASHIMOTO, T. & CONTI, S. F. (1968). Electron microscopic observations on the penetration of *Bdellovibrio bacteriovorus* into Gram-negative bacterial hosts. *J. Bact.*, **96**, 1366–1381.

BURNHAM, J. C., HASHIMOTO, T. & CONTI, S. F. (1970). Ultrastructure and cell division of a facultatively parasitic strain of *Bdellovibrio bacteriovorus*. *J. Bact.*, **101**, 997–1004.

BURNHAM, J. C. & ROBINSON, J. (1974). Bdellovibrio. In *Bergey's Manual of Determinative Bacteriology*, eighth edition, 212–214 (eds. R. E. Buchanan and N. E. Gibbons). Baltimore: Williams & Wilkins Company.

CROTHERS, S. F., FACKRELL, H. B., HUANG, J. C.-C. & ROBINSON, J. (1972). Relationship between *Bdellovibrio bacteriovorus* 6-5-S and autoclaved host bacteria. *Can. J. Microbiol.*, **18**, 1941–1948.

CROTHERS, S. F. & ROBINSON, J. (1971). Changes in the permeability of *Escherichia coli* during parasitization by *Bdellovibrio bacteriovorus*. *Can. J. Microbiol.*, **17**, 689–697.

DIEDRICH, D. L., DENNY, C. F., HASHIMOTO, T. & CONTI, S. F. (1970). Facultatively parasitic strain of *Bdellovibrio bacteriovorus*. *J. Bact.*, **101**, 989–996.

DOUTT, R. L. (1964). Biological characteristics of entomophagous adults. In *Biological Control of Insect Pests and Weeds*, 145–167 (eds. P. DeBach and E. I. Schlinger). London: Chapman & Hall.

DUNN, J. E., WINDOM, G. E., HANSEN, K. L. & SEIDLER, R. J. (1974). Isolation and characterization of temperature-sensitive mutants of host-dependent *Bdellovibrio bacteriovorus* 109D. *J. Bact.*, **117**, 1341–1349.

EHRLICH, M. A. & EHRLICH, H. G. (1971). Fine structure of the host–parasite interfaces in mycoparasitism. *A. Rev. Phytopathol.*, **9**, 155–184.

ENGELKING, H. M. & SEIDLER, R. J. (1974). The involvement of extracellular enzymes in the metabolism of *Bdellovibrio*. *Arch. Mikrobiol.*, **95**, 293–304.

FACKRELL, H. B. & ROBINSON, J. (1973). Purification and characterization of a lytic peptidase produced by *Bdellovibrio bacteriovorus* 6-5-S. *Can. J. Microbiol.*, **19**, 659–666.

GLOOR, L., KLUBEK, B. & SEIDLER, R. J. (1974). Molecular heterogeneity of the bdellovibrios: Metallo and serine proteases unique to each species. *Arch. Mikrobiol.*, **95**, 45–56.

GROMOV, B. V. & MAMKAEVA, K. A. (1972). [Electron microscope examination of *Bdellovibrio chlorellavorus* parasitism on cells of the green alga *Chlorella vulgaris*.] *Tsitologiia*, **14**, 256–260.

GUÉLIN, A. & CABIOCH, L. (1974). Caractères dynamiques de l'interaction entre le microprédateur *Bdellovibrio bacteriovorus* et la bactérie-hôte en fonction de leurs densités initiales respective. *C. r. hebd. Séanc. Acad. Sci. Paris*, **278**, 1293–1296.

HALL, R. (1974). Pathogenism and parasitism as concepts of symbiotic relationships. *Phytopathology*, **64**, 576–577.

HESPELL, R. B., ROSSON, R. A., THOMASHOW, M. F. & RITTENBERG, S. C. (1973). Respiration of *Bdellovibrio bacteriovorus* strain 109J and its energy substrates for intraperiplasmic growth. *J. Bact.*, **113**, 1280–1288.

HESPELL, R. B., THOMASHOW, M. F. & RITTENBERG, S. C. (1974). Changes in cell composition and viability of *Bdellovibrio bacteriovorus* during starvation. *Arch. Mikrobiol.*, **97**, 313–327.

HOROWITZ, A. T., KESSEL, M. & SHILO, M. (1974). Growth cycle of predacious bdellovibrios in a host-free extract system and some properties of the host extract. *J. Bact.*, **117**, 270–282.

HUANG, J. C.-C. & STARR, M. P. (1973a). Possible enzymatic bases of bacteriolysis by bdellovibrios. *Arch. Mikrobiol.*, **89**, 147–167.

(1973b). Effects of calcium and magnesium ions and host viability on growth of bdellovibrios. *Antonie van Leeuwenhoek*, **39**, 151–167.

ISHIGURO, E. E. (1973). A growth initiation factor for host-independent derivatives of *Bdellovibrio bacteriovorus*. *J. Bact.*, **115**, 243–252.

(1974). Minimal nutritional requirements for growth of host-independent derivatives of *Bdellovibrio bacteriovorus* strain 109 Davis. *Can. J. Microbiol.*, **20**, 263–265.

KESSELL, M. & VARON, M. (1973). Development of bdellophage VL-1 in parasitic and saprophytic bdellovibrios. *J. Virol.*, **12**, 1522–1533.

KEYA, S. O. & ALEXANDER, M. (1975). Regulation of parasitism by host density: The *Bdellovibrio-Rhizobium* interrelationship. *Soil Biol. Biochem.* (in press).

LAMBINA, V. A., AFINOGENOVA, A. V., KONOVALOVA, S. M., PECHNIKOV, N. V., FEDOROVA, A. M., FICHTE, B. A. & SKRYABIN, G. K. (1974). [On the character of parasitism of *Bdellovibrio bacteriovorus* Stolp et Starr gen. et sp. nov.] *Akad. Nauk. [S.S.S.R.], Institut Biochimii i Fiziologii Mikroorganizmov, Seriia Biologischeskaia*, **1**, 81–88.

LEWIS, D. H. (1973). Concepts in fungal nutrition and the origin of biotrophy. *Biol. Rev.*, **48**, 261–278.

(1974). Micro-organisms and plants: The evolution of parasitism and mutualism. *Symp. Soc. gen. Microbiol.*, **24**, 367–392.

LUTTRELL, E. S. (1974). Parasitism of fungi on vascular plants. *Mycologia*, **66**, 1–15.

MATIN, A. & RITTENBERG, S. C. (1972). Kinetics of deoxyribonucleic acid destruction and synthesis during growth of *Bdellovibrio bacteriovorus* strain 109D on *Pseudomonas putida* and *Escherichia coli*. *J. Bact.*, **111**, 664–673.

REINER, A. M. & SHILO, M. (1969). Host-independent growth of *Bdellovibrio bacteriovorus* in microbial extracts. *J. gen. Microbiol.*, **59**, 401–410.

RITTENBERG, S. C. (1973). Aspects of the physiology and biochemistry of intraperiplasmic development of *Bdellovibrio*. *Symp. First Internat. Congr. Bacteriol.* [*Jerusalem, September, 1973*], Abstracts, **1**, 108.

RITTENBERG, S. C. & SHILO, M. (1970). Early host damage in the infection cycle of *Bdellovibrio bacteriovorus*. *J. Bact.*, **102**, 149–160.

ROSS, E. J., ROBINOW, C. F. & ROBINSON, J. (1974). Intracellular growth of *Bdellovibrio bacteriovorus* 6-5-S in heat-killed *Spirillum serpens* VHL. *Can. J. Microbiol.*, **20**, 847–851.

SCHERFF, R. H., DEVAY, J. E. & CARROLL, T. W. (1966). Ultrastructure of host-parasite relationships involving reproduction of *Bdellovibrio bacteriovorus* in host bacteria. *Phytopathology*, **56**, 627–632.

SEIDLER, R. J., MANDEL, M. & BAPTIST, J. N. (1972). Molecular heterogeneity of the bdellovibrios: Evidence of two new species. *J. Bact.*, **109**, 209–217.

SEIDLER, R. J. & STARR, M. P. (1968). Structure of the flagellum of *Bdellovibrio bacteriovorus*. *J. Bact.*, **95**, 1952–1955.

(1969a). Factors affecting the intracellular parasitic growth of *Bdellovibrio bacteriovorus* developing within *Escherichia coli*. *J. Bact.*, **97**, 912–923.

(1969b). Isolation and characterization of host-independent bdellovibrios. *J. Bact.*, **100**, 769–785.

SEIDLER, R. J., STARR, M. P. & MANDEL, M. (1969). Deoxyribonucleic acid characterization of bdellovibrios. *J. Bact.*, **100**, 786–790.

SHILO, M. (1969). Morphological and physiological aspects of the interaction of *Bdellovibrio* with host bacteria. *Curr. Top. Microbiol. Immunol.*, **50**, 174–204.

(1973a). *Bdellovibrio bactereovorus* as a model for the study of bacterial endoparasitism. In *Dynamic Aspects of Host–Parasite Relationships*, vol. I, 1–12 (eds. A. Zuckerman and D. W. Weiss). New York and London: Academic Press.

(1973b). Rapports entre *Bdellovibrio* et ses hôtes. Nature de la dépendance. *Bull. Inst. Pasteur*, **71**, 21–31.

SHILO, M. & BRUFF, B. (1965). Lysis of Gram-negative bacteria by host-independent ectoparasitic *Bdellovibrio bacteriovorus* isolates. *J. gen. Microbiol.*, **40**, 317–328.

SIMPSON, F. J. & ROBINSON, J. (1968). Some energy producing systems in *Bdellovibrio bacteriovorus* strain 6-5-S. *Can. J. Biochem.*, **46**, 865–873.

SMITH, D. C. (1973). Experimental studies of lichen physiology. In *Symbiotic Associations*. *Symp. Soc. gen. Microbiol.*, **13**, 31–50.

SMITH, T. (1934; reprinted 1963). *Parasitism and Disease*. New York and London: Hafner Publishing Co. Inc.

SNELLEN, J. E. & STARR, M. P. (1974). Ultrastructural aspects of localized membrane damage in *Spirillum serpens* VHL early in its association with *Bdellovibrio bacteriovorus* 109D. *Arch. Mikrobiol.*, **100**, 179–195.

STANIER, R. Y. (1970). Some aspects of the biology of cells and their possible evolutionary significance. In *Organization and Control in Prokaryotic and Eukaryotic Cells*. *Symp. Soc. gen. Microbiol.*, **20**, 1–38.

STANIER, R. Y., DOUDORFF, M. & ADELBERG, E. A. (1970). *The Microbial World*, third edition. Englewood Cliffs, New Jersey: Prentice-Hall, Inc.

STARR, M. P. (1959). Bacteria as plant pathogens. *A. Rev. Microbiol.*, **13**, 211–238.

STARR, M. P. & BAIGENT, N. L. (1966). Parasitic interaction of *Bdellovibrio bacteriovorus* with other bacteria. *J. Bact.*, **91**, 2006–2017.

STARR, M. P. & CHATTERJEE, A. K. (1972). The genus *Erwinia*: Enterobacteria pathogenic to plants and animals. *A. Rev. Microbiol.*, **26**, 389–426.

STARR, M. P. & HUANG, J. C.-C. (1972). Physiology of the bdellovibrios. *Adv. microbial Physiol.*, **8**, 215–261.

STARR, M. P. & SEIDLER, R. J. (1971). The bdellovibrios. *A. Rev. Microbiol.*, **25**, 649–678.

STOLP, H. (1967a). *Bdellovibrio bacteriovorus* (*Pseudomonadaceae*). Parasitische Befall und Lysis von *Spirillum serpens*. Film E-1314. Göttingen: Institut für den wissenschaftlichen Film.

(1967b). Lysis von Bakterien durch den Parasiten *Bdellovibrio bacteriovorus*. Film C 972. Göttingen: Institut für den wissenschaftlichen Film. (Begleittext in *Publ. Inst. Wiss. Film*, Göttingen, Bd. A-II, 695–706, 1969.)

(1973). The bdellovibrios: Bacterial parasites of bacteria. *A. Rev. Phytopathol.*, **11**, 53–76.

STOLP, H. & PETZOLD, H. (1962). Untersuchungen über einen obligat parasitischen Mikroorganismus mit lytischer Aktivät für *Pseudomonas-Bakterien*. *Phytopathol. Z.*, **45**, 364–390.

STOLP, H. & STARR, M. P. (1963). *Bdellovibrio bacteriovorus* gen. et sp. n., a predatory, ectoparasitic, and bacteriolytic microorganism. *Antonie van Leeuwenhoek*, **29**, 217–248.

STOLP, H. & STARR, M. P. (1965). Bacteriolysis. *A. Rev. Microbiol.*, **19**, 79–104.

STORZ, J. & PAGE, L. A. (1971). Taxonomy of the chlamydiae: Reasons for classifying organisms of the genus *Chlamydia*, family *Chlamydiaceae*, in a separate order, *Chlamydiales* ord. nov. *Int. J. Syst. Bacteriol.*, **21**, 332–334.

STRALEY, S. C. & CONTI, S. F. (1974). Chemotaxis in *Bdellovibrio bacteriovorus*. *J. Bact.*, **120**, 549–551.

TAYLOR, V. I., BAUMANN, P., REICHELT, J. L. & ALLEN, R. D. (1974). Isolation, enumeration, and host range of marine bdellovibrios. *Arch. Mikrobiol.*, **98**, 101–114.

UEMATSU, T. & WAKIMOTO, S. (1970). Biological and ecological studies on *Bdellovibrio*. 1. Isolation, morphology, and parasitism of *Bdellovibrio*. *Ann. phytopathol. Soc. Japan*, **36**, 48–55.

UEMATSU, T. & WAKIMOTO, S. (1971). Biological and ecological studies on *Bdellovibrio*. 3. Growth of *B. bacteriovorus* in media composed of living and of autoclaved bacterial cells. *Ann. phytopathol. Soc. Japan*, **37**, 91–99.

VARON, M. (1974). The bdellophage three-membered parasite system. *CRC Crit. Rev. Microbiol.*, **4**, 221–241.

VARON, M., DICKBUCH, S. & SHILO, M. (1974). Isolation of host-dependent and nonparasitic mutants of the facultative parasitic *Bdellovibrio* UKi2. *J. Bact.*, **119**, 635–637.

VARON, M., DRUCKER, I. & SHILO, M. (1969). Early effects of *Bdellovibrio* infection on the synthesis of protein and RNA of host bacteria. *Biochem. biophys. Res. Comm.*, **37**, 518–525.

VARON, M. & SHILO, M. (1968). Interaction of *Bdellovibrio bacteriovorus* and host bacteria. I. Kinetic studies of attachment and invasion of *Escherichia coli* B by *Bdellovibrio bacteriovorus*. *J. Bact.*, **95**, 744–753.

(1969a). Attachment of *Bdellovibrio bacteriovorus* to cell wall mutants of *Salmonella* spp. and *Escherichia coli*. *J. Bact.*, **97**, 977–979.

(1969b). Interaction of *Bdellovibrio bacteriovorus* and host bacteria. II. Intracellular growth and development of *Bdellovibrio bacteriovorus* in liquid cultures. *J. Bact.*, **99**, 136–141.

26

Copyright © 1981 by W. H. Freeman and Company Publishers. All rights reserved
Reprinted from pages 205-213 of Symbiosis in Cell Evolution, W. H. Freeman and Company Publishers, San Francisco, 1981

MITOCHONDRIA: ACQUISITION BY WHOM?

L. Margulis

> More recently, Wallin (1922) has maintained that chondriosomes [mitochondrial] may be regarded as symbiotic bacteria whose associations with other cytoplasmic components may have arisen in the earliest stages of evolution. . . .To many, no doubt, such speculations may appear too fantastic for present mention in polite biological society; nevertheless, it is in the range of possibility that they may some day call for more serious consideration.
>
> E. B. WILSON, 1925

Do mitochondria satisfy the criteria for organelles, having originated as endosymbionts? An abundant literature on mitochondria can be brought to bear on this question (Table 8-1). Viewing mitochondria as highly integrated endosymbionts raises new questions. What were the nucleocytoplasmic hosts that acquired protomitochondria? Which aerobic bacteria became the mitochondria? Was acquisition polyphyletic? What genetic changes have integrated the mitochondrial and nucleocytoplasmic systems? Why does the mitochondrial genetic organization vary so?

Although no eukaryotes can be induced to lose mitochondria, some yeasts that obtain energy by fermentation may dedifferentiate them to protomitochondria (Figure 8-1). Although they retain the potential to redifferentiate their mitochondria, they are able to grow without respiratory metabolism. Both high concentrations of fermentable substrates, such as glucose, and the absence of oxygen repress mitochondrial development from protomitochondria, although in different ways. This unusual eukaryotic style of facultative aerobiosis permits the transmission of lethal or severely deleterious mitochondrial genes under fermenting but non-respiring conditions: such genes would not be transmitted in other eukaryotes—obligate aerobes carrying them would die. Thus, nearly all definitive information about

TABLE 8-1
The symbiotic origin of mitochondria.

What was the free-living form of the protomitochondria?	Aerobic, Gram-negative eubacteria containing the Krebs-cycle enzymes and the cytochrome system for total oxidation of carbohydrates to carbon dioxide and water. For example, *Paracoccus denitrificans* (John and Whatley, 1975a, b, 1977b; See Table 7-4) and *Bdellovibrio* (Starr, 1975a).
In what free-living host did protomitochondria become established?	Microbes able to ferment glucose to pyruvate anaerobically by the Embden-Meyerhof pathway. That is, organisms having a fermentative, heterotrophic metabolism, characteristic of the eukaryotic nucleocytoplasm. For example, *Thermoplasma acidophilum* (Searcy et al., 1978).
What environmental agents selected for and maintained the symbioses?	Atmospheric oxygen, depletion of nutrients, availability of particular food.
When did these symbioses become established?	In the Proterozoic Aeon, during or after the transition to the oxidizing atmosphere.
Is the symbiosis obligate?	Yes, in aerobic environments. Only in special cases can eukaryotic microorganisms survive dedifferentiation or loss of the organelle. For example, *Saccharomyces*, *Myxotricha*, some polymastigote flagellates (Margulis, 1976; Bloodgood and Fitzharris, 1976).
Which traits of free-living prokaryotic cells do mitochondria retain?	Circular DNA, not histone-bound (Nass, 1969; Kroon, 1966; Borst et al., 1967; Gillham, 1978; and see Table 8-4). DNA synthesized throughout life cycle and distributed equally to daughter mitochondria (Reich and Luck, 1966). "Nearest neighbor frequency" and proportion of GC nucleotide pairs higher than in eukaryotic nuclear DNA, as in bacteria (Cummins et al., 1967). "Cytoplasmic genes," non-Mendelian inheritance of mitochondria in meiotic eukaryotes (Ephrussi, 1953; Jinks, 1964; Mounolou et al., 1967; Dujon and Slonimski, 1976; Gillham, 1978). Organisms studied primarily have been yeasts, *Neurospora*, *Tetrahymena*, *Physarum*, and mammalian cells.
What mitochondrial components are coded for by nuclear genes?	Cytochrome *c* (Sherman et al., 1966), malate dehydrogenase, fumarase (Clark-Walker and Linnane, 1967). F_1 ATPase polypeptides (Bücher et al., 1976; Dujon and Slonimski, 1976).
What new syntheses have been made possible by the presence of mitochondria?	Synthesis of steroid derivatives and some polyunsaturated fatty acids (see Table 6-8), ubiquinone, and probably other secondary metabolites, especially in plants.

Why are mitochondria packaged into sperm cells, mold spores, seeds, and all other propagules?	They, or at least their genetic material, must be retained throughout life cycles because host nuclei lack the genetic potential for forming mitochondria, they are required by the hosts for ATP synthesis and calcium control.
What intracellular mechanisms maintain mitochondria in cells throughout the life cycle of the host organism?	See Table 8-6.
How can the loss of the mitochondria be induced?	Because the symbiosis is nearly always obligate, there is no way of inducing the loss of mitochondria, but in certain organisms DNA and structures can be permanently diminished by mutagens, such as acriflavin (Ephrussi, 1953; Roodyn and Wilkie, 1968), agents to which mitochondrial nucleic acid is more sensitive than host nuclear nucleic acid is (Kusel et al., 1967).
How much may mitochondria dedifferentiate?	Very little or not at all in most eukaryotes. In facultative aerobes, such as yeast and trypanosomes, they may dedifferentiate beyond the power of the electron microscope to resolve. In yeast, the entire mitochondrial differentiation system is inducible by, and sensitive to, concentration of glucose and oxygen (Roodyn and Wilkie, 1968).
What kinds of RNA do mitochondria synthesize?	Both ribosomal and transfer RNA, different from nuclear RNA (Granick and Gibor, 1967; Cummins et al., 1967; Sinclair et al., 1967; Barath and Kuntzel, 1972; Wilkie, 1973). Ribosomal RNA sequences of mitochondria are more similar to those of bacteria than to those of the cytoplasm in the same cell (Doolittle and Margulis, in press; Bücher et al., 1976).
What proteins do mitochondria synthesize?	Both ribosomal and enzyme proteins; *in vitro* protein synthesis (Kroon, 1966; Borst et al., 1967; Roodyn and Wilkie, 1968); characteristic "mitoribosomes" differing in RNA, protein, and structure from "cytoribosomes" (Linnane, 1968; Gillham, 1978). Protein synthesis sensitive to chloramphenicol, not cycloheximide (Clark-Walker and Linnane, 1967; Gillham, 1978).
Do mitochondria synthesize lipids?	Synthesis of outer mitochondrial membrane (Parsons, 1966; Clark-Walker and Linnane, 1967). The lipid fractions of daughter mitochondria receive equal amounts of labeled choline that had been fed to the parent cell (Reich and Luck, 1966; Roodyn and Wilkie, 1968).
Why are mitochondrial nucleic acids and enzymes "packaged" in all fungi, plants, and animals?	They were acquired together as an intracellular symbiont by heterotrophic ancestors.

TABLE 8-1 *(continued)*

Why are some mitochondrial functions under nuclear control?	Obligate symbionts tend to relegate redundant or dispensable metabolic functions to the host. This accounts for nuclear control of mitochondrial physiology, as in insect flight muscle (Grossman and Heitkamp, 1968), production of mitochondrial RNA polymerase by the nuclear genome (Barath and Kuntzel, 1972), and the integration of nuclear and mitochondrial functions (Gillham, 1978).
Why are there variable numbers and sizes of mitochondria per cell?	As in all host-symbiont relationships, the ratio of host to symbiont is not precisely constant.
Why are the mitochondria of plants, animals, and fungi similar?	Protomitochondria were acquired symbiotically by fermenting heterotrophs before the acquisition of photosynthetic plastids by ancestors of algae and plants; common ancestors of all eukaryotes were heterotrophic mitochondria-containing amoeboids. The similarity suggests monophyly of mitochondria, but this is not unequivocally established.
Why are mitochondrial genes in pieces?	Perhaps became so during hundreds of millions of years of endosymbiosis; or perhaps protomitochondrial genes were organized that way; not obvious.
Why are cristae found inside mitochondria?	They are adaptations that increase the surface area of oxidative enzymes, evolutionary analogues to the mesosomal membranes of many prokaryotes.
Why are yeast and regenerating rat liver mitochondria sensitive to streptomycin, chloramphenicol, spectinomycin, paromomycin, and so forth, but not to cycloheximide?	Because protomitochondria were prokaryotic cells, drugs that block prokaryote synthesis affect protein-synthesizing mitochondria. Cycloheximide blocks eukaryotic ribosomes.
If they originated as endosymbionts, why do mitochondria lack cell walls?	Walls are dispensable in the controlled, osmotically regulated, buffered conditions of the cytoplasm in which the mitochondria live; therefore, they have been selected against.
Why are there often several genomes per mitochondrion and variations on the unit size?	Mitochondria originated from intracellular symbionts that, like *Bdellovibrio*, often made several copies of their genome before forming plasma membranes and cross walls.

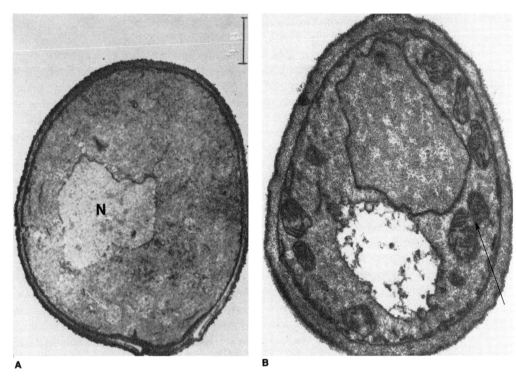

FIGURE 8-1
Yeast with and without mitochondria.
(A) Mitochondrial development is repressed in yeast grown under anaerobic conditions or in the presence of high concentrations of glucose. The capacity for the redifferentiation of the mitochondria is retained, as well as the mitochondrial DNA, even though there is no morphological evidence for the presence of mitochondria. N = nucleus. Transmission electron micrograph, ×14,000.
(B) When aerobic conditions are restored and glucose is depleted, the mitochondria reappear. Mitochondrion apparently in division can be seen in the lower right part of the cell. Transmission electron micrograph, ×20,000.
[Courtesy of A. W. Linnane.]

mitochondrial genetics and development comes from studies of yeast. No plants, animals, or fungi are known to permanently lack mitochondria. The few eukaryotes that do are protoctists—some ciliates and amoebae and a few symbionts found in the anaerobic or microaerophilic hindguts of termites and wood-eating cockroaches. Few have been cultured axenically, except for some cellulolytic forms (Yamin, 1980), and no genetic information is available about any of them.

Genuinely primitive protoeukaryotes, nucleocytoplasmic hosts that lack mitochondria, would be expected also to lack mitosis and undulipodia; unless they had already acquired respiring symbionts, they would be microaerophilic or anaerobic. Could such organisms be distinguished from other prokaryotes or from protists that lost their mitochondria when they reinvaded anaerobic environments? Perhaps not. Dennis Searcy and his co-workers have argued that *Thermoplasma acidophila,* a heat- and acid-tolerant mycoplasma is an excellent candidate for a living descendant of the protoeukaryotes (Searcy and Stein, 1980; Searcy et al., 1978). They have discovered histonelike and actinlike proteins in this wall-less prokaryote. A combination of traits—lack of cell walls, Embden-Meyerhof fermentation pathway, ease with which associations are entered, sterol requirements, acid and heat tolerance (Darland et al., 1970), as well as sensitivity to cytochalasin B (Ghosh et al., 1978)—supports Searcy's idea. It has even been suggested that *Sulfolobus,* a sulfur-oxidizing heat- and acid-resistant prokaryote related to this group is, unlike other prokaryotes, capable of particle ingestion. However, the nucleotide sequences of the 16S RNA of *Thermoplasma* and *Sulfolobus* indicate that these microbes are related more to the archaebacteria (methanogens and halobacters) than to eukaryotic nucleocytoplasm (Woese et al., 1975).

Perhaps the large anaerobic amoeba *Pelomyxa palustris* is the closest living descendant of the protoeukaryotes. *Pelomyxa* lacks mitosis and mitochondria but, unfortunately, cannot be cultured (Daniels et al., 1966; Whatley, 1975). Because it has nuclear membranes, it must be classified formally as a eukaryote, although lack of other eukaryotic features makes it anomalous. *Pelomyxa* may never have acquired mitochondria and thus a direct link to the kind of organism that acquired protomitochondria as endosymbionts. Alternatively, *Pelomyxa* may have lost its mitotic spindle centrioles, mitochondria, and the rest. It is hard to see why all of these organelles would be completely lost, but there is no rigorous basis for choosing between the alternatives.

Comparison of microaerophilic metabolism common to the cytoplasm of eukaryotes (DeDuve, 1978) with the metabolism of candidate organisms such as *Thermoplasma* would aid the search for protoeukaryotes. Some criteria that might be used are homology of DNA and ribosomal RNA sequences, organization of the genome, the presence of nucleosomes, homology of the amino-acid sequences in the proteins of the Embden-Meyerhof glycolytic pathway, presence of distinctive cytoplasmic oxidizing organelles called peroxisomes, presence in the cytoplasm and amino-acid sequence of "motile" proteins such as actins, myosins, and calcium-modulating proteins (Kretsinger, 1977), and the phospholipid-protein composition of the plasma membrane. Especially intriguing is C. DeDuve's suggestion (1969, 1978) that peroxisomes were present in microaerophilic protoeukaryotes before they acquired mitochondria.

D. F. Parsons (1966) has shown that the outer layer of the double membrane of mitochondria is more similar to endoplasmic reticular (ER) membranes than it is to inner mitochondrial membranes (Table 8-2). How do the ER and inner mitochondrial membranes compare to the plasma membranes of *Bdellovibrio, Paracoccocus,* and *Thermoplasma*? The prediction of the symbiotic theory is that the inner membranes should be homologous to those of the protomitochondria, and the outer and ER membranes to those of the protoeukaryotes.

TABLE 8-2
Mitochondrial, bacterial, and endoplasmic-reticular (ER) membranes compared.*

	MITOCHONDRIAL INNER MEMBRANE	MITOCHONDRIAL OUTER MEMBRANE	SMOOTH ER	BACTERIAL MEMBRANE†
Thickness	5.5 nm	5.5 nm	5.5 nm	—
Fine structure	"Globules," 9.0-nm membrane subunits	"Globules," 2.8-nm pits, 6.0-nm membrane subunits	"Globules," polysaccharide fringe	—
Density (g/cm^3)	1.21	1.13	1.13	—
Protein:lipid ratio	1:0.275	1:0.829	1:0.385	—
Cardiolipin (% of total phospholipids)	21.5	3.2	0.5	High
Phosphatidyl inositol (% of total phospholipids)	4.2	13.5	13.4	—
Phosphatidyl serine (% of total phospholipids)	Not detected	Not detected	4.5	—
Phosphatidyl ethanolamine	—	—	—	High
Cytochrome $a + a_3$ (μmol/g)	0.24	< 0.02	0.0	—
Cytochrome b_5 (μmol/g)	0.17	0.51	0.79	—
Permeability	Small molecules	Probably large and small molecules	Small molecules	Small molecules
Osmotic response	Responsive	Not responsive	Slightly responsive	Responsive
Effect of ATP	Contraction	No effect	No effect	—
Cholesterol	Very little	present	Large amounts	None§

*After Parsons (1966).
†From Sokatch (1969). Bacterium studied was *Serratia*
§With the exception of the mycoplasm PPLO

An important and nearly unique metabolic capability of eukaryotes is steroid synthesis. Eukaryote membranes regularly contain large quantities of these cyclic lipids. Steroid synthesis is probably the product of interaction between more than one genome—at the metabolite level (Figure 7-6). Anaerobically grown yeasts that contain dedifferentiated mitochondria must be supplied steroids, whereas these same organisms containing mature mitochondria and growing aerobically can synthesize them.

Steroids such as cholesterol are apparently indispensable for the formation of the flexible dynamic membranes of eukaryotes. In spite of their bewildering diversity, steroids can be traced back to their common biosynthetic precursor lanosterol. This compound is converted to cholesterol in animals and cycloartenol in plants. Lanosterol is formed from the universal isoprenoid precursor squalene, which is the product of a pathway from acetate through isopentenyl pyrophosphate.

Free gaseous oxygen is required to cyclize the ring in the formation of lanosterol from squalene. Atmospheric or dissolved oxygen, used in the final step of aerobic respiration, is available at mitochondrial sites.

Because they can be synthesized by the yeast nucleocytoplasm and by most prokaryotes, the diterpene alcohols probably were synthesized by protoeukaryote cells before they acquired protomitochondria. Thus they would be expected to be now under the genetic control of the nucleocytoplasm. Hence, the biosynthetic pathway from acetate to squalene is likely to be under the control of nuclear genes, whereas the enzymes in the pathway from lanosterol to cholesterol are (or were originally, at least) probably under protomitochondrial genetic control.

Complex terpenoid syntheses are talents almost exclusively of plants (Table 8-3). Biosynthesis of oils, diterpenes, and triterpenes may require the presence of more than one type of genome: the genetic control of cyclic and oxygenated terpenes, so characteristic of plants, may be due to their trigenomic nature—nucleocytoplasm, mitochondrion, and plastid. To test this concept biosynthesis must be studied during the suppression—either physiologically or by mutation of mitochondrial metabolism, plastid metabolism, and both. In yeast, mitochondrial metabolism can be entirely and reversibly turned off, and in some euglenids, photosynthetic plastid metabolism can be both irreversibly and reversibly turned off. A correlation of steroid and terpenoid synthesis with organellar functions, would help to determine both the division of organellar genomic control and the origin of these metabolic pathways. (See Table 11-2 for a summary of analogous work with carotenoids.)

TABLE 8-3
Distribution of terpenoid synthetic capability.

COMPOUNDS*	ORGANISMS
Diterpenes Phytyl, oleic acid, vitamin K	All organisms except a few bacteria
Acyclic terpenes and oxygenated derivatives Bayberry wax, oil of citronella, oil of rose, "essential oils," lycopene, rubber	Certain angiosperms
Squalene	All eukaryotes; universal precursor of steroids, through cycloartenol in plants and ergosterol in animals
Monocyclic terpenes Oil of lemon, peppermint, ginger	Certain angiosperms
Bicyclic terpenes Oil of turpentine, oil of ginger, camphor	
Sesquiterpene alcohol Farnesol	

*Terpene (by custom), C_{10}; sesquesterpenes, C_{15}; diterpenes, C_{20}; triterpenes, C_{30}; polyterpenes $(C_5H_3)_n$.

If organelles began as free-living organisms, primary metabolic pathways must have been present in both partners at the onset of the association. As natural selection reduces the inherent redundancy, the partners become progressively more interdependent. Any nutrient essential for the development and replication of a symbiont must be supplied by the environment (either the environment at large or the host cell) or it must be synthesized by the symbiont itself. A primary metabolite or enzyme made and required by both the host and symbiont will tend with time to be supplied by only one partner, usually the host. Metabolic redundancy may be selected against as long as the partners remain together, so that gene products and nutrients may be exchanged. If the host is a sexually reproducing eukaryote, it must contain precise mechanisms for the segregation of genes to daughter cells. In such a case, it is "convenient" for an endosymbiont to use the metabolic facilities of its host. If the host is a normal Mendelian diploid cell and the supply of the nutrient in question depends on the presence of dominant genes, recessive mutations may then result in the lessened reproduction, development, or function of symbionts. Of course, the phenotype of such chromosomal recessives can be expressed only if the symbiont is present in the cytoplasm—the restoration of a dominant gene that controls a metabolite required for symbiont development (for example, a required amino acid) will never be able to restore a lost symbiont. This analysis can be applied to the observation that many nuclear metabolites have crucial effects on the expression of cytoplasmically transmitted organelles.*

---

*The transmission of certain lysine auxotrophs of yeast can be interpreted in this way (Mounolou et al., 1967). For the genetic details of such effects on mitochondria and plastids, see N. W. Gillham (1978).

REFERENCES

Barath, Z., and H. Kuntzel. 1972. Cooperation of mitochondrial and nuclear genes specifying the mitochondrial genetic apparatus in *Neurospora crassa. Proc. Natl. Acad. Sci. U.S.A.* **69:**1371-1374.

Bloodgood, R. A., and T. P. Fitzharris. 1976. Specific associations of prokaryotes with symbiotic flagellate protozoa from the hindgut of the termite *Reticulitermes* and the wood-eating roach *Cryptocercus. Cytobios* **17:**103-122.

Borst, P., A. M. Kroon, and G. J. C. M. Ruttenberg. 1967. Mitochondrial DNA and other forms of cytoplasmic DNA. In *Genetic Elements: Properties and Function.* D. Shugar, ed. London: Academic Press, pp. 81-116.

Bücher, T., W. Neupert, W. Sebald, and S. Werner, eds. 1976. *Genetics and Biogenesis of Chloroplasts and Mitochondria.* Amsterdam: North-Holland.

Clark-Walker, G. C., and A. W. Linnane. 1967. The biogenesis of mitochondria in *Saccharomyces cerevisiae.* A comparison between cytoplasmic respiratory-deficient mutant yeast and chloramphenicol-inhibited wild type cells. *J. Cell Biol.* **34:**1-14.

Cummins, J. E., H. P. Rusch, and T. E. Evans. 1967. Nearest neighbor frequencies and the phylogenetic origin of mitochondrial DNA in *Physarum polycephalum. J. Mol. Biol.* **23:**281-284.

Darland, G., T. D. Brock, W. Samsenoff, and S. F. Conti. 1970. A thermophilic acidophilic mycoplasma isolated from a coal refuse pile. *Science* **170:**1416-1418.

DeDuve, C. 1969. Evolution of the peroxisome. *Ann. N.Y. Acad. Sci.* **168:** 369-381.

DeDuve, C. 1978. A re-examination of the physiological role of peroxisomes. In *Tocopherol, Oxygen and Biomembranes, Proceedings of the International Symposium.* C. DeDuve and O. Hayashi, eds. Elsevier/North-Holland Biomedical Press, pp. 351-361.

Doolittle, W. F., and L. Margulis. In press. Problems on the Prokaryotic-eukaryotic transition. *Microbiological Reviews.*

Dujon, B., and P. P. Slonimski. 1976. Mechanisms and rules for transmission, recombination and segregation of mitochondrial genes in *Saccharomyces cerevisiae.* In *Genetics and Biogenesis of Chloroplasts and Mitochondria.* T. Bücher et al., eds. Amsterdam: North-Holland, pp. 392-404.

Ephrussi, B. 1953. *Nucleo-Cytoplasmic Relations in Microorganisms.* Oxford: Oxford University Press.

Ghosh, A., J. Maniloff, and D. A. Gerling. 1978. Inhibition of mycoplasm cell division by cytochalasin B. *Cell* **13:**57-64.

Gillham, N. W. 1978. *Organelle Heredity.* New York: Raven Press.

Granick, S., and A. Gibor. 1967. The DNA of chloroplasts, mitochondria and centrioles. *Adv. Nucl. Acid Res.* **1:**143-186.

Grossman, I. W., and D. H. Heitkamp. 1968. Electron microscopy and biochemistry of developing mitochondria (abs.). *Fed. Proc.* **27:**247.

Jinks, J. L. 1964. *Extrachromosomal Inheritance.* Englewood Cliffs, N.J.: Prentice-Hall.

John, P., and F. R. Whatley. 1975a. *Paracoccus denitrificans* and the evolution of mitochondrion. *Nature* **254:**995-998.

John, P., and F. R. Whatley. 1975b. *Paracoccus denitrificans:* a present-day bacterium resembling the hypothetical free living ancestor of the mitochondrion. *Symbiosis. Proc. Soc. Exp. Biol.* **29**:39–40. Cambridge: Cambridge University Press.

Kretsinger, R. H. 1977. Evolutionary considerations of calcium pumping by biological membranes. In *The Proceedings of a Joint US-USSR Conference.* D. C. Tortenson, ed. New York: Raven Press.

Kroon, A. M. 1966. *Protein Synthesis in Mitochondria.* Ph.D. thesis, University of Amsterdam, North Holland, Netherlands.

Kusel, J. P., K. H. Moore, and M. M. Weber. 1967. Ultrastructure of *Crithidia fasiculata* and morphological changes induced by growth in acriflavin. *J. Protozool.* **14**:283–296.

Linnane, A. W. 1968. *The Nature of Mitochondrial RNA and some characteristics of the Protein-Synthesizing System of the Mitochondria Isolated from Antibiotic Sensitive and Resistant Yeasts.* E. C. Slater, J. M. Tager, S. Papa, and E. Quagliariello, eds. Bari, Italy: Adriatica Editrice.

Margulis, L. 1976. The genetic and evolutionary consequences of symbiosis. *Exp. Parasit. Rev.* **39**:277–349.

Mounolou, J. C., H. Jakob, and P. P. Slonimski. 1967. Molecular nature of hereditary cytoplasmic factors controlling gene expression in mitochondria. In *Control of Nuclear Activity.* L. Goldstein, ed. Englewood Cliffs, N.J.: Prentice-Hall, pp. 413–431.

Nass, M. M. K. 1969. Mitochondrial DNA: advances, problems and goals. *Science* **165**:25–35.

Parsons, D. F. 1966. Ultrastructure and molecular aspects of cell membranes. *Canadian Conference.* Honey Harbor, Ontario. **7**:193–246. Oxford and New York: Pergamon Press.

Reich, E., and D. J. L. Luck. 1966. Replication and inheritance of mitochondria DNA. *Proc. Natl. Acad. Sci. USA.* **55**:1600–1608.

Roodyn, D. B., and D. Wilkie. 1968. *The Biogenesis of Mitochondria.* London: Methuen.

Searcy, D. G., D. B. Stein, and G. R. Green. 1978. Phylogenetic affinities between eukaryotic cells and a thermoplastic mycoplasm. *BioSystems* **10**: 19–28.

Sinclair, J. H., B. J. Stevens, N. Gross, and M. Rabinowitz. 1967. The constant size of circular mitochondrial DNA in several organisms and different organs. *Biochim. Biophy. Acta* **145**:528–531.

Sokatch, J. R. 1969. *Bacterial Metabolism.* New York: Academic Press.

Starr, M. P. 1975a. Bdellovibrio as symbiont: the associations of Bdellovibrios with other bacteria interpreted in terms of a generalized scheme for the classifying organismic associations. *Symp. Soc. Exp. Biol.* **29**:93–124.

Whatley, F. R. 1975. Chloroplasts. In *Energy Transportation Biological Systems. CIBA Symposium* **31**:41–68. Amsterdam: Elsevier.

Wilkie, D. 1973. Cytoplasmic genetic systems of eukaryotic cells. *British Medical Bulletin* **29**:263–268.

Wilson, E. B. 1925. *The Cell in Development and Heredity.* New York: Macmillan.

Woese, C. R., G. E. Fox, L. Zablen, T. Uchida, L. Bonen, K. Pechman, B. J. Lewis, and D. Stahl. 1975. Conservation of primary structure in 16S rRNA. *Nature* **254**:83–86.

Yamin, M. A. 1980. Cellulase metabolism by the termite flagellate *Trichomitopsis termopsidis. Appl. Environ. Microbiol.* **39**:859–863.

Part VI
ORIGIN OF MOTILITY ORGANELLES

Editors' Comments
on Papers 27 through 32

27 **MAY and GOODNER**
   Cilia as Pseudo-Spirochaetes

28 **BEISSON and SONNEBORN**
   Cytoplasmic Inheritance of the Organization of the Cell Cortex in Paramecium Aurelia

29 **HEIDEMANN, SANDER, and KIRSCHNER**
   Evidence for a Functional Role of RNA in Centrioles

30 **MARGULIS and CHASE**
   Microtubules in Prokaryotes

31 **GRIMES et al.**
   Patterning and Assembly of Ciliature are Independent Processes in Hypotrich Ciliates

32 **HUANG et al.**
   Uniflagellar Mutants of Chlamydomonas: Evidence for the Role of Basal Bodies in Transmission of Positional Information

Studies of the evolutionary biochemistry and genetics of motility organelles (undulipodia) have led to exciting but inconclusive results so far. Therefore, the question of the evolutionary origin of the motility organelle (undulipodium) of eukaryotes is still at a controversial stage, and unless otherwise stated (as in Paper 30), the authors of the papers included in this section do not necessarily agree with the following editorial comments.

The motility organelles (undulipodia) of eukaryotes (the sperm tails, the cilia, etc.) are distinctive and complex organelles very different from the flagella of prokaryotes (Paper 30). For this reason we feel that the word *flagellum* should never be used for the eukaryotic organelle either as a descriptive word or as a suffix (as in dinoflagellate). Which term should be applied to the eukaryotic motility organelle? Several alternative names have been suggested, although none are yet common in the literature: *9 + 2 structure* refers to the cross-sectional arrangement of microtubules found in eukaryotic motility organelles; *undulipodium* is a revival of an old term. Throughout our comments in this book we refer to the *eukaryotic motility organelle (undulipodium),* thereby avoiding a final decision on the terminology problem but acknowledging the unacceptability of *flagellum.*

The hypothesis of the symbiotic origin of motility organelles (undulipodia) states that these organelles have their origins as formerly free-living spirochete bacteria, which established a symbiosis, conferring motility on their host.

May and Goodner noted in 1926 that eukaryotic motility organelles (undulipodia) were so similar in morphology and staining properties to spirochetes that they were often mistaken for spirochetes and sometimes named pseudo-spirochetes in the literature (Paper 27). The similarity between bacteria and organelles was among the observations leading Margulis to the idea of the symbiotic spirochete origin of motility organelles (undulipodia). The controversial hypothesis was first presented in 1967 (Paper 4) and later in Paper 30. Since 1967 biochemical and ultrastructural evidence in favor of the hypothesis has been published and is reviewed below. Although there are many in opposition to this hypothesis, there are no alternative hypotheses that take into consideration the range of evidence presented here. Therefore, this section may seem one-sided because of the lack of alternative hypothetical papers.

We concluded the introduction to this book with a list of six basic criteria for the hypothetical symbiotic origin of a given organelle. This list is applied here to current knowledge of the eukaryotic motility system:

1. a genome with characteristics more similar to that of a prokaryote than a eukaryotic genome. The base of the motility organelle (the kinetosome) contains a small amount of ribonucleic acid. Little is known about the function or coding abilities of this RNA (see Papers 29 and 32).
2. ribosomal RNA, transfer RNA, and messenger RNA more similar to that of a prokaryote than a eukaryotic cytoplasm. These are unknown in motility organelles.
3. enzymes and protein complexes more similar to prokaryotic than eukaryotic cytoplasmic analogs. The protein tubulin, which is one of the primary constituents of motility organelles, has been searched for in many prokaryotes but has not been found in any with the possible exceptions of spirochetes (Obar, 1985), spiroplasmas (long, thin helical bacteria) (Townsend, Archer, and Plaskitt, 1980),

and *Azotobacter* (Adams and Kelley, 1984). The function of tubulins or tubulin-like proteins in these bacteria is unknown.

4. an ability to replicate and a genetics separate from the nuclear genetics. The ability of the base of the motility organelle (the kinetosome) to replicate itself is well known. The structure also features a genetic system independent of the nucleus in some organisms. (Papers 28, 31, and 32).
5. a free-living prokaryotic counterpart with strong genetic, biochemical, and morphological resemblances to the organelle: an example of a similar symbiotic relationship of the counterpart organism with a host. Spirochetes have been found with tubulin-like proteins (see item 3) and with microtubules (Paper 30). Microtubules have been observed to be attached to the bacterial flagellar basal body of a spirochete suggesting a possible motility function (Hovind-Hougen, 1976). Spirochetes enter into a variety of motility symbioses with eukaryotes, for example, *Mixotricha paradoxa*, a mastigote found in the hindguts of some termites. *Mixotricha* is motile due to millions of two different types of spirochetes covering its surface and beating in synchrony (Cleveland and Grimstone, 1964).
6. an all-or-nothing phenomenon in which one either finds the organelle as a whole or does not find it at all (if it has not been acquired or was secondarily lost); one does not expect to find intermediate stages of the organelle if the organelle was acquired all at once as a symbiont. There are no obvious intermediates to the very complex eukaryotic motility organelle composed of over 200 proteins with microtubules arrayed in the distinctive 9 + 2 pattern. Typically eukaryotes either have this pattern or they do not (as in the amoebae, for example).

The tendency in close, efficient symbioses is for a streamlining of the parts, an omission of redundant functions, and a centralization of the genome (discussed in Part II). The challenge of reconstructing events in close associations has been addressed (for mitochondria) in Paper 18 and is the topic of the quotation by D. C. Smith (Paper 11) with which we dedicate our book. There is much research that remains to be done on this problem.

REFERENCES

Adams, G. M. W., and M. R. Kelley, 1984, A Tubulin-like Gene in the Bacterium *Azotobacter vinelandii*, *J. Cell Biol.* **99:**237a.

Cleveland, L. R., and A. V. Grimstone, 1964, The Fine Structure of the Flagellate *Mixtoricha Paradoxa* and Its Associated Organisms, *R. Soc. London Proc.*, series B, **157:**668–683 and unnumbered plates.

Hovind-Hougen, K., 1976, Determination by Means of Electron Microscopy of Morphology Criteria of Value for Classification of some Spirochetes, in Particular Treponemes, *Pathol. Microsc. Scand. Acta,* sec. B, **255** (suppl.):1–40.

Obar, R., 1985, Tubulin-Like Proteins from a Spirochete, Ph.D. thesis, Boston University, Boston, Mass.

Townsend, R., D. Archer, and K. Plaskitt, 1980, Purification and Preliminary Characterization of *Spiroplasma* fibrils, *J. Bacteriol.* **142:**694–700.

CILIA AS PSEUDO-SPIROCHAETES[1]

By

Henry G. May and Kenneth Goodner

It is one thing to recognize cilia when they are alive, attached to the cell and waving synchronously; but it is quite another thing to recognize them when they are detached, floating about by the thousands in a darkfield preparation or on a slide stained for spirochaetes.

Recently the author had occasion to examine material from the brains of fowls suffering from paralysis. Dark-field preparations showed curved rods, usually slightly attenuated at one end and occasionally showing an enlargement at the thicker end. The thinner end, moreover, showed a filamentous extension. These rods were easily differentiated from the ordinary pseudo-spirochaetes such as heamopodia, pieces of nerve fibers, or other tissue fibers in that they had a very sharp and clean-cut outline. This was particularly brought out in preparations stained by Fontana's method. They resembled spirochaetes in their refractive power, being practically invisible by transmitted light, but easily seen in good darkfield preparations. They also resembled spirochaetes in their staining reactions, not being stained by the ordinary dyes except after a good mordant and taking well the silver impregnation of Fontana or even that of Warthin and Starry. In all silver impregnations they appeared usually even darker than the nerve fibers that were present and could be detected even when embedded in other tissue material. Levaditi methods in tissue in bulk, however, failed to show them. With the exception of the Levaditi method, every other stain tried gave the same results as though these structures were spirochaetes.

In most cases the rods appeared to have an open spiral curvature, in some cases, however, almost regular spiral undulations were present such as one would expect in the known spirochaetes (Figs. 22, 23, 32 and 38), in other cases the spirals were quite irregular (Figs. 13 and 20).

As a rule the structures appeared rather fragile, breaking especially near the thinner end (Figs. 6 and 8); in some cases, however, stained preparations showed them to have a distinct fibrillar structure (Fig. 31). These rods were found not only in the domestic fowl but also in turkeys, pigeons, sparrows, rabbits, rats and mice.

The rods differed from all known spirocheates in that they were never observed to move and in that they were pointed only at one end, the other

[1] Contribution No. 335 From the Rhode Island Agricultural Experiment Station.

end appearing nicely rounded. They also failed to develop in cultures although they would remain intact for as many as four days in some of the media.

It was not until these structures were found in large numbers in scrapings from the mouth, trachea, ovary and oviduct that a careful study revealed their identity as cilia.

A study of the literature shows that they have already been described both as spirochaetes and as pseudo-spirochaetes without having their identity discovered. Kuhn and Steiner described them under the name of *Spirochaeta argentigensis* from animals inoculated with material from multiple sclerosis in man and others have confirmed their findings.

Adams, Blacklock and M'Cluskie also described these structures from the central nervous system of experimental animals. Although they did not recognize their true nature they classified them as pseudo-spirochaetes on account of their lack of motility and lack of growth in cultures. They were inclined to regard them as chemical secretions and attempted to dissolve them in organic solvents.

Zuelzer in discussing the results of Kunn and Steiner, is inclined to regard their spirochaetes as hematopodia or possibly other tissue fibers but does not mention the possibility of their being cilia.

It seems appropriate, therefore, to publish this short article in order to show the identity of these structures and also to figure them with sufficient clearness to aid others in avoiding a misinterpretation of these pseudo-spirochaetes.

Summary

Cilia from the brain and other tissues of birds and mammals were found to simulate spirochaetes when examined by dark-field or in stained smears.

These structures have already been described both as spirochaetes and as pseudo-spirochaetes without having their identity recognized.

Literature Cited

ADAMS. D. K., BLACKLOCK, J. S. W., AND M'CLUSKIE, J. A. W. 1925. Spirochaete-like structures in the cerebrospinal fluid of animals used for experimental purposes. J. Path. Bact., 28:115–116.

KUHN, PHILALETHES AND STEINER, GABRIEL. 1917. Ueber die Ursache der multiplen Sklerose. Med. Klin., 13:1007–1009.

1919. Forschungen ueber die Aetiologie der multiplen Sklerose. Muench. Med. Woch. No. 43:1145–1146.

ZUELZER, MARGARETE. 1925. Die Spirochaeten. Handbuch Path. Protozoen by S. von Prowazek and W. Moeller, Leipsig, No. 11:1686.

EXPLANATION OF PLATE XXV

All figures are photomicrographs and all but 1 and 2 were taken with the same camera setting and lens combination, giving a magnification of 2400 diameters. In order to facilitate measurements the marks on a very fine micrometer scale representing 10μ are given in figure 40.

Figure 1 is magnified 1200 diameters, while figure 2 has a magnification of about 3300 diameters.

FIGS. 1–13 and 41 cilia from adult fowls.
FIG. 14, embryonic nerve fibre of chick at 4 days.
FIGS. 15, 17, 18, 37 and 39 cilia from chick embryo of 4 days.
FIGS. 16 and 22–24 cilia from optic lobe of pigeon.
FIGS. 19, 25, 26, 35 and 42 cilia from the optic lobe of a sparrow.
FIGS. 20 and 21 cilia from a turkey.
FIGS. 27, 30–32 and 38 cilia from a rat.
FIGS. 28, 29, 33 and 34 cilia from a rabbit.
FIGS. 36 and 43 cilia from a mouse.

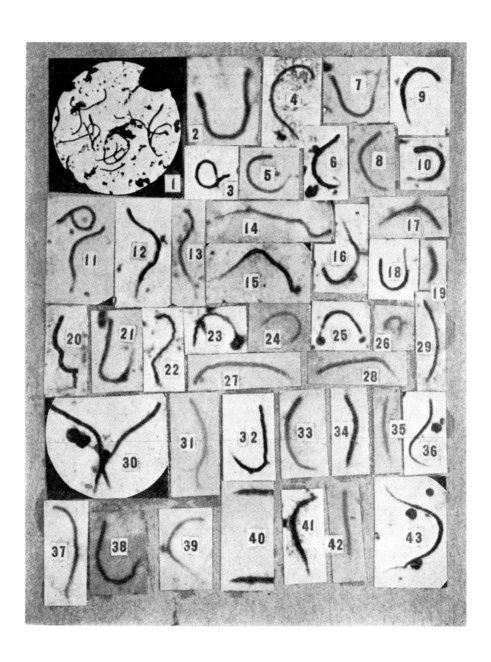

CYTOPLASMIC INHERITANCE OF THE ORGANIZATION OF THE CELL CORTEX IN PARAMECIUM AURELIA*

BY JANINE BEISSON† AND T. M. SONNEBORN

DEPARTMENT OF ZOOLOGY, INDIANA UNIVERSITY

Communicated December 14, 1964

The cortex of *Paramecium aurelia* exhibits a high degree of structural differentiation (Fig. 1). At each level of organization observable within the limits of resolution of the optical microscope, the cortical pattern is remarkably constant and reproduces faithfully through a regular cycle of changes during growth and fissions. However, this highly stable organization can be experimentally modified. Sonneborn[1] obtained several variants of the normal pattern by fusion of mates after conjugation, by partial loss of supernumerary structures, and by "cortical picking," an accident of separation after conjugation in which one of the mates acquires a piece of its partner. Sonneborn[1] showed, by all available methods of genetic analysis, that these abnormal patterns behaved like "cortical mutations." Each variation proved to be hereditary although the variants were genotypically identical to wild type, as though the existing pattern of cortical organization itself determined the pattern that arose during subsequent reproduction. This ordering of new by old cell structures, which has been called "macrocrystallinity"[2] or "cytotaxis,"[3] might be an important, although still little recognized, mechanism in cell heredity. We therefore undertook further experiments and extended the analysis to more elementary levels of structural organization.

The analysis exploited various grafts of a piece of one cell on a whole cell, using stock 51 (syngen 4) of *P. aurelia*. All of these arose from conjugating pairs which remained united by a cytoplasmic bridge instead of separating immediately after conjugation. Some pairs eventually separated spontaneously in such a way that a part of one conjugant went to the other. In other pairs most of one conjugant was cut off with a micromanipulator, the residual part persisting as a graft on the other.

The progeny of animals bearing such spontaneous and experimentally produced grafts were observed from fission to fission in order to follow the development of the grafted piece. The methods of culture and handling of the organism have been described elsewhere.[4] For detailed study of the cortical geography, specimens were prepared by a modification of the silver nitrate impregnation technique[5] and pieces of isolated cortex by a simplification of the digitonin technique.[6] Each method revealed details of organization not shown by the other.

Determination at the Level of the Repeating Unit of Cortical Structure.—Origin and nature of the mutant "twisty": This mutant was derived from an exconjugant that received a piece of its mate during spontaneous breakage of the conjugation bridge. The piece formed a little tail near the posterior end on the ventral side. Because of its posterior position, it passed to successive opisthes (posterior products of transverse fission), but became less and less obvious, disappearing by the fourth fission. Nevertheless, the clone derived from the fourth successive opisthe displayed abnormal swimming, designated as "twisty." This gives the impression of marked twisting during progression forward, but is merely an exaggeration of normal spiral progression.

Twisty swimming has a structural basis which appears from a comparison of normal (Fig. 1) with twisty (Fig. 2) cells. Unlike the regular spacing of the rows of cortical granules in normal cells (Fig. 1A,B), cells of the twisty clone (Fig. 2) show a patch of four rows bounded on one side by an unusually wide space and, on the other side, by an unusually narrow space. The intervening rows consist of cortical units with reversed polarities: instead of the normal polarity described in the Figure 1, legend the kinetosomes lie to the left (instead of the right) of the unit midline, the parasomal sacs are to the left (instead of the right) of the kinetosomes, and the kinetodesmal fibers (see Figs. 1C and 3) emerge to the left (instead of the right) of the kinetosomes and extend posteriorly (instead of anteriorly). This reversal of *both* antero-posterior and right-left polarities will be symbolized henceforth by RP; it is equivalent to 180° rotation in the plane of the body surface. Such rows of RP units constitute the first reported exception to the rule of desmodexy,[7] the rule that kinetodesmal fibers in Ciliates always lie to the right of the kinetosomes. This exception explains the abnormally wide and narrow spaces at the edges of the patch of RP rows in the twisty clone (Fig. 2, *rpr*). When a row of normal units has a row of RP units on its left, the larger clear areas of units in both rows are adjacent and the space therefore appears unusually wide. Conversely, when an RP row has a normal row on its left, the visible structures of both rows are brought into juxtaposition, leaving hardly any clear space between them.

Experimental production of cells with rows of RP cortical units: The simplest hypothesis as to the origin of the twisty clone is that the tail of the ancestral exconjugant included a piece of cortex of its mate which became assimilated in reversed orientation. This assumption is reasonable, for exconjugant pairs held together by a cytoplasmic bridge often twist into a settled heteropolar position (Fig. 4A). Moreover, it can be verified experimentally. By cutting off most of one cell from a heteropolar pair of exconjugants, one can observe that the remaining fragment becomes grafted to the intact cell and that the progeny include twisty sublines of descent. But the easiest way to obtain clones of cells with complete rows of RP cortical units was the following.

Pairs of conjugants united in heteropolar position (Fig. 4A) were isolated. Such pairs grow and divide and at each fission (Fig. 4B, C, D) give off a proter from each end while, in the central heteropolar double opisthe, the zone of union extends progressively. By about the third fission (Fig. 4D) the opisthes are fused up to or beyond the vestibular region, so that the fission furrows cut through a mixed cortex. At this stage each newly arising proter has most of its cortex derived from its "own" parent cell, but contains a little heteropolar bit in its posterior end. If such a cell is cultured, the normal process of cortical growth associated with fission brings about, in successive opisthes, the extension forward of the "grafted" rows. In the third opisthe they extend all the way to the anterior pole. All the clones derived from isolation of the third opisthe of third proters given off by a heteropolar pair displayed the twisty swimming behavior. Each clone was characterized by a distinctive number (up to 12) of rows of *RP* cortical units inserted as a single patch or as two or three patches separated by a few normal rows.[8]

Inheritance of rows of RP cortical units: The original twisty, as well as the many twisty clones obtained by the procedure just described, reproduced true to type through hundreds of fissions, dozens of autogamies, and conjugations with normals, with or without endoplasmic exchanges. Occasionally one or more RP rows could be lost from a particular subline of descent.

This persistence of rows of RP cortical units can be understood from Dippell's[9] analysis of production of new cortical units. She has shown that new structures (kinetosomes, parasomal sacs, and kinetodesmal fibers) arise within each "old" unit in locations and with polarities that are fixed in relation to the "old" components of the unit. For example, new kinetodesmal fibers grow out from kinetosomes on the same side and in the same direction as the old ones, and comparable definite spatial relationships are shown by new and old kinetosomes and parasomal sacs. Then, longitudinal growth of cortical membranes and appearance of transverse partitions yield identically organized units from each old one. In this way the units of a row are traceable to a common ancestral unit, the orientation of which is always perpetuated by the same mechanism, regardless of the orientation of the unit and of the polarity of the cell.

Determination at the Level of Differentiated Fields of Cortical Rows.—The question may now be asked: what determines the path followed by a row of cortical units and (or) what determines the differentiation of groups of rows into characteristic patterns (Fig. 1A, B): circumoral and vestibular fields around the ingestatory apparatus, right fields of rows running from anterior suture to posterior pole, left fields running from anterior to posterior sutures, and dorsal field extending from pole to pole?

First, as became evident from further observation of the twisty clones, the paths followed by the rows are *not* intrinsic properties of any of them. All *RP* rows were initially located close to the suture line on the right side. In the course of hundreds of fissions, they were observed to be located progressively further to the right, eventually reaching the left side of the cell. The mechanism of this shift relative to the oral meridian is not yet understood, but it is clearly not peculiar to *RP* rows, for the normal rows interspersed among them accompany them. As such groups

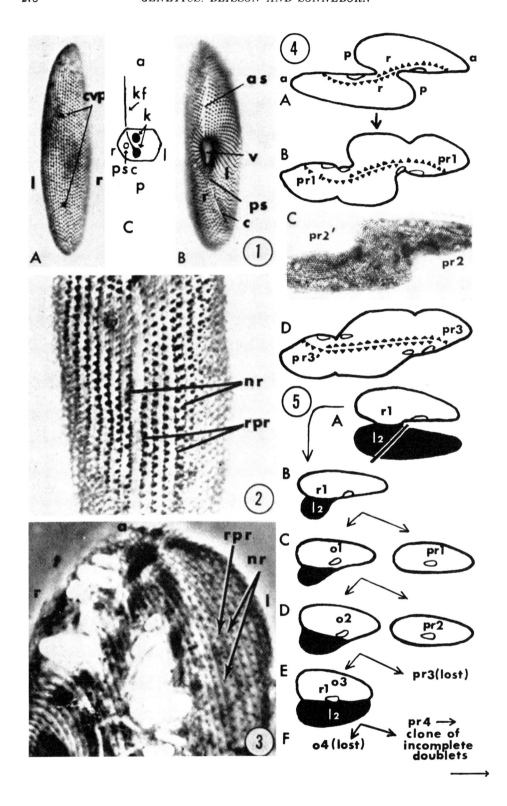

of normal and *RP* rows are followed into different fields, they fit the path characteristic of the region in which they come to lie. These paths are therefore somehow imposed on the rows by their location or their "cortical environment." This will now be demonstrated by a second type of observation.

In a pair of exconjugants united by a cytoplasmic bridge, one exconjugant was cut (Fig. 5*A*, *B*) so as to remove roughly the anterior $2/3$ of the cell including the whole ingestatory apparatus (vestibule, mouth, and gullet). What remains of the posterior left field (Fig. 5*B*, *l2*) of the amputated cell (represented in black) is fused to the posterior right field (*r1*) of the host. Conversely, 180° away (on the side not visible in the figure), what remains of the posterior right field of the amputated cell is fused to the host's posterior left field. The host's vestibule lies anterior to the graft and is surrounded on both sides by only *host* circumoral material. Figure 5*C*, *D*, *E* shows the development of the graft during the first three fissions. In the first two successive opisthes (*o1*, *2*), the cortical growth accompanying fission enlarged the graft forward to the vestibular region. Then by the third fission (Fig. 5*E*) the "daughter" vestibule, which goes to the opisthe, came to lie at the juncture of the *right* field (*r1*) of the *host* with the *left* field (*l2*) of the *graft*. At the other juncture (180° away) between host and graft cortex, the whole ingestatory apparatus being absent, no circumoral or vestibular fields developed; but at *both* junctures the other features (anterior and posterior sutures and cytopyge, see Fig. 1*B*) of the

FIG. 1.—Normal cortical geography of *Paramecium aurelia*. (*A*) Dorsal view, and (*B*) ventral view, of a silver nitrate impregnated cell. Magnification: × 416. A cell's right and left sides and anterior and posterior ends are defined in relation to the major landmarks of the ventral aspect. "To the right" means clockwise around the body from the anterior (*as*) and posterior (*ps*) sutures in (*B*) when facing the anterior end (top of fig.); the right and left sides [*r* and *l* in (*A*) and (*B*)] are to the animal's right and left of the sutures, respectively. (*C*) Diagrammatic enlargement of the unit of cortical structure, including one or two kinetosomes (*k*) or ciliary basal bodies, slightly to the right of the unit midline, one parasomal sac (*psc*) to the right of the kinetosome and anterior to it (or to the posterior kinetosome in units possessing two), and one kinetodesmal fiber (*kf*) emerging to the right and extending anteriorly (*a*) from the kinetosome (or from the posterior one when two are present). The 1 or 2 kinetosomes and parasomal sac of one unit, lying close together, appear as a single granule (silver deposit) in photographs (*A*) and (*B*), but they are partially resolvable (see Fig. 2). Kinetodesmal fibers show only in digitonin preparations (see FIG. 3).

a, anterior. *as*, anterior suture. *c*, cytopyge or cell anus. *cvp*, pores of the contractile vacuoles. *k*, kinetosome. *kf*, kinetodesmal fibre. *l*, left. *p*, posterior. *ps*, posterior suture. *psc*, parasomal sac. *r*, right. *v*, vestibule leading inward to mouth and gullet.

FIG. 2.—Part of dorsal surface of a silver impregnated typical cell of the twisty clone after about 300 cell generations. Magnification: × 1800. In the four rows with reversed polarities (*rpr*), the compound granules "point" to the observer's right, while in the normal rows (*nr*) they point to the observer's left (i.e., animal's right).

FIG. 3.—Digitonin preparation of a cell from a clone possessing a single row with reversed polarities (*rpr*). The kinetodesmal fibers in this row, unlike those in the normal rows (*nr*), emerge to the animal's left (*l*) and extend posteriorly (toward bottom of figure). Magnification: × 1800.

FIG. 4.—Origin of cells possessing some rows of cortical units with reversed polarities. (*A*) Pair of exconjugants united in heteropolar position; (*B*), (*C*), and (*D*), first, second, and third fissions of the heteropolar doublet cell. *pr1* and *pr1'*, *pr2* and *pr2'*, *pr3* and *pr3'*, are the three pairs of proters (anterior fission products) given off from the two anterior ends (*a*) of the heteropolar doublet at its first three fissions. Small ovals are the vestibules leading inward to the mouth and gullet. In the diagrams [(*A*), (*B*), and (*D*)], one row of cortical units on either side of the juncture between the two fused cells is represented by a row of triangles pointing in opposite directions. This shows the opposite polarity of the units, the points of the triangles representing the position of the parasomal sacs (see Figs. 1*C* and 2). The photograph (*C*) shows that all rows beyond the juncture are oriented like the row nearest the juncture. Note that at the third fission (*D*), the proters (*pr3* and *pr3'*) receive a mixed cortex including oppositely oriented rows on both sides of the juncture line.

FIG. 5.—Origin of clone of incomplete doublets (lacking one ingestatory apparatus) by excision of part of one conjugant of a homopolar fused pair. For description, see text.

midventral meridian were present. Between the two midventral meridians, on both sides, was a "back" with its two contractile vacuole pores (as in Fig. 1A). This incompletely double organization (lacking one ingestatory apparatus), which constitutes another type of cortical variation and has been previously described,[1] characterized the clone descended from the third posterior fission product (Fig. 5, $o3$) of the grafted cell.

The significant features of this development are (1) that the piece of grafted left field ($l2$) has acquired the vestibular and circumoral differentiations when it became associated with the host's ingestatory apparatus, and (2) that the left field of the host, originally possessing the circumoral and vestibular patterns, lost these differentiations when it developed 180° away in the absence of an ingestatory apparatus, along the right field of the graft.

In more general terms, the presence or absence of a given cortical organization (ingestatory apparatus in this case) determines whether other characteristic patterns (vestibular and circumoral fields in this case) will differentiate during growth of adjacent or nearby regions. This confirms the conclusion already drawn from the behavior of rows of RP cortical units: the cortical environment determines the path or form the rows take. Comparable observations, not reported here, on the differentiation of the cytopyge and the ingestatory apparatus show that changes in the surrounding cortical organization result in production of abnormal structures or in failure to form them altogether. In agreement, definite cortical requirements for oral differentiation have also been reported for Stentor[10] and other Ciliates.

Discussion.—The major conclusion from our results is that, at each studied level, the organization, i.e., presence, orientation, and location, of newly formed cortical structures is determined by the cortical environment existing at the time of their development. This raises the question of the mechanisms of perpetuation of both normal and variant cortical organizations and of the role of genes in this perpetuation.

The variations reported here are perpetuated during sexual and asexual reproduction in the direct cortical line of descent. As set forth previously, the variations persist in the absence of any other known differences between "mutant" and normal cells. The absence of genic and nuclear differences has been demonstrated by the genic and nuclear identities of host and graft and by breeding experiments. The absence of endoplasmic differences has been shown again by endoplasmic identity of host cell and donor cell and, in some cases, by exchange of endoplasm between normal and variants during conjugation. This, of course, does not mean that genes do not play any role in the cortical organization. They are, in all probability, necessary for production of the specific molecules that make the structures, and gene mutations are known which result in altered cortical organization. Nevertheless, the coexistence, *on the same cell*, of a normal oral meridian and of a deficient one lacking circumoral differentiations, and of normal and RP rows of cortical units, could hardly be accounted for by any noncortical difference, even by differences in local concentrations of genic products since the RP rows, for instance, persist as such as they "move" around the cell.

The mechanism of perpetuation of cortical variations is less reliable than genic mechanisms. Irreversible losses of parts and eventually of all of the variant pattern may occur with a higher order of frequency than mutations or losses of a specific gene. Nevertheless, certain variants, such as complete doublets and cells with rows of *RP* cortical units, can apparently be maintained indefinitely with periodic selection. Even those which revert 100 per cent to normal and do so rapidly (for example, incomplete doublets that revert after 30–40 cell generations) transmit the variant organization to a remarkably large number of progeny cells (10^9 to 10^{12} in the example cited). Hence, even imperfections of transmission do not greatly lessen the problem of accounting for the impressive degree of transmission that does occur in the less persistent types.

How then can the hereditary maintenance of cortical variations in the absence of relevant nuclear or endoplasmic differences be understood? Would the assumption that the cortical structures themselves, or any components of them, replicate resolve the problem? The question is most pertinent in regard to the kinetosomes which have been thought to be both self-reproducing and directive or instrumental in morphogenesis.[11] Recent evidences, at the electron microscope level, still leave the matter unsolved. However, even if they reproduced themselves or even if they contained DNA, this would not explain the hereditary maintenance of our cortical variations. These variations are not changes in the kinetosomes or associated structures *per se*. They are variations in spatial relationships among elementary structures, which themselves appear to be structurally unaltered.

The maintenance of normal or variant patterns of organization can be understood simply in terms of existing structures and the restraints they impose on the development and positioning of new structures. For example, grafted segments of rows of cortical units can be integrated only by being sandwiched between the existing rows. They are forced to grow parallel to them and their only degree of freedom is to have normal or reversed polarities. Once a row is integrated, no degree of freedom is left: as Dippell has shown,[9] new parts, as they arise, are positioned within, and in definite relations to, those of the existing unit.

We have no evidence concerning the molecular nature of such structural constraints or interactions between structures or groups of structures, but it may be worth calling attention to the absence of any necessary theoretical conflict between our results and the implications of the "self-assembly" hypothesis which ascribes the formation of organelles and subcellular structures solely to the properties of their constituent molecules. Our results merely underline a missing and probably decisive component in the system of molecular interactions which should have been recognized *a priori*, namely, the existing organization that constitutes the milieu where other subcellular organelles and structures are assembled. Newly formed molecules do not enter a vacuum, but a structured cell, the molecules of which are an essential part of the determinism for locating, orienting, and patterning new molecular formations. In this perspective it is conceivable that any subcellular structure might be affected, in the course of its formation, by changes in physical, chemical, or organizational properties of the surrounding intracellular milieu. Some changes may not allow the new structure to be formed at all; but with some probability they may induce a structural alteration in it. A modified structure might in turn, by itself constituting a change in the milieu, no longer permit the formation of the normally patterned structure, but permit reproduction of only the modified form. Such interactions at the molecular level might then result in hereditary extragenic variations.

Whatever the case may be concerning the possible mechanisms of cytoplasmic mutations, our observations on the role of existing structural patterns in the determination of new ones in the cortex of *P. aurelia* should at least focus attention on the informational potential of existing structures and stimulate explorations, at every level, of the developmental and genetic roles of cytoplasmic organization.

Summary.—Pieces of cortex of Paramecium can be grafted onto a whole cell and become integrated, yielding a modified cortical pattern which is maintained through both sexual and asexual reproduction. Two types of modified cortical patterns were studied: presence of some rows of cortical units with reversed polarities and presence or absence of whole fields of rows of cortical units. An analysis of the development and maintenance of the modified patterns shows that, both at the level of fields of rows and at the level of the cortical unit, the organization (presence, location, orientation, and shape) of newly formed structures is determined by the cortical environment existing at the time of their development.

* Contribution 756 from the Department of Zoology, Indiana University. Aided by grants from the Centre National de la Recherche Scientifique and from the Délégation Générale à la Recherche Scientifique et Technique, and by grant COO-235-13 of the Atomic Energy Commission to T. M. Sonneborn. The authors are very much indebted to Dr. Ruth V. Dippell for sharing much of the experimental work with the twisty cultures and for invaluable discussions of every aspect of the work.

† Present address: Laboratoire de Biologie Expérimentale, Faculté des Sciences, Orsay (Seine-et-Oise) France.

[1] Sonneborn, T. M., *The Nature of Biological Diversity*, ed. J. M. Allen (New York: McGraw-Hill Book Co., 1963), p. 165.

[2] Weiss, P., *The Molecular Control of Cellular Activity*, ed. J. M. Allen (New York: McGraw-Hill Book Co., 1962), p. 1.

[3] Sonneborn, T. M., these PROCEEDINGS, **51**, 915 (1964).

[4] Sonneborn, T. M., *J. Exptl. Zool.*, **113**, 87 (1950).

[5] Chatton, E., and A. Lwoff, *Compt. Rend. Soc. Biol.*, **104**, 834 (1930).

[6] Child, F., and D. Mazia, *Experientia*, **12**, 161 (1956).

[7] Chatton, E., and A. Lwoff, *op. cit.*, **118**, 1068 (1935).

[8] The possession of rows of *RP* cortical units is doubtless the basis of the twisty swimming behavior, but the relation of the orientation of cortical units to the direction of the ciliary beat is still obscure.

[9] Dippell, Ruth V., *Excerpta Medica, Intern. Congr. Ser.* 77, 16 (1964).

[10] Tartar, V., *J. Exptl. Zool.*, **144**, 187 (1960).

[11] Lwoff, A., *Problems of Morphogenesis in Ciliates* (New York: John Wiley & Sons, Inc., 1950).

Evidence for a Functional Role of RNA in Centrioles

Steven R. Heidemann, Greta Sander, and
Marc W. Kirschner
Department of Biochemical Sciences
Moffett Laboratories
Princeton University
Princeton, New Jersey 08540

Introduction

The centriole and its immediate surroundings were presumed by early cytologists to organize the mitotic apparatus in all animal cells. Recent experiments by Weisenberg and Rosenfeld (1975), McGill and Brinkley (1975), Snyder and McIntosh (1975), and Osborn and Weber (1976) have supported the view that the centriole is one of the principle microtubule-organizing centers in both mitotic and interphase animal cells. The ubiquity of the centriole, its unusual mechanism of formation in association with preexisting centrioles, and its role as a microtubule organizing center have fueled much speculation on the function and possible autonomy of this organelle (Stubblefield and Brinkley, 1967; Pickett-Heaps, 1971; Fulton, 1971; Wolfe, 1972; Hartman, 1975).

The centriole is typically a cylinder of dimensions $0.2 \times 0.5 \mu$ of nine triplet microtubules arranged around some internal structure. The identical structure has been found at the base of all flagella and cilia, and has been termed the basal body. It is probable that the basal body and centriole are functionally interchangeable as well as structurally identical (Sorokin, 1968; Fulton, 1971; Anderson and Brenner, 1971), and therefore both terms will be used to refer to the same structure. It is clear from ultrastructural studies of the basal body that it serves as a template for the growth of the 9 fold doublet array of microtubules in the cilia or flagellum (see Wolfe, 1972). What is not apparent is how the cylindrical array of microtubules in the centriole could organize the roughly spherical array of microtubules at the poles of the mitotic spindle. In fact, numerous ultrastructural studies have concluded that the microtubules of the mitotic spindle do not arise directly from the centriole, but arise from a poorly characterized dense material surrounding the centriole known as the centrosphere (Robbins and Gonatas, 1964; Porter, 1966). It has even been suggested that the centriole has no function in organizing the spindle (Pickett-Heaps, 1969; Friedlander and Wahrman, 1970), but recent experimental studies of aster induction have strongly implicated the centriole in this process (Heidemann and Kirschner, 1975; Maller et al., 1976; Weisenberg and Rosenfeld, 1975).

One approach to understanding the nature of the centriole has been to determine the macromolecular composition of isolated basal bodies (Seaman, 1960; Gould, 1975) or of basal bodies in situ by histochemical methods (Randall and Disbrey, 1965; Smith-Sonneborn and Plaut, 1967; Dippell, 1976). These studies have shown that the centriole and basal body are composed largely of tubulin, but have produced conflicting evidence for the presence of nucleic acid (reviewed by Fulton, 1971).

Recently, we reported that purified basal bodies from Chlamydomonas and Tetrahymena induce the formation of asters when injected into unfertilized eggs of the African frog, Xenopus laevis (Heidemann and Kirschner, 1975). The aster-inducing activity seemed to be unique to basal body. Maller et al. (1976), using the same system, assayed parts of sea urchin sperm for parthenogenetic activity and found the centriole to be the active component of sperm for inducing both spindle and cleavage formation. Using the Xenopus egg system as a functional assay for the aster-inducing activity, we have tried to determine which components of the basal body are required for aster induction. We report here on the effect of various chemical and enzy-

338 S. R. Heidemann, G. Sander, and M. W. Kirschner

matic treatments on the aster-inducing activity of basal bodies and suggest a possible role for RNA in centriolar function.

Results

Preparation of Basal Bodies

In the experiments reported here, purified basal bodies from Chlamydomonas reinhardtii and Tetrahymena pyriformis were studied as aster-inducing agents. Basal bodies were purified from Chlamydomonas by the method of Snell et al. (1974). Quantitative electron microscopy of such preparations (Figure 1) indicate that basal bodies are the major structures in these preparations with small spherical vesicles (0.1–0.3 μ) as the principle contaminant. These basal bodies have the typical paired structure of Chlamydomonas basal bodies in situ.

We have developed a new procedure for the purification of basal bodies from Tetrahymena using the deciliation technique of Thompson, Baugh, and Walker, (1974) and the procedure for cell disruption of Wolfe (1970). Earlier procedures for isolating basal bodies from Tetrahymena were time-consuming, produced poor yields, involved partial fixation of the cells, and used harsh procedures for disruption (Rubin and Cunningham, 1973). Our procedure is easier and more rapid than existing methods for basal body isolation, and produces preparations whose purity and morphology are similar to those prepared by previous methods (Figure 2). The principle contaminants of the Tetrahymena preparations are membranous skirts surrounding the proximal end of the basal bodies. These skirts seem to be a result of the deciliation procedure. All deciliation procedures involving calcium "mobilization" (Watson and Hopkins, 1962; Rosenbaum and Carlson, 1969; Thompson et al. 1974) left identical skirts around the isolated basal body. Other deciliation procedures were either ineffective on Tetrahymena, such as pH shock of Witman et al. (1972), or left behind a number of contaminating cilia, although they produced basal bodies free of membranous skirts (Child, 1959). Basal bodies isolated by the above procedure were as effective at inducing the formation of asters in unfertilized eggs as previous preparations (Heidemann and Kirschner, 1975).

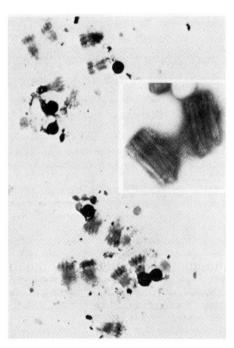

Figure 1. Quantitative Electron Micrograph of a Basal Body Preparation from Chlamydomonas reinhardtii

This quantitative electron micrograph shows the typical degree of purity of Chlamydomonas basal bodies. The very dark circles are vinyl beads 0.312 μ in diameter. The less densely stained circles are the small spherical vesicles which are the major contaminant in Chlamydomona preparations. The paired structure of the basal bodies is shown at higher magnification in the inset.

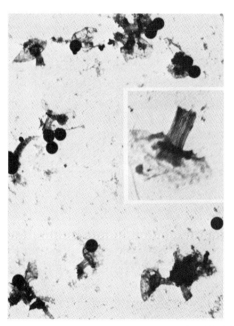

Figure 2. Quantitative Electron Micrograph of a Basal Body Preparation from Tetrahymena pyriformis

This micrograph shows the purity of basal bodies obtained from Tetrahymena. The dark circles are vinyl beads 0.312 μ in diameter. The membranous skirts surrounding the basal body are typical of Tetrahymena basal bodies obtained by this procedure and one is shown at higher magnification in the inset.

Assay for Aster-Inducing Activity of Basal Bodies

Aster-inducing activity was assayed using eggs which had been stratified by centrifugation at 2500 × g for 30 min on a cushion of 60% Ficoll in Barth's saline (Masui, 1972). These eggs stratified into four layers: yolk platelets, pigment granules, clear cytoplasm, and an oil cap (Figure 3). Basal bodies were injected into the cytoplasmic layer of these eggs, the only layer in which asters were found to form. Asters induced in stratified eggs were more symmetric and more delicately fibrous than those induced by an equivalent injection into undisturbed eggs (Figure 4).

The number of asters induced in an egg appears to depend in a poorly understood way on the quality of eggs from a given female, the concentration and activity of the basal bodies injected, whether there is leakage during the particular injection, and possibly other undefined variables. These same factors probably account for the fraction of eggs injected with untreated basal bodies which do not produce asters. Quantitative differences in aster formation are therefore hard to interpret; however, our experience with the assay indicates that treated basal bodies reproducibly do or do not induce asters on an all or nothing basis.

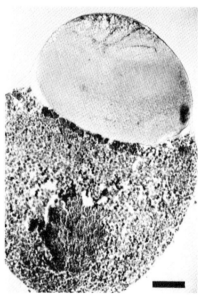

Figure 3. Light Microscope Section through a Stratified Egg of Xenopus laevis

Unfertilized eggs stratified by centrifugation on a cushion of Ficoll show the four layers seen in this photograph. The dark granules which compose the bottom half of the egg are yolk platelets; the pigment layer is seen as an area of increased density between the yolk platelets and the lightly stained cytoplasm. The area of lowest density near the top of the photograph above the cytoplasm is the oil cap. Bar = 100 μ.

The Effect of Enzyme Treatments on the Aster-Inducing Activity of Basal Bodies

The effects of various enzyme treatments on the aster-inducing activity of the basal body are summarized in Table 1. Digestion of basal bodies with DNAase I, alkaline phosphotase, wheat germ lipase, and lysozyme had no effect on aster formation. However, digestion with proteolytic enzymes and ribonuclease eliminated aster-inducing activity. Boiling basal bodies for 5 min also completely destroyed their activity.

Proteolytic enzymes affected both the structure and the activity of basal bodies. Treatment of basal bodies with trypsin or S. aureus protease at 5 μg/ml or 10 μg/ml resulted in substantial degradation of basal bodies as shown in Figure 5a. These degraded basal bodies were ineffective aster inducers. Lower concentrations of trypsin, however, affected the basal body to a much lesser extent, and aster-inducing activity persisted. The only apparent ultrastructural effect of treating Chlamydomonas basal bodies with 0.5 μg/ml trypsin was to digest the cross-linking elements between the two-paired basal bodies giving rise to singlets as seen in Figure 5b. These basal bodies were still capable of inducing asters in Xenopus eggs. Figure 5c shows that no ultrastructural changes were observed in basal bodies treated with 0.1 μg/ml trypsin. These preparations induced asters to the same extent as untreated basal bodies.

On the other hand, ribonuclease-treated basal bodies showed no alterations in overall structure at any of the concentrations used. Ribonuclease A-treated basal bodies at 10 μg/ml retained the array of nine triplet microtubules as well as internal structures such as the cartwheel and were generally indistinguishable from untreated basal bodies (Figure 6).

The effect of RNAase on aster-inducing activity was quite pronounced, as shown in Figures 7 and 8, and highly reproducible. In eleven separate experiments, ribonuclease A-treated (5 μg/ml) basal bodies failed to induce asters in approximately 100 eggs from 20 females. In one exceptional egg in one experiment, basal bodies treated with ribonuclease A produced two asters. In approximately 100 eggs injected with untreated basal bodies, some 2000 asters were formed. The aster-inducing activity of the basal body was sensitive to low concentrations (0.5 μg/ml) of ribonuclease A, under rather mild conditions of treatment, room temperature for 30 min. Basal bodies treated with lower concentrations of RNAase A, 0.1 μg/ml and 0.01 μg/ml, retain some ability to induce asters in Xenopus eggs. Control basal bodies allowed to remain at room temperature for 30 min without treatment lost none of their effectiveness.

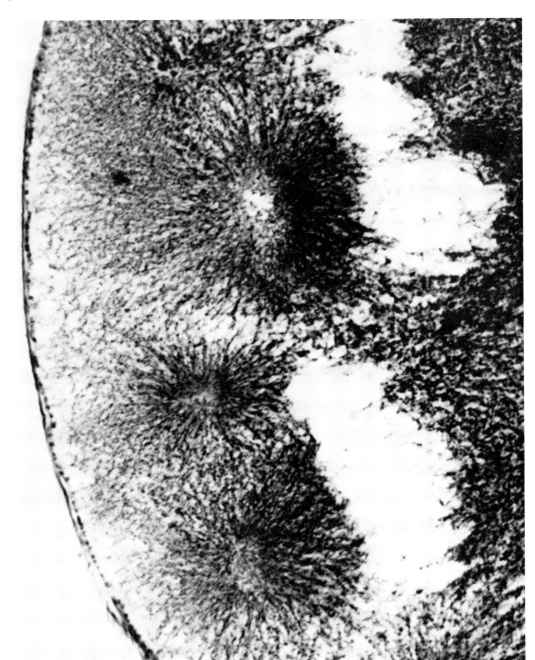

Figure 4. Induced Asters Seen in a Section through the Cytoplasmic Layer of a Stratified Egg
Typical asters induced by the injection of basal bodies into the cytoplasmic layer of a stratified egg. Bar = 30 μ.

Table 1. Effect of Enzymatic Treatments on Aster-Inducing Activity of Basal Bodies

Treatment of Basal Bodies	Number of Eggs Injected	Number Showing Asters	Percentage Showing Asters
Untreated	97	77	79%
10 μg/ml DNAase I	18	16	88%
10 μg/ml Lysozyme	6	6	100%
5 μg/ml Alkaline Phosphatase	8	6	75%
5 μg/ml Wheat Germ Lipase	8	7	87%
Boiling	12	0	0%
Trypsin			
10 μg/ml	6	0	0%
5 μg/ml	18	0	0%
0.5 μg/ml	8	7	87%
0.1 μg/ml	8	8	100%
5 μg/ml S. aureus Protease	8	0	0%
RNAase A			
10 μg/ml	24	0	0%
5 μg/ml	66	1	1.5%
5 μg/ml Unsedimented	6	0	0%
1 μg/ml	37	0	0%
0.5 μg/ml	8	0	0%
0.1 μg/ml	8	3	37%
0.01 μg/ml	8	8	100%
RNAase T1			
10 μg/ml	6	0	0%
5 μg/ml	15	0	0%
5 μg/ml S1 Nuclease	8	0	0%
5 μg/ml Boiled S1 Nuclease	8	4	50%
5 μg/ml Diethylpyrocarbonate-Treated RNAase A	8	5	67%
0.1 μg RNAase A Injected Followed by Untreated Basal Bodies	6	4	66%
Untreated Basal Bodies + 100 μg/ml Yeast RNA	8	7	87%
10 μg/ml RNAase A Basal Bodies + 100 μg/ml Yeast RNA	8	0	0%
1 μg/ml RNAase A Basal Bodies + 100 μg/ml Yeast RNA	8	0	0%
100 μg/ml Yeast RNA Injected	8	0	0%

Four lots of ribonuclease A from two sources, Sigma Chemical (Type 1A) and Worthington (Type R and Type RAF), proved equally effective at elimination of the aster-inducing activity of basal bodies. Boiling 1 mg/ml stock solutions of these enzymes for 3 min had no effect on their ability to eliminate aster-inducing activity. The possibility that RNAase A was acting by some mechanism based on its physical properties rather than its enzymatic activity was tested by the following experiments. Ribonuclease A which was inactivated with diethylpyrocarbonate was added to basal body preparations. The inactivated enzyme had no effect on aster-inducing activity. Lysozyme, like RNAase A, is a basic protein. Treatment of basal bodies with 10 μg/ml of lysozyme had no effect on aster induction. Another ribonuclease, ribonuclease T1, which is an acidic protein, was found to eliminate the aster-inducing activity of basal bodies at 5 μg/ml. A third nuclease, S1 nuclease, specific for single-stranded nucleic acids, also eliminated aster formation. Heat-inactivated S1 nuclease did not eliminate the ability of basal bodies to induce asters.

We also studied the effect of injected RNAase A on the egg cytoplasm. An aliquot of 0.1 μg of RNAase A was injected into the egg cytoplasm giving an internal concentration of approximately 50 μg/ml, followed 5–10 min later by an injection of untreated basal bodies. The injection of ribonuclease A into the egg did not eliminate aster formation, but seemed to inhibit it to some extent.

It was possible that during treatment with RNAase, some active component was released by digestion. When the basal bodies were sedimented, the activity would have then been discarded with the supernatant. To test this possibility, basal bodies treated with 5 μg/ml RNAase A were injected into stratified eggs without pelleting. No asters were formed in eggs injected with this material. This amount of ribonuclease injected into

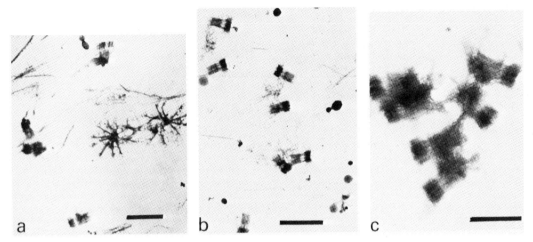

Figure 5. The Effect of Trypsin on Basal Body Structure
(a) Shows the effect of 5 μg/ml of trypsin on Chlamydomonas basal bodies. Bar = 1 μ.
(b) Shows the effect of 0.5 μg/ml of trypsin on the same preparation as in (a). Note that the basal bodies are no longer linked as pairs. Only singlet basal bodies were seen under these conditions. Bar = 1 μ.
(c) Shows the effect of 0.1 μg/ml of trypsin on this basal body preparation. The basal bodies remain paired and no other structural change is apparent. Bar = 0.5 μ.

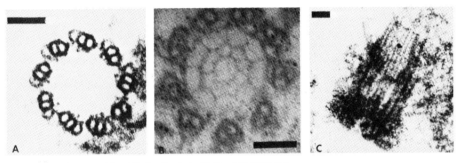

Figure 6. Thin Sections of Basal Bodies Treated with 10 μg/ml RNAase A
(A) section through Chlamydomonas basal body showing intact triplet microtubules. Bar = 0.1 μ.
(B) section through Tetrahymena basal body showing internal "cartwheel" structure. Bar = 0.1 μ.
(C) longitudinal glancing section through Chlamydomonas basal body. Bar = 0.1 μ.

the egg does not itself inhibit aster formation by subsequently injected basal bodies as described above. We wished to see if yeast RNA would complement the loss of RNA from basal body preparations treated with RNAase. To 50 μl of RNAase-treated, pelleted basal bodies were added 5 μg of yeast RNA, and the mixture was injected into Xenopus eggs. No asters formed.

Effect of Ribonuclease on the Ability of Basal Bodies to Nucleate Microtubule Assembly in Vitro

As shown above, the aster-inducing activity of the basal body injected into Xenopus eggs was eliminated by ribonuclease digestion. We wished to determine whether another microtubule-organizing function of the basal body more analogous to its role in flagellar growth, that of a nucleation site for microtubule growth in vitro (Snell et al. 1974), was affected by ribonuclease digestion. Purified basal bodies from Chlamydomonas were treated with 5 μg/ml RNAase A at room temperature for 30 min or allowed to remain untreated at room temperature for 30 min. These basal bodies were used as initiating sites for tubulin preparations which show poor spontaneous assembly into microtubules but which will readily elongate existing structures (Allen and Borisy, 1974; Binder, Dentler, and Rosenbaum, 1975). We used both high speed supernatants of depolymerized microtubules and phosphocellulose-purified tubulin with small amounts of added tau protein as described by Witman et al. (1976). As shown in Figure 9, RNAase-treated basal bodies retained their capacity to serve as seeds for the growth of microtubules. There was no apparent difference between untreated and RNAase A-treated basal bodies in their capacity to serve as initiating sites for microtubule assembly in vitro.

Presence of RNA in Basal Bodies

We wished to show whether purified basal bodies, whose aster-inducing activity is RNAase-sensitive, actually contain RNA. Tetrahymena and Chlamydomonas were labeled with $^{32}PO_4$ or ^3H-uridine as described in Experimental Procedures. Basal bodies were then purified by the procedures above. To fractionate further any possible RNA contamination from the basal body, preparations were banded to equilibrium in a density gradient of 0–75% sucrose in 99% D_2O, which gives a density range of 1.13–1.35 g/cc. The gradient was centrifuged at 100,000 × g for 2 hr at 4°C. The position of the basal body band remained unchanged at 1.30 g/cc from 1.5–2.5 hr for Tetrahymena basal bodies and at 1.33 g/cc for Chlamydomonas. Small particulate contaminants remained at or near the gradient-sample interface. The few contaminating axoneme fragments seemed to band a slightly lower density than basal bodies, at approximately 1.28 g/cc. In addition to giving the proper density range, D_2O sucrose did not affect the basal body structure and might even be expected to stabilize the microtubule structure (Inoué, 1964; Olmsted and Borisy, 1973). Renograffin (Tamir and Gilvarg, 1966), which gave the proper density range, disrupted the basal bodies.

Ten fractions were collected from these gradients, and each was analyzed for aster-inducing activity, number of basal bodies, and ribonuclease-sensitive, TCA-precipitable counts. The result of such an experiment with Tetrahymena basal bodies is depicted in Figure 10. Similar results were obtained with Chlamydomonas basal bodies and Tetrahymena labeled with ^3H-uridine. The ribonuclease-sensitive counts follow quite precisely the number of basal bodies in the fraction. A plot of total $^{32}PO_4$ counts in this gradient was similar to that of the ribonuclease-sensitive counts, but about twice as great. The difference was released by trypsin. The amount of RNA per basal body, based on the data obtained from the gradient and on the specific activity of the $^{32}PO_4$ in the culture medium, ranged from 2×10^{-16} g RNA per basal body to 8×10^{-16} g RNA per basal body for both Tetrahymena and Chlamydomonas. A micro-injection assay of these fractions repeatedly indicated that only fractions containing basal bodies induced the formation of asters in stratified eggs.

Discussion

We report here on an assay for centriolar function based on the capacity of purified basal bodies to induce the formation of asters when injected into unfertilized eggs of Xenopus laevis. The use of eggs stratified by centrifugation to remove yolk and lipid from the cytoplasm improves the clarity of the assay. Asters induced in stratified eggs are larger, more delicately fibrous, and more symmetric than those induced in unstratified eggs. The increased clarity of the asters in stratified eggs suggests that in unstratified eggs, the yolk platelets can obscure the radial structure at the limits of the aster, retard fixation, or actually limit the extent of microtubule growth. The use of stratified eggs also eliminates the problem of interpreting yolk-free, aster-like areas, which we surmised in a previous paper (Heidemann and Kirschner, 1975) were caused by poor fixation in the interior of the egg. All cytoplasmic patterns in stratified eggs can be clearly distinguished as being astral or not.

The Xenopus egg offers a clear-cut assay for the aster-inducing function of the centriole or basal body. No other microtubule structure, such as flagella, flagellar axonemes, cilia, or brain microtubules, induced asters in Xenopus eggs (Heidemann and Kirschner, 1975). However, flagellar microtubules from sea urchin sperm tails have been used to initiate cleavage in fish eggs (Iwamatsu, Miki-

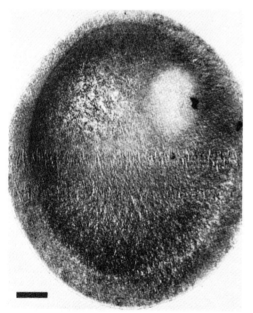

Figure 7. Section through the Cytoplasmic Layer of a Stratified Egg Injected with Basal Bodies Treated with 5 μg/ml RNAase A
No astral structures are observed. Bar = 100 μ.

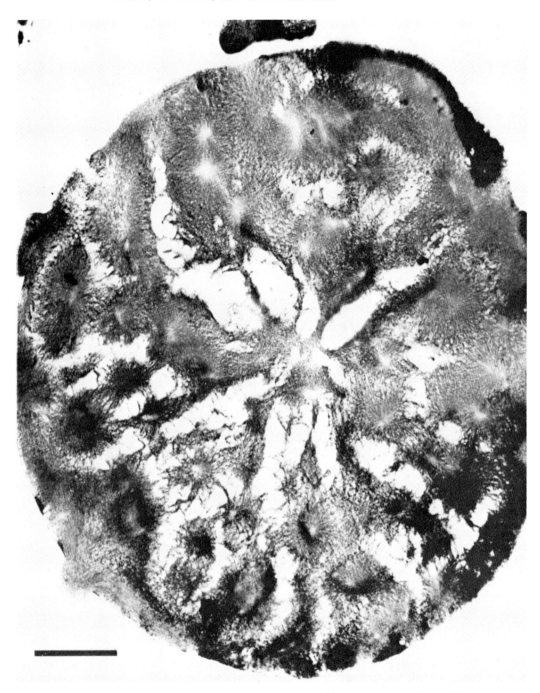

Figure 8. Section through Cytoplasmic Layer of a Stratified Egg Injected with Basal Bodies Treated with 10 µg/ml DNAase I

This micrograph shows the appearance of the cytoplasmic layer of a stratified egg injected with active basal bodies in this case treated with DNAase I. The appearance of eggs injected with untreated basal bodies is identical. The open space within the cytoplasmic layer is a characteristic of aster formation and seems to be due to the concentration of cytoplasm around the asters producing the observed gaps. This is an egg from the same female injected with the same basal body preparation as that shown in Figure 7.

Functional Role of RNA in Centrioles 345

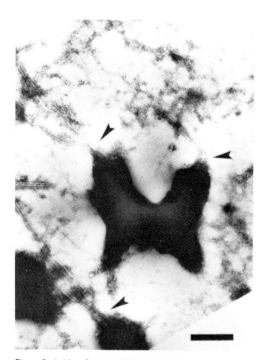

Figure 9. In Vitro Growth of Microtubules off Basal Bodies after Treatment with 10 μg/ml RNAase A

Basal bodies were treated with 10 μg/ml RNAase A for 30 min at 23°C. These basal bodies were then mixed with high speed supernatant tubulin such that the final tubulin concentration was 1.5 mg/ml. GTP was added to 1 mM, and the mixture was incubated for 15 min at 37°C. Electron microscope grids were prepared by staining with 2% uranyl acetate. Microtubules can be clearly seen growing from the distal end of these basal bodies (arrows). To visualize the microtubules, it was necessary to overexpose the basal body. Bar = 0.3 μ.

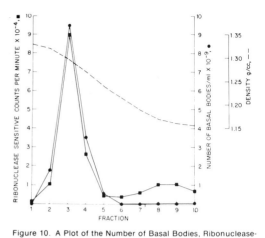

Figure 10. A Plot of the Number of Basal Bodies, Ribonuclease-Sensitive ^{32}P Counts, and Solvent Density as a Function of Fraction Number in a D_2O-sucrose Equilibrium Density Gradient

A 5 ml, linear, 0–75% sucrose in D_2O gradient was overlaid with 0.3 ml of a Tetrahymena basal body preparation. The gradient was spun at 100,000 × g for 2 hr at 4°C. Ten fractions were collected and analysed for density (----), number of basal bodies by quantitative electron microscopy (●——●), and ^{32}P counts sensitive to RNAase A (■——■).

Nomura, and Ohta, 1976) and in mutant axolotl eggs (Raff, Brothers, and Raff, 1976), although in these experiments, no attempt was made to assay for aster formation. The evidence suggests that these cleavage-initiating activities may be different from the aster induction which is assayed here. In the experiments of Iwamatsu et al. (1976), 6% bovine serum albumin initiated an appreciable amount of cleavage, whereas injections of material not containing centrioles into hundreds of Xenopus eggs never once produced a single aster. In any case, fractionated sea urchin sperm has been assayed for its ability to induce parthenogenesis, cleavage, and asters in unfertilized Xenopus eggs by Maller et al. (1976). These investigators found that sperm flagellar fractions were essentially inactive at inducing parthenogenesis, cleavage, or asters, and that these activities were associated with the sperm centrioles.

For assays reported here, basal bodies were isolated from either the green algae Chlamydomonas reinhardtii or the ciliate protozoan Tetrahymena pyriformis. Basal bodies from Chlamydomonas were isolated by the complete procedure of Snell et al. (1974). In our previous report, too few basal bodies remained to carry out the final step of purification. The overall yield of basal bodies is about 10–15% when the final number of basal body pairs, as determined by quantitative electron microscopy, is compared to the initial number of cells. The preparations contain, in addition to basal bodies, about 25% by number small spherical particles of unknown composition which are not sensitive to trypsin or ribonuclease. We have also presented a new procedure for purifying basal bodies from Tetrahymena. It is more rapid, less harsh, and gives a higher yield of basal bodies than previous methods. The overall yield is approximately 25% if it is assumed that Tetrahymena has 2000 basal bodies per cell. Tetrahymena basal body preparations contain a few ciliary axonemes, and the basal bodies have attached to them membranous skirts, neither of which are found in basal bodies isolated from Chlamydomonas.

As shown in Table 1, the effects of various enzymatic treatments on the aster-inducing activity of basal bodies fall into two categories: either the treatment had no effect or the treatment completely eliminated aster formation. In a typical single experiment, untreated basal bodies induced asters in all eight eggs sectioned, producing a total of approximately 250 asters. Basal bodies treated with 5 µg/ml of RNAase produced no asters in eight eggs. The only enzymatic treatments which eliminated aster-inducing activity were proteolytic or ribonucleolytic digestion. Treatment of basal bodies with DNAase, alkaline phosphatase, wheat germ lipase, and lysozyme had no effect on aster formation.

The effect of proteolytic enzymes on the aster-inducing activity of the basal body directly correlates with observable structural changes caused by the treatment. At 5 µg/ml of trypsin or of S. aureus protease, aster-inducing activity was eliminated, and the basic architecture of the basal body was disrupted as shown in Figure 5a. At 0.5 µg/ml of trypsin, the cross-links between paired basal bodies in Chlamydomonas were completely digested, but no effects on basal body structure proper were observed (Figure 5b). At that concentration, there was little or no effect on aster-inducing activity. The simplest interpretation of these experiments is that the structural integrity of the basal body is required for its activity to induce asters. This may be because some specific protein which takes part in aster induction is destroyed by proteolytic digestion. It is also possible, however, that some nonprotein component, which is packaged in the basal body, is released or inactivated by the breakdown in the basal body architecture.

The most striking and unexpected conclusion from the experiments is that the aster-inducing activity of basal bodies is sensitive to ribonuclease. Treatment of basal bodies with 0.5 µg/ml or more of RNAase A completely eliminates aster-inducing activity. At 0.1 µg/ml, the aster-inducing activity is retained, although qualitatively there seem to be fewer asters formed. At 0.01 µg/ml of RNAase, there is no effect. The loss of aster-inducing activity was not accompanied by any observable change in the basal body structure, as shown by Figure 6. Treatment of basal bodies with even 10 µg/ml RNAase also had no observable effect on the ultrastructure. The simplest interpretation of these experiments is that RNA is associated with the basal body and is required for aster-inducing activity. However, sensitivity to treatment, even with low concentrations of pancreatic ribonuclease A, could have other explanations.

The most obvious alternative explanation is that the sensitivity to ribonuclease is due to the physical rather than the enzymatic properties of pancreatic ribonuclease A. Pancreatic ribonuclease is a basic protein (pI = 7.5) which could bind to the acidic tubulin subunits (pI = 5.3) of the basal body and inhibit its function. This interpretation is refuted by several experiments. Treatment with lysozyme, a much more basic protein (pI = 11), at a concentration 20 times the minimum effective ribonuclease concentration had no effect on the capacity of basal bodies to induce asters. Pancreatic RNAase A which had been inactivated with diethylpyrocarbonate did not inhibit basal body activity. Most importantly, treatment with two other unrelated ribonucleases, ribonuclease T1 and S1 nuclease, completely inhibited aster-inducing activity. Ribonuclease T1 is a highly acidic protein (pI = 2).

Another alternative explanation is that some contaminating enzymatic activity common to all preparations of ribonuclease is responsible for the elimination of aster-inducing action by basal bodies. Although it is impossible to refute this argument completely, several experiments render it improbable. First, ribonuclease A, ribonuclease T1, and S1 nuclease are obtained from quite different biological sources and are warranted to be free of DNAase and protease contamination, yet all eliminated aster-inducing activity at low concentrations. Second, ribonuclease A solutions which had been boiled for 3 min retained their activity. Many enzymatic activities, unlike RNAase A, are destroyed by boiling. Since proteolytic enzymes were the only other class of enzymes effective at eliminating aster-inducing activity, the most probable contaminant which would be effective is proteolytic activity. However, purified trypsin at a concentration of 0.5 µg/ml is ineffective at eliminating aster-inducing activity, while RNAase A at the same concentration is effective. Moreover, effective concentrations of trypsin and of S. aureus protease cause disintegration of basal body structure, while no such disintegration is observed with any of the ribonuclease treatments. Finally, the basal bodies show ultrastructural effects of the proteases (digestion of cross-links) at concentrations which are too low to inhibit aster-forming activity, whereas basal bodies treated with 10 µg/ml of RNAase (20 times the minimum effective concentration) show no discernable ultrastructural changes. These data indicate that the activity of the ribonuclease preparations is almost certainly the result of their ability to digest RNA and not a general physical interaction or some contaminating proteolytic activity.

It is possible, however, that some ribonuclease might be carried along in the injection procedure and have its effect on the cytoplasm of the egg rather than on the basal body. Although treated basal bodies were washed and pelleted before injection, some residual ribonuclease activity might remain and be injected into the egg. However, when a much larger amount of ribonuclease is first injected into the egg to give a final internal concentration of 50 µg/ml and is followed by an injection of basal bodies, there is still some aster formation. It is difficult to imagine why ribonuclease did not completely eliminate aster formation in the egg. It may be that the RNAase binds to highly acidic yolk protein or is inactivated in some other way in vivo. In any case, small quantities of ribonuclease should certainly have no effect on the capacity of the egg cytoplasm to respond to injected basal bodies.

Although ribonuclease abolishes the aster-inducing activity of the basal body, it has no effect on the capacity of the basal body to serve as a nucleation site for the assembly of microtubules in vitro. As shown in Figure 9, RNAase-treated basal bodies serve as nuclei for the growth of microtubules from tubulin preparations which have reduced ability to assemble spontaneously. The two different microtubule-organizing functions of this organelle thus appear to be separable. The "centriole" function where it acts as a nucleus for aster formation is RNAase-sensitive. The "basal body" function where it acts as a simple template for microtubule growth, analogous to flagellar growth, is not sensitive to RNAase, as is to be expected since microtubule assembly in vitro is not inhibited by RNAase (G. Sander and M. W. Kirschner, unpublished data).

Analysis and characterization of the RNA species present in purified basal bodies or centrioles is beyond the scope of the present work. However, for the interpretation of the experiments presented above, it would be important to know whether purified basal bodies in fact contain RNA. As shown in Figure 10, Tetrahymena basal bodies labeled with $^{32}PO_4$ for more than ten generations contain ribonuclease-sensitive $^{32}PO_4$ which fractionates with the basal bodies on an equilibrium density gradient. Similar banding patterns are obtained with ^{32}P-labeled Chlamydomonas and ^3H–uridine-labeled Tetrahymena. The aster-inducing activity, the number of basal bodies, and the ribonuclease-sensitive counts co-band at 1.3 g/cc. The amount of RNA calculated per basal body is about 5×10^{-16} g for both Tetrahymena and Chlamydomonas. This agrees fairly well with the estimate of 2×10^{-16} g of nucleic acid per basal body by Randall and Disbray (1965) from acridine staining of Tetrahymena pellicle preparations.

The subject of nucleic acids and basal bodies has a confusing and, at present, an unresolved history. The impetus for looking for nucleic acids, principally DNA, has come from the curious way in which most centrioles seem to arise from preexisting centrioles or from intermediate structures (Sorokin, 1968; Lwoff, 1950). For example, Sagan (1967) included centrioles along with mitochondria and chloroplasts as organelles which probably arose endosymbiotically. As was mentioned, Randall and Disbrey (1965) used acridine staining and thymidine incorporation to show histochemically that Tetrahymena basal bodies contain DNA. This was extended and confirmed by Smith-Sonneborn and Plaut (1967) on Paramecium. However, chemical studies of subcellular fractions containing basal bodies showed little if any DNA (Argetsinger, 1965; Hoffman, 1965), and more recent studies (Flavell and Jones, 1971; Younger et al. 1972) attributed the small remaining DNA to contamination from chromatin or mitochondria. Hartman, Puma, and Gurney (1974) showed that acridine staining of pellicle fractions was RNAase-sensitive and not DNAase-sensitive, although this contradicted the experiments of Randall and Disbrey (1965), who arrived at the opposite conclusions. Hartman et al. (1974) have also shown that some of the RNA in the Tetrahymena pellicle would not be competed by ribosomal RNA in hybridization experiments against total cell DNA. Dippel (1976) has shown from ultrastructural studies that part of the inner core of Paramecium basal bodies is digested by RNAase. Stubblefield and Brinkley (1967) found a RNAase-sensitive structure at the foot of the A tubule in centrioles.

Our present analysis of purified basal bodies supports experiments which show RNA in basal bodies. It is interesting that we find that the amount of RNA is approximately the same in purified Chlamydomonas and Tetrahymena basal bodies, and that the RNA remains associated with basal bodies through both velocity and density gradient sedimentation. However, arguments in favor of centriolar RNA based on chemical analysis of partially purified preparations are inherently weak. What is required is either absolute purity of the basal bodies and, hopefully, chemical uniqueness of the RNA or evidence that centriole function depends upon RNA.

Our experiments on ribonuclease treatment of basal bodies indicate that some RNA species associated with the basal body is required for aster-inducing activity. There have been a few recent preliminary reports suggesting that RNA may be involved in aster formation. Zackroff, Rosenfeld, and Weisenberg (1975) found that asters formed in homogenates of surf clam eggs were decreased in size after treatment with RNAase. A similar finding has been reported by Snyder and McIntosh (1976) on asters formed in gently lysed mammalian cells. Berns and Rattner (1975) have suggested, on the basis of selective sensitivity of the centrosomal region to laser irradiation in living cells, that nucleic acid might be involved in the mitotic organizing function of the centriole. However, in none of these experiments could the effect be localized in the centriole, and the problems in interpreting ablation and digestion experiments carried out on complex cell homogenates are formidable. Hartman (1975) has also proposed, on the basis of speculative arguments, that RNA may be involved in the microtubule-organizing function of centrioles.

The question which remains is what is the possible role of RNA in the centriole? The ability of basal bodies to induce asters when injected into frog eggs strongly supports an active role for the centriole in aster formation. It is intriguing, however, that although the centriole initiates aster formation, it is itself not the nucleation point for astral microtubules. Instead, a centrospheric cloud, of unknown composition, surrounding the centriole appears to nucleate microtubule assembly. It therefore seems reasonable that the centriole contains part of the information necessary to set up the centrospheric cloud. RNA may take part in the transfer of this information. RNA has three known functions: as a source of genetic information in certain viruses, as a template for protein synthesis, and as a structural component with a role in protein and nucleic acid recognition in tRNA and ribosomes. One might imagine, therefore, that the RNA in the centriole could code for an mRNA which produces for a specific protein required for aster formation; it could itself be that mRNA, or it could have some unique structural role in nucleating aster assembly.

Experimental Procedures

Isolation of Basal Bodies

Basal bodies were isolated from Chlamydomonas reinhardtii (strain 137C) by the method of Snell et al. (1974). They were suspended after the final pellet in 10% sucrose on 10 mM Tris–HCl, 1 mM EDTA (pH 7.5).

Basal bodies were isolated from Tetrahymena pyriformis (Wards Natural Science) by the following method. Tetrahymena pyriformis were grown at room temperature to a density of 5×10^5 cells per ml in a medium containing 1.5% Bacto Peptone and 0.1% yeast extract (Difco). Cells were harvested by centrifugation at $1000 \times g$ and resuspended in half the original culture volume of fresh sterile culture medium. These cells were again harvested at $1000 \times g$, resuspended in 10 ml of fresh culture medium for each 10^8 cells, and deciliated according to the procedure of Thompson et al. (1974). Deciliated cells were spun at 4°C at $1500 \times g$ for 2 min. The pellet was resuspended in 50 ml of 0.25 M sucrose in 10 mM Tris–HCl, 1 mM EDTA, 1 mM EGTA (pH 9.3), and spun again at $1500 \times g$ for 2 min. The washed, deciliated cells in the pellet were then lysed by resuspending the pellet in 30 ml for each 10^9 cells in 1 M sucrose in the above buffer containing 1% Triton X-100 and 0.1% mercaptoethanol at 0°C as described by Wolfe (1970) for isolating oral apparatus of Tetrahymena. The lysed cells were further disrupted by homogenizing at 0°C with a glass tissue grinder with a motor-driven teflon pestle. This homogenate was layered over a discontinuous sucrose gradient of 10 ml 60% sucrose and 10 ml 50% sucrose in 10 mM Tris–HCl, 1 mM EDTA (pH 7.5), and spun for 1 hr at $14,000 \times g$ at 4°C in a Sorvall HB4 (swinging bucket) rotor. The 50% sucrose layer was recovered, diluted 1:1 with cold buffer, and spun $35,000 \times g$ for 40 min at 4°C in a Sorvall SS34 rotor. This pellet material may be purified further by sedimentation to equilibrium in a D_2O-sucrose gradient described below.

Quantitative Electron Microscopy

The number of basal bodies and the purity of basal body preparations were determined by a quantitative electron microscopic technique based on the centrifugation technique of Sharp (1949). 1 µl of the purified basal body preparation was diluted into 0.5 ml of 10 mM Tris–HCl, 1 mM EDTA (pH 7.5) which contained vinyl beads (Ted Pella Co.) 0.312 µ in diameter at a concentration of 2.4×10^7 beads per ml. Formvar-coated copper grids which had not been carbon-shadowed were affixed to the bottom of a 4°, 12 mm path length, Kel F, single-sector analytical ultracentrifuge cell (Beckman) by double-stick tape. The sample was added to the cell and spun at 24,630 rpm for 40 min at 25°C in the model E analytical ultracentrifuge. Glassware used throughout the procedure was siliconized by submersion in a solution of Siliclad (Clay Adams, Inc.) according to the directions supplied with the product. After centrifugation, the grids were removed from the cell, negatively stained with 2% uranyl acetate, and examined in a JEOL 100C electron microscope (Jeolco Electronics). In one preparation, the number of basal bodies and vinyl beads per μ^2 were determined on samples which were rotary-shadowed with tungsten and compared with those which were negatively stained, and the number was found to be the same. This indicated that no basal bodies were removed by the negative staining procedure. For nonquantitative analysis of basal bodies, negatively stained samples were prepared as described by Witman et al. (1976).

Treatment of Basal Bodies

Basal bodies were treated in 10 mM Tris–HCl, 1 mM EDTA (pH 7.5), except for DNAase I treatment, in which $MgCl_2$ was added to the treated sample to give a final concentration of 3 mM. All enzyme treatments unless otherwise specified were for 30 min at 25°C, after which the treated sample was cooled on ice and then spun at $16,000 \times g$ for 30 min. The pellet was then gently washed once with 10 mM Tris, 1 mM EDTA (pH 7.5) buffer and resuspended by vortexing in the original volume of buffer. DNAase I (Worthington), Trypsin (Worthington), RNAase T1 (Boehringer Mannheim) were added to basal body preparations at concentrations of 5 µg/ml, 10 µg/ml. All RNAase A stock solutions were routinely heated to 100°C for 3 min before use. In one experiment, basal bodies were treated with 5 µg/ml of RNAase A that had been inactivated with diethylpyrocarbonate (Solymosy et al., 1968). The ribonuclease was allowed to incubate overnight in a polysyrene tube to eliminate any excess diethylpyrocarbonate before addition to the basal body preparations. In two experiments, 5 µg of yeast total RNA (Sigma) was added to 50 µl of pelleted basal bodies after treatment with 5 µg/ml RNAase A. In all experiments, control basal bodies were allowed to remain at room temperature for 30 min in buffer with no treatment, then centrifuged and resuspended as described above.

Injections

Unfertilized eggs of Xenopus laevis were microinjected as described by Heidemann and Kirschner (1975), except that eggs were stratified before injection. Dejellied eggs were suspended at the interface of 5 ml 60% Ficoll (w/v) in Barth's saline and 5 ml Barth's saline (Barth and Barth, 1959) and then centrifuged for 30 min at $2500 \times g$ at 4°C in a Sorvall HB4 rotor (Masui, 1972). Eggs stratified in this manner were then injected with 0.1 µl of material into the cytoplasmic layer. Usually each egg was injected with 10^4 basal bodies, an amount which saturates the egg response. Similar results were obtained using 10^2 basal bodies. Eggs were allowed to incubate for 1 hr, fixed, serial sectioned, and stained as previously described (Heidemann and Kirschner, 1975).

Labeling of Tetrahymena and Chlamydomonas

Chlamydomonas reinhardtii was labeled with ^{32}P as described by Piperno and Luck (1976). Labeled vegetative cells were allowed to differentiate into gametes preparatory to basal body preparation in unlabeled nitrogen-free medium as described by Snell et al. (1974).

Tetrahymena pyriformis was labeled with ^{32}P or ^{3}H-uridine (both 1–2 µCi/ml) (New England Nuclear) in medium containing 0.5% proteose peptone, 0.05% yeast extract. Cultures were grown at room temperature until the cell density reached approximately 4×10^5 cells per ml. Basal bodies were prepared from these cells as described.

The specific activity of the $^{32}PO_4$ in the culture medium was determined by calculating the total PO_4 concentration in an aliquot of the sterile medium by the method of Ames (1966) and by measuring the total radioactivity by liquid scintillation counting.

Assay for Nucleation of Microtubule Assembly in Vitro

Basal bodies purified from Chlamydomonas were either treated with 5 µg/ml of RNAase A or with buffer as described above, and incubated at room temperature for 30 min. These basal bodies were then used as nucleating centers for microtubule protein polymerization by the procedures described by Witman et al. (1976) for assembly of flagellar axonemes.

Equilibrium Density Centrifugation of Basal Bodies

Purified basal bodies labeled with $^{32}PO_4$ or ^3H-uridine from Chlamydomonas and Tetrahymena were layered over a 5 ml linear sucrose gradient, 0–75% sucrose in heavy water (D_2O) containing 10 mM Tris, 2 mM EDTA (pH 7.5). The gradients were centrifuged in a Beckman SW39 rotor at 4°C at 100,000 × g for 2 hr. Ten fractions were collected from each gradient, and the density was analyzed by refractometry. Electron microscope grids were made from each fraction, and those fractions found to contain any basal bodies were analyzed by quantitative electron microscopy. To determine the amount of RNA in each fraction, duplicate aliquots were removed from each fraction and one set was treated with 10 µg/ml RNAase A for 30 min at 37°C, while the other was incubated at 37°C for 30 min with no enzymatic treatment. 50 µg of bovine serum albumin were added to each fraction, which was then cooled on ice and precipitated with an equal volume of ice-cold 20% TCA. Precipitates were collected on nitrocellulose filters (0.45 µ; Millipore Corp.) and were rinsed with seven 3 ml washes of 10% TCA and two 3 ml washes of ethanol. The filters were then dried and added to 5 ml Aquasol LSC cocktail (New England Nuclear) in shell vials and counted. Aliquots from the same gradient were analyzed for aster-inducing activity by microinjection into stratified Xenopus eggs after the fractions had been centrifuged 100,000 × g for 1 hr, and the pellet was resuspended for injection in 10 mM Tris–HCl, 1 mM EDTA (pH 7.5).

Acknowledgments

We would like to thank Dr. Victor Bruce and Nancy Bruce for kindly supplying the Chlamydomonas strains used in these experiments and for the use of their growth facilities. We thank Dr. Marjorie Marsden for doing some sectioning for the electron microscope. This work was suppported by a USPHS grant and a grant from the American Cancer Society. S. R. H. was supported by a USPHS grant. We would like to thank the Dreyfus Foundation for their support. M. W. K. is the recipient of a USPHS research career development award.

Received November 5, 1976

References

Allen, C., and G. G. Borisy (1974). Structural polarity and directional growth of microtubules of Chlamydomonas flagella. J. Mol. Biol. 90, 381–402.

Ames, B. N. (1966). Assay of inorganic phosphate. Total phosphate and phosphatases. In Methods in Enzymology, 8, (New York: Academic Press), pp. 115–118.

Anderson, R. G. W., and R. M. Brenner (1971). The formation of basal bodies in the Rhesus monkey oviduct. J. Cell Biol. 50, 10–34.

Argetsinger, J. (1965). The isolation of ciliary basal bodies from Tetrahymena pyriformis. J. Cell Biol. 24, 154–157.

Barth, L. G., and L. J. Barth (1959). Differentiation of cells of Rana pipiens gastrula in unconditioned medium. J. Embryol. Exp. Morphol. 7, 210–222.

Berns, M. W., and J. B. Ratner (1975). Irradiation of the centriolar region in mitotic potoporous cells with a laser microbeam. J. Cell Biol. 67, 30a.

Binder, L. I., W. L. Dentler, and J. L. Rosenbaum (1975). Assembly of chick brain tubulin onto flagellar microtubules from Chlamydomonas and sea urchin sperm. Proc. Nat. Acad. Sci. USA 72, 1122–1126.

Child, T. M. (1959). Characterization of the cilia of Tetrahymena pyriformis. Exp. Cell Res. 18, 258–267.

Dippell, R. V. (1976). Effects of nuclease and protease on the ultrastructure of Paramecium basal bodies. J. Cell Biol. 69, 622–637.

Flavell, R. A., and I. G. Jones (1971). DNA from isolated pellicles of Tetrahymena. J. Cell Biol. 9, 719–726.

Friedlander, M., and J. Wahrman (1970). The spindle as a basal body distributor. J. Cell Sci. 1, 65–89.

Fulton, C. (1971). Centrioles. In Origin and Continuity of Cell Organelles, W. Beerman, J. Reinert, and H. Ursprung, eds. (New York: Springer-Verlag).

Gould, R. R. (1975). The basal bodies of Chlamydomonas reinhardtii. J. Cell Biol. 65, 65–74.

Hartman, H. (1975). The centriole and the cell. J. Theoretical Biol. 51, 501–509.

Hartman, H., Puma, J. D. and Gurney, T., Jr. (1974). Evidence for the association of RNA with the ciliary basal bodies of Tetrahymena. J. Cell Sci. 16, 241–259.

Heidemann, S. R., and Kirschner, M. W. (1975). Aster formation in eggs of Xenopus laevis: induction by isolated basal bodies. J. Cell Biol. 67, 105–117.

Hoffmann, E. J. (1965). The nucleic acids of basal bodies isolated from Tetrahymena pyriformis. J. Cell Biol. 25, 217–228.

Inoué, S. (1964). Organization and function of the mitotic spindle. In Primitive Motile Systems in Cell Biology, R. D. Allen and N. Kamiya eds. (New York: Academic Press).

Iwamatsu, T., Miki-Nomura, T., and Ohta, T. (1976). Cleavage initiation activities of microtubules and in vitro reassembled tubulins of sperm flagella. J. Exp. Zool. 195, 97–106.

Lwoff, A. (1950). Problems of morphogenesis in ciliates. In the Kinetosomes in Development, Reproduction, and Evolution (New York: J. Wiley and Sons).

McGill, M., and B. R. Brinkley (1975). Human chromosomes and centrioles as nucleation sites for the in vitro assembly of microtubules from bovine brain tubulin. J. Cell Biol. 67, 189–199.

Maller, T., Poccia, D., Nishioka, D., Kidd, P., Gerhart, T., and Hartman, H. (1976). Spindle formation and cleavage in Xenopus eggs injected with centriole containing fractions from sperm. Exp. Cell Res. 99, 285–294.

Masui, Y. (1972). Distribution of the cytoplasmic activity inducing germinal vesicle breakdown in frog oocytes., J. Exp. Zool. 179, 365–378.

Olmsted, J. B., and Borisy, G. G. (1973). Characterization of microtubule assembly in porcine brain extracts by viscometry. Biochemistry 12, 4782–4789.

Osborn, M., and Weber, K. (1976). Cytoplasmic microtubules in tissue culture cells appear to grow from an organizing center towards the plasma membrane. Proc. Nat. Acad. Sci. USA 73, 867–871.

Pickett-Heaps, J. D. (1969). The evolution of the mitotic apparatus: an attempt at comparative ultrastructural cytology in dividing plant cells. Cytobios 1(3), 257–280.

Pickett-Heaps, J. D. (1971). The autonomy of the centriole: fact or fallacy. Cytobios 3, 205–214.

Piperno, G., and Luck, J. L. (1976). Phosphorylation of axonemal proteins in *Chlamydomonas reinhardtii*. J. Biol. Chem. 251, 2161-2167.

Porter, K. R. (1966). Cytoplasmic microtubules and their functions. In Principles of Biomolecular Organization, G. E. W. Wolstenholme and M. O'Connor, eds. (London: J. A. Churchill).

Raff, E. C., Brothers, A. J., and Raff, R. A. (1976). Microtubule assembly mutant. Nature 260, 615-617.

Randall, J. T., and Disbrey, C. (1965). Evidence for the presence of DNA in basal body sites in *Tetrahymena pyriformis*. Proc. Roy. Soc. Ser. B 162, 473-491.

Robbins, E., and Gonatas, N. K. (1964). The ultrastructure of a mammalian cell during the mitotic cycle. J. Cell Biol. 21, 429-463.

Rosenbaum, J. L., and Carlson, K. (1969). Cilia regeneration in *Tetrahymena* and its inhibition by colchicine. J. Cell Biol. 40, 415-425.

Rubin, R. W., and Cunningham, W. P. (1973). Partial purification and phosphotungstate solubilization of basal bodies and kinetodesmal fibers from Tetrahymena pyriformis. J. Cell Biol. 57, 601-612.

Sagan, L. (1967). On the origin of mitosing cells. J. Theoretical Biol. 14, 225.

Seaman, G. R. (1960). Large scale isolation of kinetosomes from the ciliated protozoan *Tetrahymena pyriformis*. Exp. Cell Res. 21, 292-302.

Sharp, D. G. (1949). Enumeration of virus particles by electron microscopy. Proc. Soc. Exp. Biol. Med. 70, 54-49.

Smith-Sonneborn, J., and Plaut, W. (1967). Evidence for the presence of DNA in the pellicle of *Paramecium*. J. Cell Sci. 2, 225-234.

Snell, W. J., Dentler, W. L., Haimo, L. T., Binder, L. I., and Rosenbaum, J. L. (1974). Assembly of chick brain tubulin onto isolated basal bodies of *Chlamydomonas reinhardtii*. Science 185, 357-359.

Snyder, J. A., and McIntosh, J. R. (1975). Initiation and growth of microtubules from mitotic centers in lysed mammalian cells. J. Cell Biol. 67, 744-760.

Snyder, J. A., and McIntosh, J. R. (1976). Studies on the centriolar region of mammalian cells at the onset of mitosis. J. Cell Biol. 70, 368a.

Solymosy, F., Fedorcsak, I., Gulyas, A., Farkas, G. L., and Ehrenberg, L. (1968). A new method based on the use of diethylpyrocarbonate as a nuclease inhibitor for the extraction of undegraded nucleic acid from plant tissues. Eur. J. Biochem. 5, 520-527.

Sorokin, S. P. (1968). Reconstruction of centriole formation and ciliogenesis in mammalian lungs. J. Cell Sci. 3, 207.

Stubblefield, E., and Brinkley, B. R. (1967). Architecture and function of the mammalian centriole. Symp. Int. Soc. Cell Biol. 6, 175-218.

Tamir, H., and Gilvarg, C. (1966). Density gradient centrifugation for the separation of sporulation forms of bacteria. J. Biol. Chem. 241, 1085-1090.

Thompson, G. A., Jr., Baugh, L. C., and Walker, L. F. (1974). Nonlethal deciliation of *Tetrahymena* by local anesthetic and its utility as a tool for studying cilia regeneration. J. Cell Biol. 61, 253-257.

Watson, M. R., and Hopkins, J. M. (1962). Isolated cilia from *Tetrahymena pyriformis*. Exp. Cell Res. 28, 280-295.

Weisenberg, R. C., and Rosenfeld, A. C. (1975). *In vitro* polymerization of microtubules into asters, and spindles in homogenates of surf clam eggs. J. Cell Biol. 64, 146-158.

Witman, G. B., Carlson, K., Berliner, J., and Rosenbaum, J. L. (1972). *Chlamydomonas* flagella. I. Isolation and electrophoretic analysis of microtubules, matrix, membranes, and mastigonemes. J. Cell Biol. 54, 507-539.

Witman, G. B., Cleveland, D. W., Weingarten, M. D., and Kirschner, M. W. (1976). Tubulin requires tau for growth onto microtubule initiation sites. Proc. Nat. Acad. Sci. USA 73, 4070-4074.

Wolfe, J. (1970). Structural analysis of basal bodies of the isolated oral apparatus of *Tetrahymena pyriformis*. J. Cell Sci. 6, 679-700.

Wolfe, J. (1972). Basal body fine structure and chemistry. Adv. Cell Mol. Biol. 2, 151-192.

Younger, K. B., Banerjee, S., Kelleher, J. K., Winston, M., and Margulis, L. (1972). Evidence that the synchronized production of new basal bodies is not associated with DNA synthesis in *Stentor coerulus*. J. Cell Sci. 11, 621-637.

Zackroff, R., Rosenfeld, A., and Weisenberg, R. (1975). Effect of RNAase on aster formation *in vitro*. J. Cell Biol. 67, 469a.

Microtubules in Prokaryotes

Universally involved in mitosis and motility in eukaryotes, microtubules are seen in spirochetes.

Lynn Margulis, Leleng To, David Chase

Microtubules, 250 angstroms in diameter, composed of tubulin proteins are universal constituents of eukaryotic cells. They take part in flagella movement (1), and in intracellular transport such as that in nerve cells (2), protozoans (3), hydras (4), and fungi (5). Mi-

---

Summary. Longitudinally aligned microtubules, about 220 Å in diameter, have been seen in the protoplasmic cylinders of the following spirochetes (symbiotic in the hindguts of dry-wood and subterranean termites): *Pillotina* sp., *Diplocalyx* sp., *Hollandina* sp. They are also present in a gliding bacterium from *Pterotermes occidentis*. These microtubules are probably composed of tubulin, as determined by staining with fluorescent antibodies to tubulin and comigration with authentic tubulin on acrylamide gels. *Treponema reiteri* lack tubulin by these same criteria. These observations support the hypothesis of the symbiotic origin of cilia and flagella from certain spirochetes.

---

crotubules have a role in regenerative morphogenesis (6) and underlie many cell structures, especially those in protists and animals (7). They comprise the mitotic spindle (8, 9) and are intimately involved, in still incompletely known ways, in the segregation of the chromosomes to the poles in mitosis in nearly all eukaryotes (8–11). Tubulin proteins from very different sources show a great deal of homology (12, 13); most microtubules are sensitive to cold and most bind alkaloids such as colchicine, vinblastine, and lignans such as the podophyllotoxin derivatives (12–14). But some microtubules are insensitive to these agents (15).

In general, neither microtubules nor microtubule proteins are known to occur in prokaryotic organisms such as *Escherichia coli* or *Bacillus*. However, hollow tubular structures have been reported in cell walls in certain blue-green algae (16). Smaller tubular structures have also been reported in *Proteus mirabilis* and in *Treponema*, but not in *Borrelia* (17, 18). None of these prokaryotic tubules have been studied chemically. Microtubules have not been observed to be involved in prokaryotic cell division.

There are two classes of hypotheses for the origin of eukaryotic microtubules. The exogenous (19) hypothesis suggests the origin of tubules in eukaryotic cells by symbiotic acquisition of tubule-containing spirochetes. Originally selected because associated spirochetes conferred motility on their hosts, as time went on the spriochetes evolved into the ubiquitous (9 + 2) flagella or cilia (undulapodia) (20) of eukaryotes. If such ancestral spirochetes have not become extinct, it is expected that free-living descendant spirochetes, ancestors to the (9 + 2) flagellar organellar system, will be found and that these will contain microtubules. These hypothetical tubules are predicted, therefore, to be composed of tubulin proteins and to be homologous to those in eukaryotic cells. The endogenous hypotheses of Pickett-Heaps (11) and Taylor (21) of the origin of tubules suggests that tubulin differentiated in primitive photosynthetic organisms ancestral to modern red algae.

Electron Microscopy

The observations (22, 23) of intracytoplasmic microtubules, 250 Å in diameter, in large spirochetes—such as those in the hindguts of termites—did not come to our attention until 1972. Hollande and Gharagozlou placed the microtubule-containing spirochetes in the genus *Pillotina* and in a new family, the Pillotaceae. The difficulty of obtaining the termite host endemic to the island of Madeira limited further work until these same pillotinas were recognized in electron microscopic (EM) preparations of *Reticulitermes flavipes* (24). Since 1974 we have found the large distinctive pillotina spirochetes and their smaller hollandina relatives present in hindgut microbiota from 21 out of 21 species of dry-wood termites [family Kalotermitidae (Fig. 1)], from five out of five species of subterranean termites (family Rhinotermitidae, for example, *R. flavipes*, *R. flavicollis*, *R. hesperus*, and *Heterotermes aureus*), but absent in two out of

L. Margulis and L. To are professor and graduate student, respectively, at the Department of Biology, Boston University, Boston, Massachusetts 02215. D. Chase is a senior research investigator in the Cell Biology Laboratory at Veterans Administration Hospital in Sepulveda, California.

two damp-wood termites (Hodotermitidae) and one unidentified species of nasutitermitid from Paquera, Puerto Rico (25). The abundance of these large spirochetes in the paunch (modified hindgut) (26) in dry-wood termites made biochemical studies possible although no attempts at in vitro growth of pillotinas have yet succeeded [motile spirochete maintenance of up to 5 days in culture under anaerobic conditions is described in (25)].

In this article, we present evidence for tubulin-containing microtubules in three types of related prokaryotic microbes. Our evidence is based on three different techniques: (i) observation of microtubules by transmission electron microscopy (TEM); (ii) isolation and separation of a soluble protein that comigrates electrophoretically on acrylamide gels with authentic tubulin from a hindgut fraction that contained only prokaryotes, primarily those spirochetes in which microtubules had been seen; and (iii) specific antitubulin immunofluorescent staining with fluorescent antibody to various authentic eukaryotic tubulins.

Micrographs of the basic structure of these extremely obscure symbionts are

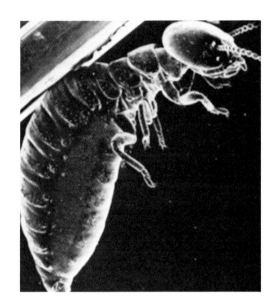

Fig. 1 (left). The Sonoran desert termite, *Pterotermes occidentis* (Walker), has been the best source for spirochetes and related bacteria in quantity that apparently contain microtubule protein. Tubulin protein previously has been thought to be limited to eukaryotes. The swollen abdomen of this pseudergate (wood-eating larva, or worker) is typical: within it lies the paunch or hypertrophied hindgut packed with many species of cellulolytic and nitrogen-fixing microorganisms which permit the termite colony to survive on an exclusive diet of wood, in this case, of the Palo verde (*Cercidium*) tree. Fig. 2 (below). Microorganisms symbiotic in termite hindguts in which microtubules have been found. (a) *Pillotina* sp. from *P. occidentis*. Phase contrast, live (bar, 1.0 μm). (b) Transverse section of *Pillotina* sp. from *Kalotermes schwarzi*. Transmission electron micrograph (bar, 0.1 μm). (c) Longitudinal section of *Pillotina* sp. from *Reticulitermes hesperus*. Transmission electron micrograph (bar, 1.0 μm). (d) *Hollandina pterotermitidis* (h), *Pillotina* sp., and unidentified gliding bacteria (g), from *Pterotermes occidentis* (bar, 10 μm). (e) Transverse section of *Hollandina* sp. Transmission electron micrograph (bar, 0.1 μm). (f) Longitudinal section of *Hollandina* sp. Transmission electron micrograph (bar, 0.1 μm).

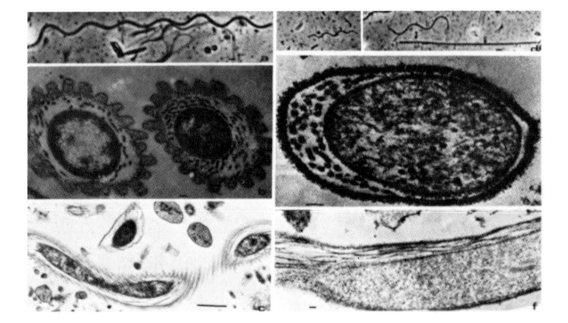

presented in Fig. 2. All three—*Pillotina* sp. (Fig. 2, a to c), *Hollandina* sp. (Fig. 2, d to f), and the long skinny unidentified gliding rod [probably one of the organisms referred to as *Bacillus flexillus* in 1927 (*27*)] (Fig. 2d, labeled *g*)—are always extremely motile when healthy. Thus, the original observations of microtubules in pillotinas (*22, 23*) have been confirmed by us and extended to include a third spirochete and a gliding bacterium.

Termites were collected and identified as indicated in the tables; they were maintained as described (*25*). Electron microscopic techniques (*24*) were used, except that 5 percent tannic acid was added in some cases to the glutaraldehyde fixative to enhance the probability that the tubules would be seen, and impregnation times were lengthened to as much as 36 hours, depending on the size of the gut.

Occasional sections containing microtubules were seen in pillotinas from subterranean termites *R. flavipes* and *R. hesperus*. They were even more frequently seen in the medium-sized (0.75 to 0.90 micrometer in diameter) hindgut spirochetes described as *Hollandina calotermitidis*, from *Kalotermes schwarzi* (*25*). The microtubules are longitudinally aligned, and often look as though they follow the contours of the spirochetes' helical bodies, but they are not seen in all sections. Similar, if not identical, microtubules have also been seen in long skinny "gliding bacteria" unidentified but under study (Fig. 3c) and in hollandinas (Fig. 3d); both microbes come from the hindguts of the Sonoran desert dry-wood termite *Pterotermes occidentis* (Fig. 1). The outer diameters of the microtubules measured 240 ± 15 nanometers as determined on glossy prints (8 by 10 inches), with the aid of calipers and a dissecting microscope. The microtubules were always intracytoplasmic and were always larger than the atypically large axial filaments (spirochete endoflagella, 200 to 210 Å in diameter) of these microbes. In one case, tubules were seen in a gliding bacterium also infected with phage (Fig. 3e). No comparable tubules were seen in the many other termite hindgut prokaryotes studied (for example, spirilla; *Coleomitus* sp., the large filamentous endospore-forming bacterium, or the tiny treponeme-like spirochetes) nor have they been seen in *Treponema reiteri* (*18*).

Specific Immunofluorescence of Antibody to Tubulin

A specific antibody to microtubule protein has been developed (*8, 9, 28*). This antibody can be labeled with fluorescein and used in combination with fluorescence microscopy as a cytological stain; tubulin can be directly visualized, as can be seen in the micrographs of Fuller *et al.* (*9*). A Zeiss microscope was fitted with an AO xenon ultraviolet light

Table 1. Antibody tubulin; sources and results; SAA, same as above.

Antigen	Antibody originally tested against and result	F*	Microbe termites†	F	Comments
Source: G. M. Fuller and B. R. Brinkley					
Bovine brain 6S tubulin	Dividing and interphase human, kangaroo, mouse, monkey, rat, and hamster embryo cells (9)	+	*K. schwarzi* Pillotinas Hollandinas	+ +	Indirect technique‡
Rabbit serum			Tiny (treponeme-like) spirochetes Flagellates Spirilla *Coleomitus* sp. Skinny gliders	− + − − +	
			K. schwarzi Fluorescein alone IgG goat-fluorescein alone Buffer preparations alone	− − −	
Source: K. Fujiwara					
Strongylo- centrotus (sea urchin) un- fertilized eggs vinblastine precipitated tubulin	Egg homogenate Tetrahymena cilia Sperm (cranefly testes) flagella Sea urchin sperm tail axonemes Rabbit muscle actin Myosin Bovine serum albumin Chick, rat, mouse, fish frog brain tubulin HeLa cells, mitotic spindle	+ + + + − − − + +	*K. schwarzi* SAA *Pterotermes occidentis* Hollandinas Skinny gliders Miscellaneous small bacteria and spirochetes *Coleomitus* (two sp.) *Treponema reiteri* Pelleted, pure culture	SAA + + − − −	Direct antitubulin labeled with fluorescein. Best preparations were with this antibody‡
Source: V. Kalnins					
Porcine brain tubulin	Embryonic chick and brain tubulin Porcine brain Actin	+ + −	*K. schwarzi* Hindgut microbes, results identical to those with Fuller antibody listed above		Indirect technique, same as with Fuller antibody

*Fluorescence. †*Kalotermes schwarzi* was collected by R. Syren and W. Ormerod in South Florida and identified by R. Syren, University of Miami. *Pterotermes occidentis* was collected by W. Nutting and L. Margulis in the Sonoran desert 20 miles south of Tucson, Arizona, and identified by W. Nutting, University of Arizona. ‡The best technique for indirect staining was as follows. Hindguts were removed with fine forceps on alcohol cleaned and flamed slides fixed with 5 percent formaldehyde in Trager's solution. The gut contents were drawn out into fixative with a fine-needle syringe. The preparation was washed with Trager's, placed in absolute acetone at −8°C for 7 minutes, washed two to three times again with Trager's solution, and dried in air. It was incubated for 45 minutes at 35° ± 1°C with rabbit antibody to tubulin; the particles were then washed three times. When just a film of Trager's solution was left, goat antibody to normal rabbit IgG conjugated with fluorescein was added dropwise, and the preparation was incubated for 30 minutes in a humid chamber. It was washed thoroughly with Trager's and observed, and was then stored in a refrigerator with a desiccant. Direct techniques are the same except that only one 45-minute incubation with fluorescein-bound rabbit antibody to tubulin made against vinblastine precipitated sea urchin tubulin. Slides mounted for observation in 90 percent glycerin and 5 percent phosphate-buffered saline.

source and appropriate heat and interference filters to permit absorption of light at wavelengths from 490 to 495 nm. Guts were dissected, and the contents were removed to slides cleaned with alcohol. Formaldehyde (1 to 5 percent, or occasionally ethanol) was used as fixative; the hindgut preparations were stained for 45 minutes at 30°C with fluorescein-labeled antibody to tubulin directly or indirectly (Table 1). This experiment was repeated with slight variations five times, and the same observations were consistently made. In hindgut preparations of microbes from *Kalotermes schwarzi* or *Pterotermes occidentis*, the following organisms showed bright fluorescence: polymastigotes and hypermastigotes (particularly the rostral region although these fixatives were very poor for flagellates) and their flagella, pillotina and hollandina spirochetes, and the long, thin "gliding bacteria". In the same preparations the large *Coleomitus* sp., the smaller treponeme, spirilla, and other bacteria did not fluoresce; wood particles in the gut lumen and within the flagellates tended to fluoresce orange, whereas the spirochetes and flagella fluoresced green (Fig. 3, f and g). The consistent fluorescence of the thin gliding bacteria (0.45 by 15 μm) prompted our search for microtubules in them. These observations on the structure, behavior, and movement patterns of the gliders have led us to agree with Canale-Parola (29) that the "phylogenetic relationship between spirochetes and gliding bacteria may be closer than previously believed."

Significant and regular fluorescence was seen down the entire length of those spirochetes and gliding bacteria that gave positive signals. In fact, the fluorescent stain in the spirochetes was comparable in intensity to that of the flagella of neighboring polymastigotes and hypermastigotes. Preparations, even if refrigerated, aged in a day or so. The sharp image was replaced by a zone of fluorescence surrounding their contours or collecting toward the periphery of the spirochetes or flagella.

To verify the specificity of the fluorescence observations the following experiments were done with absorbed antiserum. Antiserum to tubulin (from K. Fujiwara, see Table 1) was absorbed with mouse brain tubulin for 18 hours at 4°C. It was also absorbed with a spirochete-enriched hindgut preparation free of flagellates, and washed three times with phosphate buffered saline. The indirect immunofluorescence, checked on slide preparations of hindgut microbes, treated with serum, yielded the following results: (i) antiserum to tubulin and fluo-

rescein-conjugated immunoglobulin G (IgG)–stained preparations gave fluorescent spirochetes and flagellates as above and in Fig. 2, f and g; (ii) antitubulin absorbed with brain tubulin gave no fluorescence for either spirochetes or flagellates; (iii) antitubulin absorbed with the sonicated spirochete-enriched fraction gave positive but diminished fluorescence for both spirochetes and flagellates; (iv) normal rabbit serum post-stained with fluorescein-conjugated IgG; and (v) fluorescein isothiocyanate alone gave no fluorescence of either spirochetes or flagellates. These results are consistent with our hypothesis that protein with antigenic sites that cross react with authentic tubulin is present in hollandina and pillotina spirochetes, as well as, of course, in the hypermastigote and polymastigote flagellates of the termite hindgut.

We feel that it is improbable that the fluorescing material is due to nonspecific adherence to the surface of the spirochetes and gliding bacteria, especially since aged preparations of flagella of hypermastigotes and polymastigotes behaved the same way, whereas filamentous spore-forming bacteria took up no stain whatsoever. It is more likely that the soluble complex of tubulin with the fluorescein-conjugated antibody to tubulin diffuses, over time, out of fixed material. In fact, the intensity of the spirochete fluorescence suggests that pillotinas, hollandinas, and gliding bacteria contain a large tubulin fraction that is not

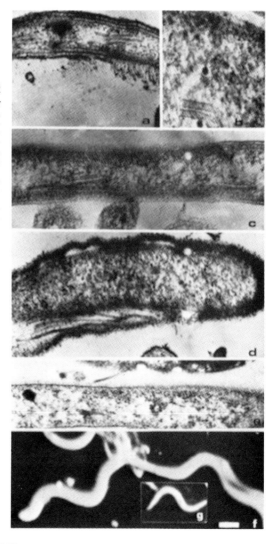

Fig. 3. Microtubules in the cytoplasm of the three types of prokaryotes shown in Fig. 2 and fluorescence micrographs of hindgut spirochetes stained with fluorescein - labeled antibody to tubulin. All microtubules are 240 ± 15 Å in diameter. (a) *Hollandina pterotermitidis*, longitudinal section. (b) *Hollandia pterotermitidis*, higher magnification, tannic acid added to fixative. Walls of tubules from 35 to 50 Å. (c) Unidentified gliding bacterium from *P. occidentis*. (d) *Hollandia pterotermitidis* showing microtubules and sheath bound flagella. (e) Unidentified gliding bacterium. Note both microtubules and bacteriophage with tail present in section. (f) Fluorescing *Pillotina* sp. from *Kalotermes schwarzi*. (g) Fluorescing *Hollandina* sp. from *K. schwarzi* (bar, 1.0 μm.).

Fig. 4. Sample of gel electrophoresis from *Pterotermes occidentis*. 1, Tubulin standard, pig brain; 2, spirochete fraction separated by centrifugation, only prokaryotes; 3, polymastigote fraction unavoidably containing attached and associated spirochetes; 4, hypermastigote fraction unavoidably containing some polymastigotes, and attached and associated spirochetes; 5, termite hindgut wall fraction; 6, tubulin standard (1 and 6 contained 2 ml per well; 2 to 5 contained 30 ml per well).

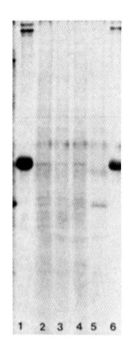

necessarily always organized into classical tubules. The stained preparations were made and observed at 22°C. The ambient temperature of the termite hindgut is 28° to 36°C. Whether the organized tubules are sensitive to lower temperatures is not known.

Gel Electrophoresis

Although many controls were used (Table 1) to verify that the fluorescence was due to tubulin immunofluorescence, such techniques are intrinsically limited. Direct biochemical techniques comparing properties of the putative spirochete tubulin with those of well-known eukaryotic tubulin are potentially more definitive; thus, sodium dodecyl sulfate gel electrophoresis studies were undertaken.

Although they could not be grown in culture, large quantities of spirochetes could be separated from the other host termite microbiota by differential centrifugation (24). The hindguts were removed intact from the termites and collected in Trager's A solution (30). When enough pseudergates were available (60 to 70), they were used for the *R. hesperus* gel electrophoresis, whereas only eight to ten *Pterotermes* were required because of the much larger size of the their gut. Tissue homogenization was used to release the microbes all at once into a small volume of Trager's solution. The flagellates were easily sedimented by desk centrifugation at low speed. The resulting supernatants were collected and centrifuged at high speed for approximately 5 minutes. The pellets were centrifuged and washed several times in Trager's solution until microscopic examination showed that they contained large numbers of live spirochetes, both pillotinas and hollandinas, skinny gliding bacteria, and other smaller bacteria. Centrifugation was continued, if necessary, until screening for flagellates revealed none. Nearly all the visible flagellates were actively swimming even after they were collected in the pellet fraction, and all the flagellates appeared to be intact even if not motile. Therefore, we believe that the prokaryotic fractions could have only been minimally contaminated by tubulin of flagellate orgin.

The total cell protein was then solubilized in sodium dodecyl sulfate buffer, and boiled for 5 minutes; for electrophoresis, samples from various fractions were placed on slab gels made with 12.5 percent acrylamide running gels and 1 percent bis-acrylamide stacking gels (31). From 10 to 50 milliliters of total soluble protein was placed in each gel well. This experiment was repeated with minor variation three times. Electrophoresis of the proteins was conducted at least twice with the cleanest possible spirochete fraction from each species of termite host. In all cases, except when a pure culture of Reiter treponemes was used and when a slice of termite gut tissue was the sole protein source, a conspicuous band comigrating with authentic brain tubulin as standard was seen (Fig. 4). In one test, a sample of purified spirochete fraction was divided in half, and egg-white lysozyme at a final

Table 2. Summary of electrophoresis experiments.

Sample	Type of gel	Bands		Gel facilities, sources, comments
		Conspicuous (No.)	Identical with tubulin*	
Treponema reiteri pure culture	Tube, SDS–7.5 percent acrylamide	6 to 8	No	K. Fujiwara, 7-day cultures
Pterotermes occidentis	Slab, 12.5 percent acrylamide			E. Lazarides, 7 to 10 worker (pseudergates) termites per run; ran twice; same results
Hindgut microbes mixed bacterial spirochete fraction (no flagellates)		14 to 18	Yes	
Mixed bacterial spirochete flagellates—mainly hypermastigotes		8 to 12	Yes	
Mixed bacterial spirochete flagellates—mainly polymastigotes		12 to 14	Yes	
Termite gut wall tissue		30	Faint or no	
Reticulitermes hesperus	Slab 12.5 percent acrylamide			E. Lazarides, one actin-like band
Mainly pillotina and hollandina spirochetes		10 to 12	Yes†	
Mixed flagellates and spirochetes		6 to 8	Yes	Two actin-like bands; 70 to 80 worker termites per run; rabbit muscle actin run as internal standard

*That is, comigrating with standard brain tubulin. †Increased quantity of protein with lysozyme treatment.

concentration of 0.1 milligram per milliliter in Trager's solution was added to one member of the paired sample. To the other, Trager's solution alone was added. After a 15-minute incubation with lysozyme at room temperature, the proteins were solubilized in sodium dodecyl sulfate. After electrophoresis, it was obvious that lysozyme had increased the yield of all of the major protein bands, including the putative tubulin band. It seems unlikely that such a mild enzyme treatment with lysozyme, which is known to dissolve bacterial cell walls, could have released tubulin from contaminating flagellates, especially when no contaminating flagellates had been seen in the preparation before protein solubilization (Table 2).

Since *Treponema reiteri* could be axenically cultured (*32*) we attempted to find tubulin in these small spirochetes by the same methods: fluorescence of specific antibody to tubulin and gel electrophoresis. We were also able to test the effects of agents known to inhibit the polymerization of microtubule protein into tubules, such as podophyllotoxin, β-peltatin, and vinblastine (Table 3). The results of these experiments were consistent. On gel electrophoresis the treponemes have no protein band corresponding to tubulin, and they do not fluoresce when stained with fluorescein-conjugated antibody to tubulin. Their growth is not inhibited by any antimitotic drugs. These characteristics are consistent with the electron microscopic observations of Hovind-Hougen (*19*), who did not find 250 Å microtubules in *T. reiteri*.

There are many possible sources of error in our results. The mixed culture of bacteria may contain a nontubulin protein of molecular weight similar enough to that of tubulin to comigrate on gels; the fluorescense may be due to an unidentified artifact; and the microtubular structures may be convergent—similar to those of eukaryotes by chance alone. Furthermore, tubulin may be present in these spirochetes and gliding bacteria through phage transfer from flagellates in the crowded gut. Nothing short of detailed biochemistry on pure cultures of these large spirochetes will permit the definitive solution to this problem. However, if, as has been hypothesized, the cilia or flagellar system of eukaryotes was symbiotically acquired and later deployed in the origin of mitosis (*19*), the prediction is unambiguous. Tubulin protein of these spirochetes ought to be homologous to that of eukaryotic microtubules, and the nucleic acid responsible for the replication of microtubule organizing centers (*33*) should be homologous

to nucleic acid of the relevant tubulin-containing spirochetes. Tubulin should not be found in cyanophyte tubules (*16*). Eventually, perhaps, colchicine-binding tubulin-containing spirochetes with a ninefold symmetry may even be discovered. At any rate, on the basis of our data (Table 3), the probability that tubulin-containing intracytoplasmic microtubules are present in the large symbiotic spirochetes and their relatives has risen significantly (*34*).

Conclusions

Microtubules, 240 ± 15 Å in diameter, have been found in the cytoplasm of large spirochetes (*Pillotina* sp. and *Hollandina* sp.) symbiotic in the hindguts of subterranean and dry-wood termites. Microtubules, aligned longitudinally down the long axes of these spirochetes, were also found in unidentified long, skinny, gliding bacteria. Large quantities of these prokaryotes were separated by centrifugation; their soluble proteins were released and studied by sodium dodecyl sulfate gel electrophoresis. Protein bands that comigrate with authentic brain tubulin were observed in preparations from these prokaryotes. The quantities of "tubulin" and other protein in solution were increased with gentle lysozyme treatment. Furthermore, specific antibodies to tubulin made from various antigenic sources (such as vinblastine-precipitated sea urchin egg tubulin and bovine brain tubulin) labeled with fluorescein was used as a stain on the hindgut microbiota. In such in situ cytological preparations, specific fluorescence was associated with hypermastigote and polymastigote flagellates, pillotinas, hollandinas, and the unidentified gliders (but none of the many other prokaryotes showed fluorescence). Cultivable treponemes (*Treponema reiteri*) do not contain microtubules, do not fluoresce in comparable cytological preparations, are not sensitive to antimitotic compounds such as vinblastine and podophyllotoxin, and show no tubulin-like band in sodium dodecyl sulfate gel electrophoretic preparations. Although alernative interpretations have not rigorously been excluded, we consider this presumptive evidence that there are bona fide tubulin microtubules in certain, but not all, spirochetes.

Note added in proof: We wish to make an important qualification concerning our claims of microtubule size in prokaryotes.

Hollande and Gharagozlou (*23*) reported microtubules 250 Å in diameter in *Diplocalyx* and *Pillotina*. We originally measured microtubules in fixed preparations in tannic acid. Subsequent work showed (i) that there was an error in magnification of microtubules in the slender rod and (ii) that tannic acid artificially inflates the diameter of microtubules so fixed. We then remeasured

Table 3. Summary of evidence for microtubules and microtubule protein in prokaryotes. The following antimitotic agents were tested and found to have no effect on growth of *T. reiteri* at concentrations between 10^{-7} to $10^{-3}M$: podophyllotoxin, β-peltatin, vinblastine, vincristine, and Colcemid. We have no comparable data for the other organisms. Symbols: +, observation positive; −, observation negative.

Termite and microbe organisms	TEM	Gel electrophoresis	Antibody to tubulin fluorescence
Kalotermes praecox (Madeira)*			
Pillotina calotermitidis	+ (*23*)		
Calotermes flavicollis (France)	+ (*22*)		
Diplocalyx calotermitidis			
Reticulitermes flavipes (Mississippi, Massachusetts)			
Pillotina sp.	+†		
Reticulitermes hesperus (San Diego)			
Skinny, gliding bacteria (unidentified)	+	+	
Pillotina sp.		+	
Hollandina sp.		+	
Pterotermes occidentis (southern Arizona)			
Pillotina sp.		+	+
Hollandina sp.	+	+	+
Gliding bacteria (unidentified)	+	+	+
Coleomitus sp.	−		−
Kalotermes schwarzi (Miami)‡			
Pillotina sp.		+	+
Gliding bacteria (unidentified)	+		+
Coleomitus sp.	−		−
Cultivable spirochetes			
Treponema reiteri	− (*19*)	−	−

*The same as *Postelectrotermes praecox*.   †Only occasionally.   ‡The same as *Incisitermes schwarzi*.

many prokaryotic and eukaryotic microtubules in the same preparations. We found that, in spirochetes, the microtubule size varied from 150 to 210 Å without tannic acid and from 150 to 250 Å with tannic acid. In both preparations eukaryotic microtubules were consistently larger. These data are taken to mean that both the size and size range of eukaryotic microtubules are larger than those of prokaryote microtubules. Most eukaryotic microtubules are 240 ± 20 Å in diameter, but in some protists a larger range has been observed. The published range for protist microtubules is from 150 to 300 Å (35). In order to make detailed morphological comparisons between prokaryotic and eukaryotic microtubules the presence, number, and dimensions of the microtubules must be determined.

References and Notes

1. S. Inoué and R. Stephens, Eds., *Molecules and Cell Movement* (Raven, New York, 1975); P. Satir, in *ibid.*, p. 143.
2. S. Ochs, *Ann. N.Y. Acad. Sci.* **253**, 470 (1975).
3. C. Bardele, *Sym. Soc. Exp. Biol.* **28**, 191 (1975); L. H. Tilney, in *Origin and Continuity of Cell Organelles*, J. Reinhert and H. Ursprung, Eds. (Springer-Verlag, New York, 1971), vol. 2, p. 222; M. O. Soyer, *C. R. Acad. Sci.*, in press.
4. G. Cooper and L. Margulis, *Cytobios*, in press.
5. W. Ormerod, S. Francis, L. Margulis, *Microbios* **17**, 189 (1976); M. Raudaskoski, *Arch. Mikrobiol.* **86**, 91 (1972).
6. S. Banerjee and L. Margulis, *Exp. Cell Res.* **78**, 314 (1973); J. K. Kelleher, *Mol. Pharmacol.* **13** 232 (1977); L. Margulis, *Int. Rev. Cytol.* **34**, 333 (1973).
7. R. A. Bloodgood and J. K. Kelleher, in *Biochemistry and Physiology of Protozoa*, S. Hutner and L. Provasoli, Eds. (Academic Press, New York, 1978); R. Goldman, T. Pollard, J. Rosenbaum, Eds., *Cell Motility* (Cold Spring Harbor Laboratory, Cold Spring Harbor, N.Y. 1976), vols. A to C.
8. J. R. McIntosh, Z. Cande, J. Snyder, K. Vanderslice, *Ann. N.Y. Acad. Sci.* **253**, 407 (1975).
9. G. M. Fuller, B. R. Brinkley, J. M. Boughter, *Science* **187**, 948 (1975).
10. M. O. Soyer, *BioSystems* **7**, 306 (1975).
11. J. Pickett-Heaps, *ibid.* **6**, 37 (1974).
12. B. R. Brinkley, G. M. Fuller, D. P. Highfield, in *International Symposium on Microtubules and Microtubule Inhibitors*, M. DeBrabander and M. Borgers, Eds. (North-Holland, Amsterdam, 1975), p. 297.
13. R. F. Luduena and D. O. Woodward, *Proc. Natl. Acad. Sci. U.S.A.* **70**, 3594 (1973).
14. G. G. Borisy and E. W. Taylor, *J. Cell Biol.* **34**, 535 (1967); J. K. Kelleher, *BioSystems* **9**, (1977); L. Wilson, K. Anderson, L. Grisham, D. Chin, in *Microtubules and Microtubule Inhibitors*, M. Borgers and M. DeBrabander, Eds. (North-Holland, Amsterdam, 1975), p. 103.
15. I. B. Heath, *Protoplasma* **85**, 147 (1975).
16. T. Bisalputra, B. R. Oakley, D. C. Walker, S. M. Shields, *ibid.* **86**, 19 (1975).
17. L. To and L. Margulis, *Int. Rev. Cytol.*, in press; W. Van Iterson, J. F. Hoeniger, E. N. VanVanten, *J. Cell Biol.* **32**, 1 (1967).
18. K. Hovind-Hougen, *Acta Pathol. Microbiol. Scand. Sec B* **255** (Suppl.), 1 (1976).
19. L. Margulis, *Origin of Eukaryotic Cells* (Yale Univ. Press, New Haven, Conn., 1970); in *International Symposium on Microtubules and Microtubule Inhibitors*, M. DeBrabander and M. Borgers, Eds. (North-Holland, Amsterdam, 1975), p. 3.
20. L. Kuznicki, L. Jahn, J. R. Fonesca, *J. Protozool.* **17**, 16 (1970).
21. F. J. R. Taylor, *Taxon* **25**, 377 (1976).
22. I. Gharagozlou [*C. R. Acad. Sci.* **266**, 494 (1968)] also discovered microtubules in a large spirochete, *Diplocalyx calotermitidis* (found in *Calotermes flavicollis* from southern France), an organism similar to the hollandinas.
23. A. Hollande and I. Gharagozlou, *C. R. Acad. Sci.* **265**, 1309 (1967).
24. R. A. Bloodgood, thesis, University of Colorado, Boulder (1974); ———, K. R. Miller, T. P. Fitzharris, J. R. McIntosh, *J. Morphol.* **143**, 77 (1974).
25. L. To, L. Margulis, A. T. W. Cheung, *Microbios*, in press.
26. J. Breznak and H. S. Pankratz, *J. Appl. Env. Microbiol.* **33**, 406 (1977).
27. O. Duboscq and P. Grassé, *Arch. Zool. Exp. Gen.* **66**, 451 (1927).
28. K. Weber, R. Pollack, T. Bibring, *Proc. Natl. Acad. Sci. U.S.A.* **72**, 459 (1975).
29. E. Canale-Parola, *Bacteriol. Rev.* **41**, 181 (1977).
30. W. Trager, *Biol. Bull.* (Woods Hole, Mass.) **66**, 182 (1934).
31. U. K. Laemmli, *Nature (London)* **227**, 680 (1970).
32. L. V. Holdeman and W. E. Moore, *Anaerobe Laboratory Manual* (Virginia Polytechnic Institute, Blacksburg, 1972).
33. C. F. Bardele, *J. Protozool.* **24**, 9 (1977); R. V. Dippell, *J. Cell Biol.* **69**, 622 (1976); S. R. Heidemann, G. Sander, M. W. Kirschner, *Cell* **10**, 337 (1977).
34. We recently have observed large free living spirochetes in the anaerobic zones of algal mats of Baja California, in an environment locally devoid of flagellates and ciliates.
35. M. Cachon and J. G. Cachon, *Ann. Biol.* **13**, 536 (1974).
36. We thank K. Fujiwara, V. Kalnins, and G. M. Fuller for antibody to tubulin; K. Fujiwara, E. Lazarides, C. Spada, and B. Hubbard for aid in the electrophoresis and D. C. Smith for advice; R. Smibert for advice and cultures of *Treponema reiteri*; W. Nutting for aid in all aspects of the termite biology and C. Spada, W. Ormerod, and L. Ozin for laboratory assistance. Supported by the Sherman Fairchild Distinguished Scholarship Program of the California Institute of Technology and by the Geology Division of NASA as well, by NASA (NGR-004-025 to L.M.) and (for L.T.) the American Association of University Women.

Patterning and Assembly of Ciliature Are Independent Processes in Hypotrich Ciliates

G. W. Grimes, M. E. McKenna, C. M. Goldsmith-Spoegler, and E. A. Knaupp

Abstract. *Mirror-imaged doublets of the hypotrich ciliate* Pleurotricha lanceolata *were induced and analyzed with respect to the overall patterning (structural asymmetry and polarity) of the individual components of the ciliature. The overall pattern is arranged as a mirror image, but the individual components in the two halves of the doublet show the same organizational asymmetry. These data demonstrate the independence of the mechanisms for this kind of large-scale (global) patterning and control of assembly of the individual ciliary components.*

Among unicells, the mechanism of cell patterning has been studied most extensively in the ciliated protozoa in which complex arrays of cilia form the cell pattern. The demonstration by Sonneborn of the lack of genic differences, genic activity differences, or differences in the fluid cytoplasm in cells possessing different cortical phenotypes (*1, 2*) emphasizes the value of these organisms in studies of cell patterning and intracellular localizations. A question remaining from such studies is whether or not the overall pattern of the ciliature is exclusively a reflection of the assembly processes of the individual components of the ciliature. We report that the overall pattern of the ciliature is determined independently from the detailed structure and assembly of the component ciliature.

The hypotrich ciliates, including *Pleurotricha lanceolata*, are well suited for studies of cell patterning because of the specific localization of ciliary units, structural polarity and asymmetry of each ciliary unit, overall cellular polarity and asymmetry, and developmental flexibility (*3-5*). The typical morphostatic cell (Fig. 1a) has an oral apparatus composed of parallel arrays of rectangularly packed cilia (each array is a membranelle) occupying approximately the antero left cell quadrant. Furthermore, clusters of hexagonally packed cilia are located regularly elsewhere on the ventral surface (the ventral ciliature and the marginal rows of ciliature, one on the left side, and two on the right). In addition to this overall polarization and asymmetry of the cell, each component of the ciliature likewise is polarized and asymmetric. Each membranelle is composed of four rows of cilia; the two postero-most the longest, the antero-most composed of only three cilia at the antero right edge of each membranelle (Figs. 1e and 2b).

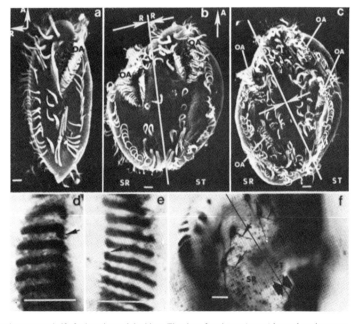

Fig. 1. Micrographs of the ventral surface of *Pleurotricha lanceolata* visualized from outside the cells. (a to c) Scanning electron micrographs; (d to f) light micrographs of silver-stained preparations. A-P, cellular polarity; L-R, axis of asymmetry of the entire cell or halves of the cell; *OA*, oral apparatus. Each bar represents 10 micrometers. (a) Typical morphostatic singlet cell illustrating the standard asymmetric array of ventral ciliature as well as the position and curvature of the oral apparatus (*OA*) in the antero left quadrant. (b) Morphostatic mirror-imaged doublet, the common polarity of both halves (approximate line of bilateral symmetry marked by vertical line) and mirrored patterning of ciliature, including curvature of the oral apparatus are shown. The lateral arrows indicate L-R asymmetry of each half. *ST*, standard symmetry half of the doublet [see (a)]; *SR*, symmetry reversal half of the doublet. (c) Predivision mirror-imaged doublet. Lateral arrows indicate L-R asymmetry of each half as above. There are four oral apparatuses (*OA*), and ventral ciliature is present on both sides of the line of bilateral symmetry (vertical line). (d) Organization of membranelles within the symmetry reversed oral apparatus (*OA* of *SR*) [see (b)]. The short fourth row (at arrow) is on the postero left margin of each membranelle; that is, it is an *inversion* of the standard arrangement [*OA* of *ST* illustrated in (e)] rather than a *mirror image* of the standard arrangement (Fig. 2). (e) Structure of membranelles in the oral apparatus of typical singlet cells and the standard symmetry half of mirror-imaged doublets. The short fourth row (arrow) located on the antero right margin of each membranelle (see also Fig. 2). (f) Morphostatic mirror-imaged doublet illustrating the same position of the lateral fiber bundles (at narrow arrows) on the ventral ciliature of both the standard symmetry half of the cell (*ST*) as well as on the symmetry reversed half of the cell (*SR*).

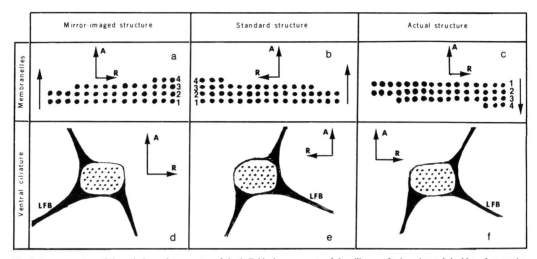

Fig. 2. Representations of the polarity and asymmetry of the individual components of the ciliature of mirror-imaged doublets. Intersecting arrows indicate polarity and asymmetry of the overall ciliary pattern of the halves of the cell on which the structures are located. (a to c) Single arrows indicate structural polarity of individual membranelles; each membranelle row is correspondingly numbered (*LFB*, lateral fiber bundle). (b and e) The standard organization of components of ciliary organelles typical of singlet cells and the standard symmetry half of mirror-imaged doublets. Row 4 is on the anatomical antero right of each membranelle and the lateral fiber bundle is on the left side of the ventral ciliary aggregates. (a and d) The expected organization of ciliary organelles in the symmetry reversal half of mirror-imaged doublets if patterning and assembly were coordinately determined. The polarity of individual membranelles would be typical but with reversed asymmetry (a). Likewise, the lateral fiber bundle would originate on the opposite side of the ventral ciliature (d). (c and f) The actual organization of structures in the symmetry reversal half of mirror-imaged doublets. The membranelle (c) is organized as an inversion of the standard arrangement and the ventral ciliature (f) is identical to the standard ventral ciliature.

Each ventral cluster of cilia also has an asymmetric array of associated microtubular bundles (Figs. 1f and 2e). Thus from observation of an individual ciliary organelle in a typical morphostatic cell, the cellular axes and the overall orientation of organellar pattern can be ascertained.

To investigate the relation of assembly processes and patterning, an axial conflict was created in the organism which yielded asexually propagative mirror-imaged doublets. Two lines of these cells were isolated from independent experiments under two different experimental conditions (6). However, the cells from both lines are similar.

The mirror-imaged doublet (Fig. 1b) has two oral apparatuses and approximately a double complement of ventral ciliature. The polarity of each half is normal; however, the asymmetry of the right half is reversed (the symmetry reversal half); thus the overall pattern of the right half is a mirror image of the left half. The two distinct and independent developmental fields that are expressed during prefission morphogenesis are arranged as mirror images (Fig. 1c).

If pattern and assembly are coordinately determined, then the detailed structure of individual components of the ventral ciliature on the symmetry reversal half would also be arranged in mirror-imaged symmetry as predicted in Fig. 2d. The fibers attached to the ventral clumps of cilia would be the mirror image of those of the typical half; that is, the lateral fiber would be attached to the opposite side of the ciliary aggregate. The membranellar structure in the symmetry reversal half would be a mirror image (Fig. 2a). Stained preparations of these cells, however, reveal that none of these structures is a mirror image with respect to details of their individual structural components (Figs. 1, d to f, and 2, c and f). The ventral ciliary aggregates would have the lateral microtubular bundles on the anatomical right if complete mirror imagery existed, but the lateral bundles are on the left side (as in the typical morphostatic singlet cell) (Figs. 1f and 2f). Furthermore, the structure of each membranelle in the symmetry reversal half is not mirror-imaged, but rather inverted (the short row of cilia is on the postero left of each membranelle instead of the antero left, Figs. 1d and 2c). The assembly of membranelles within the primordium in the symmetry reversal half occurs from anterior to posterior, the same as in the typical symmetry half of the mirror-imaged doublet. However, assembly occurs from left to right instead of the typical right to left (3); thus the pattern of development is also mirror-imaged. Nevertheless, each basal body is positioned typically relative to its adjacent one, thus constructing a membranelle whose internal organization is normal. The overall mirror-imaged pattern plus the typical internal assembly mechanism result in membranelles in the symmetry reversed half that are inverted relative to cellular polarity. These data show that the patterning of ciliary organelles as well as the pattern of development of these organelles on the ventral surface of hypotrich ciliates does not represent the sum of the individual assembly events of those organelles.

Other types of asexually propagated cortical anomalies have been analyzed with respect to the control of their inheritance. Previous accounts of homopolar doublets have demonstrated that the inheritability of the doublet phenotype is independent of changes in nuclear genotype or fluid cytoplasm (1, 2, 7, 8). In these studies doublets in which both halves showed a common polarity and asymmetry were used. Another extensively investigated cortical anomaly is the inverted ciliary row, which is propagated in the inverted fashion during both sexual and asexual reproduction of the cell, and the individual components of the rows have the same asymmetry (2, 7). Common to both of these experimentally derived anomalies is the typical asymmetry of overall pattern and of the component ciliature, as well as the propagation of those anomalies under the direct influence of the existing ciliature. In *Pleurotricha*, most primordial fields develop without direct structural continuity with ciliature of a like kind. Indeed, when these hypotrichs encyst, all ciliature is broken down and an entirely new set is formed during excystment (9).

The mirror-imaged doublets reported here also are capable of encystment and excystment, and they do so true to type. Thus, redevelopment of the entire mirror-imaged pattern occurs without the direction of any existing ciliature, and illustrates that information for pattern asymmetry is retained in the cyst even in the absence of ciliature. If the overall pattern of the ciliature were the sum of individual events of the assembly of the ciliary components and there were no directive influence of existing ciliature, then the prediction would be that mirror-imaged doublets would not pass through the cyst true to type but rather revert to a typical symmetry. The fact that mirror-imagery is retained upon excystment further substantiates the conclusion that the mechanism of global patterning of the ciliature is independent of the mechanism of assembly of the individual ciliary components.

Although Tchang Tso-run and coworkers (10) have studied this type of mirror-imaged doublet and we have studied a different type of mirror-imaged doublet (11), none of those studies presented data on the internal organization of the ciliary components. Presumably, this is why the conclusions that we present have not been reached previously.

References and Notes

1. T. M. Sonneborn, in *The Nature of Biological Diversity*, J. M. Allen, Ed. (McGraw-Hill, New York, 1963), pp. 165–221.
2. ———, *Proc. R. Soc. London Ser. B* **176**, 347 (1970).
3. G. W. Grimes, *J. Protozool.* **19**, 428 (1972).
4. ——— and J. A. Adler, *J. Exp. Zool.* **204**, 57 (1978).
5. The details of structure, standard development, and regenerative morphogenesis of *Pleurotricha lanceolata* have been studied (M. E. McKenna and G. W. Grimes, in preparation).
6. Mirror-imaged doublets were obtained either by heat-shocking (41°C; 17 minutes) random cultures of cells in the log phase of growth and subcloning or by subcloning surgically induced longitudinal fragments of cells in the log phase of growth. The exact mechanism of mirror-image induction remains to be elucidated.
7. S. Ng and J. Frankel, *Proc. Natl. Acad. Sci. U.S.A.* **74**, 1115 (1977).
8. G. W. Grimes, *Genet. Res.* **21**, 57 (1973).
9. ———, *J. Cell Biol.* **57**, 229 (1973).
10. W. A. Dembowska, *Arch. Protistenkd.* **91**, 89 (1938); Tchang Tso-run and Pang Yan-bin, *Sci. Sin.* **20**, 235 (1977); *J. Protozool.* **26**, part 1, 31a (1979).
11. G. W. Grimes and S. W. L'Hernault, *Dev. Biol.* **70**, 372 (1979).
12. We thank Drs. G. E. Dearlove, R. L. Hammersmith, and C. H. Harris for their comments on the manuscript. Supported by NSF grant PCM 79-08992 to G.W.G.

22 January 1980; revised 30 March 1980

32

Copyright © 1982 by MIT Press
Reprinted from Cell **29**:745-753 (1982)

Uniflagellar Mutants of Chlamydomonas: Evidence for the Role of Basal Bodies in Transmission of Positional Information

Bessie Huang,* Zenta Ramanis, Susan K. Dutcher and David J. L. Luck*
The Rockefeller University
New York, New York 10021

Introduction

Chlamydomonas, a unicellular alga, has been used extensively as a model system for analyzing the eucaryotic flagellum. Most experimental observations reveal a high degree of coordination of assembly and function between the paired flagella carried on the anterior surface of each cell. Coordinate behavior is seen in the flagellar wave form during forward and backward swimming (Schmidt and Eckert, 1976; Hyams and Borisy, 1978) and in flagellar shortening or regeneration (Rosenbaum et al., 1969; Lefebvre et al., 1978). Almost all mutations for the more than 40 genes identified as affecting flagellar function or assembly (Harris, 1980) appear to be coordinately expressed (Randall and Starling, 1972; Huang et al., 1977; Witman et al., 1978; Luck et al., 1982). One exception is a group of mutants with defects in flagellar length control. These strains, carrying short or long flagella, may show differences in length between flagella of a single pair (McVittie, 1972a).

During mutagenesis experiments with ICR-191 (Creech et al., 1960), we recovered among a large number of motility-defective strains of Chlamydomonas reinhardtii a series of mutants showing an unusual flagellar phenotype. In cultures from each of the independently isolated mutants, spinning cells bearing only a single flagellum were present. A striking feature of these strains, designated *uni* mutants, was that the position of the formed flagellum was constant relative to the position of the asymmetrically located eyespot, a chloroplast organelle located just below the plasma membrane that appears to function in phototaxis (see Foster and Smyth, 1980). If the flagella of wild-type cells are identified as being either trans or cis relative to the eyespot, the assembled flagellum in uniflagellate cells of the *uni* mutants was found to be in the trans position with a frequency of >95%. The finding that this positional phenotype was stably transmitted through both mitotic and meiotic cell divisions led us to study in detail the phenotypic expression and genetics of the mutations.

Data derived from analysis of four of the *uni* mutants are presented in this report. The four mutants define either a single gene or a set of closely linked genes that appear to influence basal body maturation and, secondarily, flagellar assembly. The positional effect of the mutations may reveal a closely regulated segregation pattern for parental and daughter basal bodies during cell divisions in wild-type cells.

Results

Isolation and Evidence for Genetic Linkage of the uni Mutants

The four mutants described in this report, *uni1*, *uni2*, *uni3* and *uni4*, were recovered as motility-defective strains following mutagenesis of wild-type strain 137c mt^+ with ICR-191. Each mutant was derived independently, and its flagellar phenotype was found to segregate two mutant colonies to two wild-type colonies in backcrosses to 137c mt^-. Tetrad analysis was performed on pairwise crosses between the *uni* mutants to test for linkage. The results of this analysis are summarized in Table 1. In a total of 471 tetrads analyzed, no recombination events were observed. The absence of detected recombination indicates that the four *uni* mutants define either a single gene or a set of closely linked genes (≤0.06 map units).

Positional Uniflagellar Phenotypes of the Mutants

In liquid cultures the *uni* mutants are readily distinguished from wild-type cells by the high proportion of spinning cells bearing only a single flagellum. Figure 1 illustrates the light microscopic appearance of the uniflagellate cells and their rotational movement. The flagellar bending cycle for rotational movement in *uni* cells has been analyzed; it is indistinguishable from the asymmetric ciliary-type bending that propels wild-

---

* Present address: Department of Cell Biology, Baylor College of Medicine, Houston, Texas 77030.
** Correspondence should be sent to: David J. L. Luck, The Rockefeller University, 1230 York Avenue, New York, New York 10021

276

Table 1. Tetrad Analysis of Four Independent *uni* Mutations to Determine Linkage

Parent 1/Parent 2	Total No. Tetrads Given as PD:NPD:T for Mutant Pairs			
	uni 1	uni 2	uni 3	uni 4
uni1	86:0:0	46:0:0	130:0:0	68:0:0
uni2			40:0:0	33:0:0
uni3				68:0:0

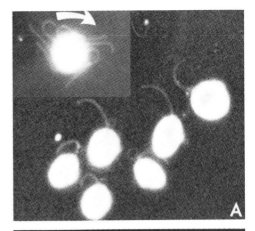

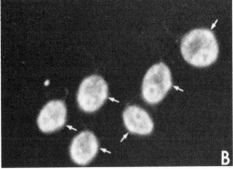

Figure 1. Dark-Field Microscopy of *uni1*
(Inset) Stroboscopic dark-field light micrograph of a uniflagellate *uni1* cell illustrating the rotational movement of the *uni* mutants and the wild-type bending pattern of the assembled flagellum. The image was obtained with a strobe frequency (60 Hz) slightly below the beat frequency (70 Hz); therefore, successive elements of single beat cycles are seen as a result of phasing. The arrow points in the direction of rotation. Exposure time, 18 sec. A single dark-field light micrograph of fixed *uni1* cells is printed to show the flagellum (A) and the location of the eyespot (B) in the same cells. In all cells except that on the far right, the assembled flagellum is trans to the eyespot.

type cells forward (Brokaw et al., 1982).

As illustrated in Figure 1, a striking correlation was found between the position of the formed flagellum in *uni* mutants and the location of the eyespot. In dark-field microscopic preparations of wild-type cells, more than 80% of the cells show clearly the points of emergence of both flagella. The brightly refractive single eyespot is found in a lateral position at the midline or slightly anteriorly. In mutant cells, too, the eyespot is in a lateral position. This makes it possible to distinguish the two flagella of a cell as being trans or cis relative to the eyespot. A common feature of the *uni* mutants is that the genetic lesions appear to affect specifically and preferentially the assembly of the flagellum normally found in the cis position. This biased effect of the mutations is evident in Figure 1. In five of the six uniflagellate *uni1* cells seen in Figures 1A and 1B, the formed flagellum is clearly in the trans position relative to the eyespot. In the other uniflagellate cell (on the far right), the assembled flagellum would be scored in the cis position. In an examination of a large number of cells from each of the four mutants (including different clones and different meiotic segregants), we observed that in more than 95% of the uniflagellate cells, the formed flagellum was in the trans-eyespot position (Figure 2).

Included in the analysis summarized in Figure 2 are the results obtained from an examination of another mutant, *lf2*. Mutant *lf2* was previously isolated, mapped to chromosome XII and described as being defective in flagellar length control (McVittie, 1972b). McVittie noted that the mutant was characterized not only by long flagella, but also by flagella of distinctly unequal lengths. In our analysis of *lf2*, we observed that more than 50% of the mutant cells showed differences between the lengths of the paired flagella, and that in approximately half of these cases only a single long flagellum was visible at the light microscopic level. The position of the flagellum in the uniflagellate *lf2* cells relative to the eyespot was examined in three different *lf2* meiotic segregants. In contrast to the *uni* mutants, the flagellum in *lf2* cells showed little or no positional effect. The flagellum in the uniflagellate *lf2* cells was found in the trans position at a frequency of 0.60. In contrast, among uniflagellated cells in *lf2 uni1*, positional assembly was detected. More than 95% of the uniflagellate double mutant cells showed a flagellum in the trans configuration. Thus the four *uni* mutations define a unique positional phenotype, and the *uni* positional effect is epistatic.

Analysis of the Expression of the uni Mutations in Haploid and Diploid Strains

During the normal cell cycle in Chlamydomonas, cell division is preceded by the resorption of the interphase flagellar pair. At the end of mitosis new pairs of flagella are rapidly assembled in the daughter cells, and the cells emerge from the wall of the mother cell with equal, full-length flagella (Cavalier-Smith, 1974).

Since we had observed variation in the mutant phenotype of cells grown at high temperatures, we characterized the occurrence of cells bearing no flagella, one flagellum or two flagella under three conditions of culture (Figure 3). These included nonsynchronous,

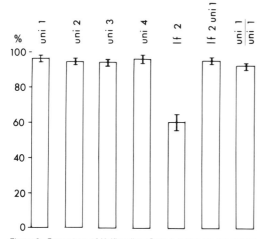

Figure 2. Percentage of Uniflagellate Cells in Which an Assembled Flagellum was Observed in a Position Trans to the Eyespot

The cells were observed with dark-field optics as seen in Figure 1. Bar indicates the standard error of the measurements at a 95% confidence limit. The number of cells of each mutant analyzed ranged from 200 to 600.

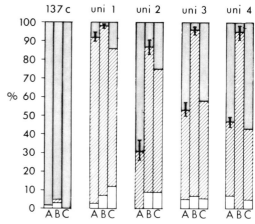

Figure 3. Expression of the Uniflagellar Phenotypes of the Mutants under Different Growth Conditions

(A) Exponential vegetative cultures grown at 22°C; (B) exponential vegetative cultures grown at 32°C; (C) stationary, gametic cultures at 32°C. The percentage of cells with no flagella (□), with one flagellum (▨) and with two flagella (▩) are shown. The results in each lane were based on light microscopic counts of at least 200 cells. In the case of the mutants grown exponentially at 22°C and 32°C, the data were derived from an analysis of 100 cells from each of ten different clones. The standard error at a 95% confidence level for the average frequency of uniflagellate cells in the different clones is indicated by the bars (⊢⊣).

vegetative cells growing exponentially at 22°C (A) and 32°C (B). In addition, we observed stationary-phase gametic cells held at 32°C (C). These conditions were chosen to identify a possible direct effect of high temperature and to distinguish it from the effect of high temperature in enhancing the cellular growth rate. For both wild-type and *uni* strains the doubling time at 32°C was approximately 9 hr, while at 22°C the doubling time was approximately 14 hr.

As is shown in Figure 3, under all three conditions more than 95% of wild-type, 137c cells were biflagellate. The small fraction of cells without flagella in vegetative cultures appeared to be cells undergoing mitosis. In *uni1* cells under all growth conditions, 75%–90% of the cells were uniflagellate. The *uni2* cells showed more than a doubling in the uniflagellate cell population when exponential growth at 32°C was compared to exponential growth at 22°C. The uniflagellate population was also high in stationary-phase gametic cultures at 32°C, suggesting that the *uni2* gene product may be thermolabile.

The situation is different in *uni3* and *uni4* cultures. In both cases there is close to a doubling of the exponential growth of the uniflagellate population at 32°C over that at 22°C. However, in nongrowing gametic cultures at 32°C, the *uni* flagellar population is reduced, and the level of uniflagellate and biflagellate cells resembles that at 22°C in vegetative growth. This result was also obtained in *uni3* and *uni4* gametic cells at 32°C when gametogenesis was allowed to take place at 32°C (data not shown). It appears that in these two mutant strains the effect of high temperature may be indirect. Since flagellar reassembly must occur with each mitotic cycle, *uni3* and *uni4* may represent cases where increasing the growth rate at 32°C reveals a rate-limiting mutant function in the reassembly of the flagellum in the cis-eyespot position.

The extreme variation in *uni2*, *uni3* and *uni4* phenotype during growth at low and high temperatures suggests that in each case the mutant produces an altered gene product affecting flagellar assembly. A similar conclusion concerning *uni1* comes from analysis of stable diploids.

Heterozygous *uni1/UNI1* and homozygous *uni1/uni1* diploids were constructed, with the *arg2* and *arg7* markers used for selection of prototrophs (Ebersold, 1967). Figure 4 compares the flagellar phenotypes of the mutant diploids with that of *UNI1/UNI1* diploids. Diploid strains of Chlamydomonas are predominantly biflagellate, with approximately 3% of the cells showing more than two flagella (Ebersold, 1967). In the heterozygous *uni1/UNI1* diploids constructed with the *uni1* mutation contributed by either an mt^+ or an mt^- parent, the majority of cells was biflagellate; only 5%–10% of the cells expressed the uniflagellar phenotype. The homozygous *uni1/uni1* diploids had a weaker mutant phenotype than the mutant haploid cells. Only 40% of the diploid cells were uniflagellate, as compared with the 75%–90% of cells in the haploid mutant. This observation that gene dosage can affect the expression of the phenotype of the *uni1* mutant

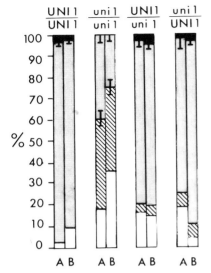

Figure 4. Analysis of Mutant Diploids

Comparison of the percentage of no flagella (□), one flagellum (▨), two flagella (▩) and more than two flagella (■) cells in diploids constructed from crosses of mt^+ arg2 and mt^- arg7. (A) Exponential vegetative cultures at 22°C; (B) exponential vegetative cultures at 32°C. The genotype at the uni1 locus is given above each lane. Ten different diploid clones from each cross were analyzed; 100 cells from each clone were counted. Significant clonal variation was not observed.

suggests that the altered gene product of uni1 may be functional in flagellar formation, but that there is a quantitative deficiency in the mutant for the functionally active form of the product. Since only a single pair of flagella is normally assembled in the diploid cells, the uni1 mutation is compensated by the 2N size of the gene product pool.

Although the frequency of uniflagellate cells was reduced in the homozygous uni1 diploids, the positional effect of the mutation was expressed. As in haploid cells, only a single eyespot is assembled in diploid cells. As seen in Figure 2, the flagellum in the uniflagellate uni1/uni1 diploids was observed in the trans-eyespot position in more than 95% of the cells.

An important feature of the biflagellate cells observed in the cultures of each of the haploid mutants and in the uni1 mutant diploids was that the paired flagella appeared equal and wild-type in length. The fact that differences in length of a pair of flagella were rarely observed suggests that the uni mutations are more likely to affect the initiation rather than the growth or elongation of the flagellum in the cis-eyespot position.

Morphological Evidence That the uni Mutations Affect Basal Body Development

In Chlamydomonas, as in other cells, the flagella are assembled onto basal bodies. With each cell division in Chlamydomonas new basal bodies are generated.

The evidence suggested that the uni mutations specifically affected the initiation of flagellar formation cis to the eyspot rather than the assembly of the flagellum itself. This led us to examine in detail the ultrastructure of the basal body complex in the mutants.

Figure 5A illustrates the morphology of the basal body complex in a wild-type interphase cell. Detailed descriptions of the ultrastructure of this region have been published (Ringo, 1967; Johnson and Porter, 1968; Cavalier-Smith, 1974). Two features of the fine structure of the basal body apparatus are most relevant to our analysis of the uni mutants. First, the paired basal bodies, each bearing a flagellum, lie at oblique angles to one another and are connected by dense striated fibers. Second, between the basal body and flagellum proper is a region known as the transition zone. This region characteristically contains two densely staining central cylinders, which appear either closely apposed or separated by approximately 25 μm. The transition zone, like the flagellum, is resorbed during mitosis and reassembled at the end of mitosis in the daughter cells prior to flagellar outgrowth.

Vegetative cultures of each of the uni mutants were grown under conditions in which more than 80% of the cells were uniflagellate, and the cells were studied by thin-section electron microscopy. A total of 51 midsagittal images of basal bodies from random sections of the four uni mutants was obtained. Two consistent morphological defects were common to all the mutants. These are illustrated in the micrograph of a uni1 cell shown in Figure 5B. In this section a pair of basal bodies in the normal orientation and association are seen. The basal body on the left bears a flagellum distal to its transition zone. The basal body on the right lies just below the plasma membrane, shows no assembly of a transition zone, no flagellar structure and no tunnel in the overlying cell wall through which the flagellum would normally emerge. Closer examination of the left basal body of Figure 5B reveals an abnormality in the formation of the transitional zone. Distal to the normal transition from microtubule triplet to microtubule doublet structures and distal to transition zone cylinders with normal appearance, additional dense material resembling cylinder wall is visible asterisk in Figure 5B).

Of 51 mutant midsagittal basal body sections examined, 12 lacked a flagellum. Since these basal bodies were not indexed by an attached flagellum, they were difficult to identify in thin sections. It is likely that this problem in identifying such images accounts for their relatively low representation in our sample. Five of the basal body images lacking a flagellum resembled the right basal body of Figure 5B, in which no distal differentiation was observed. The length of these structures closely corresponded to the length of normal, proximal, basal body structures, and it is unlikely that they were images of probasal bodies (Gould, 1975). An additional seven images showed

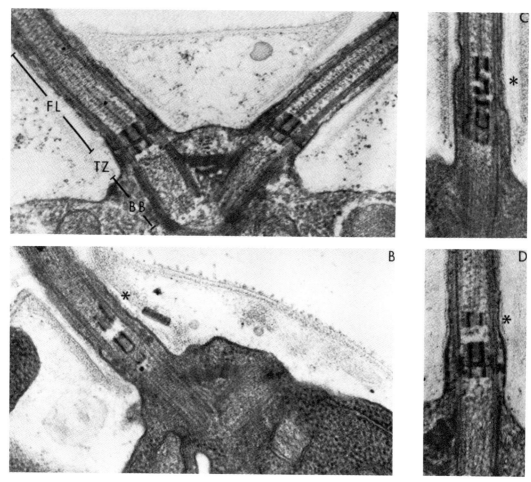

Figure 5. Longitudinal Thin-Section Electron Micrographs Illustrating Morphology of the Basal Body Complex in Interphase Wild-type and uni Mutant Cells

(A) Shows a relatively rare view in which the plane of the section cuts midsagittally through both basal bodies of a wild-type cell. The transition zone (TZ), which lies between the basal body (BB) and flagellum (FL) proper, contains two densely staining central cylinders. The central pair microtubules of the flagellum appear just distal to the upper closed cylinder. The flagella extend through channels in the cell wall. (B) A uni1 mutant cell from the same view as in (A). A pair of basal bodies in their normal orientation and association is seen. The basal body on the right approaches the cell membrane. Transition zone structures and emergent flagellum are absent. Assembled onto the basal body on the left is an apparently normal transition zone and flagellum; however, distal to the transition zone is additional dense material (*) similar in morphology to the walls of the transition zone cylinders. (C and D) Images of basal bodies bearing flagella from uni1 and uni3, respectively, illustrating the various forms of aberrant accumulation of transition zone material (*) that have been observed in the uni mutants. A, B, C and D: 75,000×.

some assembly of the transition zone region, but neither a flagellum nor a flagellar tunnel was present.

Thirty-nine of the midsagittal images of mutant basal bodies that were obtained carried a flagellum. Of these, 25 (64%) showed assembly of abnormal transition zone material. There was considerable variation in its configuration. Figures 5C and 5D illustrate some of the forms encountered.

Two images of uniflagellate cells were obtained that included the paired basal bodies as well as the eyespot. In both cases the missing flagellum was cis to the eyespot. In one case the trans basal body showed an alteration in transition zone resembling the alteration shown in Figure 5B, and in the other case the transition zone resembled sections from wild-type cultures.

Linkage Analysis of the uni Mutants and Evidence of an Aberrant Frequency of Second Division Segregation of the uni Mutations

The uni1 and uni3 mutants were studied by recombination analysis to 33 markers mapping to different

750 B. Huang et al.

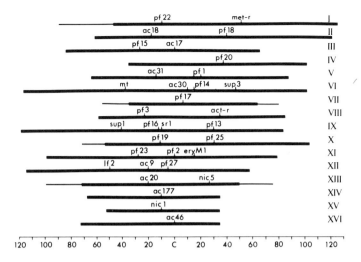

Figure 6. Linkage Map of Chlamydomonas

The map positions and designations of 33 markers studied by recombination analysis to *uni1* and *uni3* are indicated. On the basis of their map locations a chi-square analysis of the observed frequency of parental ditype, nonparental ditype and tetratype tetrads identified those regions on the linkage groups where *uni1* and *uni3* do not map within 95% confidence limits. These regions are indicated by the heavy lines. Data were derived from analysis of 2300 dissected tetrads; sample size per cross ranged from 13 to 240 tetrads. Six flanking regions on the left or right arms of linkage groups I, VII, X and XIII for which markers are available but have not been tested are indicated by thin lines.

positions on each of the 16 nuclear linkage groups in C. reinhardtii. In no case was evidence of linkage detected. On the basis of the map positions of known markers and the observed frequency of parental ditype, nonparental ditype and tetratype tetrads, a chi-square analysis identified those regions on the 16 linkage groups where *uni1* and *uni3* do not map within 95% confidence limits.

Figure 6 summarizes the results of this analysis. The map locations of the tested markers are indicated, and those regions of the linkage groups where *uni1* and *uni3* do not map are indicated by the heavy lines. In many cases our analysis excluded regions extending far beyond 50 map units on either side of the centromere. Six flanking regions on the left or right arms of linkage groups I, VII, X and XIII, for which markers are available but have not been tested, are indicated by the thin lines.

Preliminary experiments do reveal, however, that the *uni* mutations are linked and separable by recombinations to four loci that were previously unmapped to any nuclear linkage group: *pf7*, *pf8* (McVittie, 1972b), *fla9* (Adams et al., 1982) and an extragenic suppressor $sup_{uni}1$. Linkage was indicated by a high frequency of parental ditypes in single factor crosses with *uni1*. The following parental ditype frequencies were observed among the total number of tetrads as shown in parentheses: *pf7*, 0.91 (55); *pf8*, 0.95 (21); $fla_{ts}9$, 0.74 (111); $sup_{uni}1$, 0.50 (58). All of these mutants share with *uni* the property that they affect assembly of flagella rather than motility of flagella.

In our linkage analysis we observed that both *uni1* and *uni3* showed elevated frequencies of second-division segregation. In crosses of *uni1* and *uni3* to *ac17*, a centromere marker on linkage group III, the frequency of tetratype tetrads was greater than the

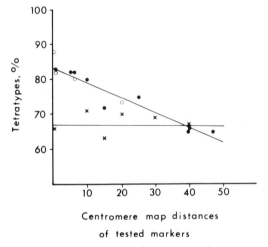

Figure 7. Unusual Second-Division Segregation of *uni* Locus

The percentage of tetratype tetrads in crosses of *uni1* (●), *uni3* (○) or *msr1* (×) to nuclear markers located from 0–51 map units from their centromere. Each point is based on 50–240 tetrads, with an average of 119 tetrads per point for *uni1* or *uni3*, and 78 tetrads per point for *msr1*. The crosses, identified from left to right were as follows:
uni1 × *ac17*, *pf27*, *pf17*, *sr1*, *pf14*, *pf23*, *pf18*, $sup_{pf}1$, $sup_{pf}3$,
uni3 × *ac17*, *pf2*, *pf17*, *pf1*
msr1 × *ac17*, *sr1*, *ery1*, *pf1*, *pf13A*, *pf25*.

maximum theoretical expectation of 67% with no interference (Perkins, 1952). Of the 169 tetrads examined, 85% had a tetratype configuration. In additional crosses of *uni1* and *uni3* to 14 unlinked loci, the frequency of tetratype tetrads varied systematically with the distance to the centromere of these markers (Figure 7). When the second marker was at or close to its centromere, the frequency of tetratypes was between 83% and 88%. When the second marker

281

Table 2. Analysis of Recombination Frequencies in the Presence or Absence of uni1

Centromere Distances Locus (Linkage Group)	Calculated Distances in centiMorgans (PD:NPD:T)	
	uni1 × UNI1	UNI1 × UNI1
sr1 (IX)	14 (7:14:8)	12 (25:31:18)
ery1 (XI)	20 (12:5:12)	16 (20:30:24)
msr1 (I)	31 (7:4:18)	33 (15:10:49)
Recombination Distances Gene Pair		
$sup_{pf}1-sr1$	18 (17:1:7)	18 (42:0:24)
$sup_{pf}1-pf16$	15 (25:0:11)	9 (30:0:7)

The effect of the uni1 allele on recombination in the genome was monitored by two parallel crosses. The marker sr1 confers resistance to streptomycin, ery1 to erythromycin, and msr1 to methionine sulfoximine. The markers $sup_{pf}1$ and pf16 affect motility and code for structural components of the axoneme (see Harris, 1980). Centromere distances were obtained from crosses between ac17 and an unlinked locus (sr1, ery1 or msr1). Distances were calculated by the formula $\frac{P(T)}{2} = a - 3ab + b$, where $P(T)$ is the percentage of tetratypes, a is the distance of one locus from its centromere and b is the distance of the other locus from its centromere (Gowans, 1965). Recombination distances were measured along the left arm of linkage group IX. Distances (D) were calculated by the formula $D = \frac{N(T)}{2} + N(NPD)$/total tetrads, where $N(T)$ is the number of tetratypes, and $N(NPD)$ is the number of nonparental ditypes (Perkins, 1952).

was unlinked to its centromere (more than 38 map units), the frequency of tetratype tetrads fell to an average value of 67%. A similar analysis of msr1, which maps at 51 map units on the right arm of linkage group I, showed no systematic variation in tetratype frequencies in pairwise crosses with six unlinked markers with centromere distances of 0-40 map units (Figure 7). A distal location in Chlamydomonas does not always lead to aberrant frequencies of second division segregation.

We tested for the possibility that the presence of uni1 might alter recombination distances for genomic markers unlinked to uni1. In crosses shown in Table 2 we demonstrated that measurements for distance from the centromere for three markers on linkage groups I, IX and XI were no different in the presence or absence of uni. Furthermore, the presence of uni did not alter recombination distances for three linked markers on linkage group IX.

Discussion

We have described four uniflagellar mutants of Chlamydomonas reinhardtii in which the formation of one specific flagellum of the wild-type pair is uniquely affected. The unilateral expression of the uni mutations on the formation of the flagellum cis to the eyespot has demonstrated that although the postmitotic assembly of the paired flagella is normally under coordinate control, at some level in their development, the two flagella can be differentiated. Moreover, the mutant phenotypes indicate that each flagellum of the pair bears a constant positional relationship to the asymmetrically distributed eyespot that is reproduced with each cell division.

Genetic analysis of the uni mutations suggest that they affect nuclear genes. In all heterozygous zygotes the segregation pattern is two wild-type to two mutant daughters. The mutations are stable through at least four sequential meiotic divisions, as well as through repeated mitotic divisions. In test crosses to the 33 markers we were unable to demonstrate linkage to any of the 16 established linkage groups. This failure to establish linkage to known nuclear loci may result simply from the relative paucity of mapped genes in Chlamydomonas. The uni mutants, however, did show linkage to four other mutations. These mutations segregate 2:2 and are also unmapped with respect to a nuclear linkage group. Strikingly, these mutants, pf7, pf8, fla9 and $sup_{uni}1$, share the property with the uni mutants that they affect flagellar assembly. The mutants pf7 and pf8 result in cells with short flagella or without flagella. The mutant fla9 is a temperature-sensitive flagellar assembly mutant, and $sup_{uni}1$ is an ultraviolet-light-induced extragenic suppressor of the uni mutants. The functional significance of the linkage of these five loci is not known.

An aberrantly high frequency of second-division segregation was documented for the uni1 and uni3 mutations in the course of linkage analysis. High frequencies of second-division segregation are not a general feature of Chlamydomonas chromosomes, as confirmed by our analysis of msr1, which maps at 51 map units on the right arm of linkage group I, and by analyses of various other linkage groups (Ebersold et al., 1962). Furthermore, no evidence was obtained to indicate that the uni1 mutation affects recombination of nuclear markers unlinked to it. The unusual segregation pattern of uni1 and uni3 can be explained by invoking either chiasma interference or chromatid interference in which two-strand and four-strand double exchanges are reduced as a feature of the uni linkage group. With the availability of linked markers, it will be possible to demonstrate the presence or absence of interference by analyzing various three-point crosses. These mutants may reveal the existence of a region with high interference on one of the established linkage groups or on a new linkage group in Chlamydomonas.

Our only clue to understanding the basis of the uni phenotype comes from morphological analysis of the site of the flagellum that fails to be assembled. We obtained 12 midsagittal images of mutant basal bodies lacking flagella. Five showed the assembly only of a basal body proper, and seven showed partial assembly of the transition zone. Although limited in number, these images suggest that the rate-limiting steps in the *uni* mutants appear to affect development of the basal body and transition zone. Observations of *uni2*, *uni3* and *uni4* cells at 22°C and 32°C and of *uni1/uni1* diploids indicate that once development at the basal body is complete, flagellar outgrowth is rapid. Biflagellate cells with flagellar stumps or unequal flagella are rarely seen.

Flagella are disassembled in preparation for mitotic cell division. Each of the paired basal bodies, lacking a transition zone, migrates to a more lateral position in the cell (Johnson and Porter, 1968). At the new site, the basal body appears to function as an organizing center for the mitotic spindle (Coss, 1974). There is evidence to suggest that in the subsequent events of cytokinesis each daughter cell receives a mature (parental) basal body with its associated probasal body (Gould, 1975). Thus when flagellar outgrowth occurs in newly formed daughter cells, the two basal bodies involved may differ initially in their stage of development. This difference in maturity could be the basis of the unilateral expression and the variable phenotypes of the *uni* mutants. If *uni* gene products were required for development of the basal body, deficiency in their function might preferentially affect maturation of the least mature of the pair. Under these circumstances, during basal body maturation in the mutants there might be a time when the parental basal body would be competent for flagellar assembly, while its probasal body counterpart would not. Prolonging the cell division cycle, as in *uni3* and *uni4*, or increasing the *uni* gene dosage, as in *uni1/uni1* diploids, might facilitate maturation and the acquisition by the progeny basal body of the competence to initiate flagellar assembly.

Another morphological alteration was encountered in approximately 65% of the images of assembled flagella—namely, the accumulation of additional transition zone cylinders (Figure 5B). Based on the high frequency of their occurrence, and in some cases on direct observation of paired basal bodies in uniflagellate cells, we conclude that most, if not all, of these abnormal transition zones are associated with the assembled flagellum (trans to the eyespot) in uniflagellate cells. According to the model just presented, in daughter cells the altered transition zone would be at the site of the mature basal body derived from the parental cell. The significance or mechanism of this alteration is not clear. It could reflect disruption of the normally coordinated postmitotic development of the flagellar pair. Similar abnormal transition zone cylinders have been described in several flagellaless strains (McVittie, 1972a).

The positional effect of the *uni* mutations is fascinating, but not easily explained. Clearly, in dividing cells some coordination exists between flagellar development and the positioning of the eyespot. In interphase cells four bands of microtubules originate at the proximal region of basal body pairs and extend in different directions laterally just below the cell surface (Ringo, 1967). These microtubule bands provide a possible structural link between basal bodies and more distant cell structures. Within 6 hr after treatment of cells with colchicine, at a time before flagella are lost, the eyespot has been observed to move from its usual position and to appear in the posterior region of the cell (Walne, 1967). In untreated cells, microtubules have been shown to lie in near association with the eyespot (Gruber and Rosario, 1974).

A highly speculative model to explain positional effects would emphasize the role of basal bodies in organizing microtubular frameworks that function to position other cell structures. Following cell division, the time course of development or the form of organization of the microtubular framework might reflect differences in maturity of the two basal bodies. There is little experimental support for this model. However, it is of potential interest as a general explanation for how asymmetric distribution of basal bodies and probasal bodies, or centrioles and procentrioles, to daughter cells might provide a basis for cellular asymmetry.

Experimental Procedures

The methods used for mutagenesis with ICR-191 (Terochem Laboratories) were as previously described (Huang et al., 1981). All genetic analysis was performed according to standard techniques (Levine and Ebersold, 1960). Cultures of *lf2* and several marker strains used in linkage analysis were provided by E. Harris. The procedures for culturing cells and preparing samples for electron microscopy were essentially as previously described (Huang et al., 1977, 1979).

Acknowledgments

We are grateful to Judy Fleming for her excellent assistance in the electron microscopic studies, and we thank Dr. Nicholas Gillham for advice and useful discussion during the early phase of these studies.

This work was supported by grants from the National Institute of General Medical Science.

The costs of publication of this article were defrayed in part by the payment of page charges. This article must therefore be hereby marked "*advertisement*" in accordance with 18 U.S.C. Section 1734 solely to indicate this fact.

Received January 27, 1982; revised April 14, 1982

References

Adams, G. M. W., Huang, B. and Luck, D. J. L. (1982). Temperature-sensitive, assembly defective flagella mutants of *Chlamydomonas reinhardtii*. Genetics, in press.

Brokaw, C. J., Luck, D. J. L. and Huang, B. (1982). Analysis of the movement of *Chlamydomonas* flagella: the function of the radial-spoke system is revealed by comparison of wild-type and mutant flagella. J. Cell Biol. 92, 722–732.

Cavalier-Smith, T. (1974). Basal body and flagellar development during the vegetative cell cycle and the sexual cycle of *Chlamydomonas reinhardtii*. J. Cell Sci. 16, 529–556.

Coss, R. A. (1974). Mitosis in *Chlamydomonas reinhardtii* basal bodies and the mitotic apparatus. J. Cell Biol. 63, 325–329.

Creech, H. J., Breuninger, E., Hankwitz, R. F., Polsky, G. and Wilson, M. L. (1960). Quantitative studies of the effects of nitrogen mustard analogs and other alkylating agents on ascites tumors in mice. Cancer Res. 20, 471–494.

Ebersold, W. T. (1967). *Chlamydomonas reinhardtii* heterozygous diploid strains. Science 157, 447–449.

Ebersold, W. T., Levine, R. P., Levine, E. E. and Olmsted, M. A. (1962). Linkage maps in *Chlamydomonas reinhardtii*. Genetics 47, 531–543.

Foster, K. W. and Smyth, R. D. (1980). Light antennas in phototactic algae. Microbiol. Rev. 44, 572–630.

Gould, R. R. (1975). The basal bodies of *Chlamydomonas reinhardtii*. Formation from probasal bodies, isolation and partial characterization. J. Cell Biol. 65, 65–74.

Gowans, C. S. (1965). Tetrad analysis. Taiwania 11, 1–19.

Gruber, H. E. and Rosario, B. (1974). Variation in eyespot ultrastructure in *Chlamydomonas reinhardtii*(ac-31). J. Cell Sci. 15, 481–494.

Harris, E. H. (1980). *Chlamydomonas reinhardtii*. In Genetic Maps, 1, S. J. O'Brien, ed. (Washington, D.C.: National Cancer Institute), pp. 142–146.

Huang, B., Rifkin, M. and Luck, D. J. L. (1977). Temperature-sensitive mutations affecting flagellar assembly and function in *Chlamydomonas reinhardtii*. J. Cell Biol. 72, 67–85.

Huang, B., Piperno, G. and Luck, D. J. L. (1979). Paralyzed flagella mutants in *Chlamydomonas reinhardtii*: defective for axonemal doublet microtubule arms. J. Biol. Chem. 254, 3091–3099.

Huang, B., Piperno, G., Ramanis, Z. and Luck, D. J. L. (1981). Radial spokes of *Chlamydomonas* flagella: genetic analysis of assembly and function. J. Cell Biol. 88, 80–88.

Hyams, J. S. and Borisy, G. G. (1978). Isolated flagellar apparatus of *Chlamydomonas*: characterization of forward swimming and alteration of waveform and reversal of motion by calcium ions *in vitro*. J. Cell Sci. 33, 235–253.

Johnson, U. G. and Porter, K. R. (1968). Fine structure of cell division in *Chlamydomonas*. Basal bodies and microtubules. J. Cell Biol. 38, 403–425.

Lefebvre, P. A., Nordstrom, S. A., Moulder, J. E. and Rosenbaum, J. L. (1978). Flagellar elongation and shortening in *Chlamydomonas*. J. Cell Biol. 78, 8–27.

Levine, R. P. and Ebersold, W. T. (1960). The genetics and cytology of *Chlamydomonas*. Ann. Rev. Microbiol. 14, 197–216.

Luck, D. J. L., Huang, B. and Piperno, G. (1982). Genetic and biochemical analysis of the eukaryotic flagellum. In Society for Experimental Biology Symposium 35. Eukaryotic and Prokaryotic Flagella, W. B. Amos and J. G. Duckett, eds. (New York: Cambridge University Press), in press.

McVittie, A. (1972a). Flagellum mutants of *Chlamydomonas reinhardtii*. J. Gen. Microbiol. 71, 525–540.

McVittie, A. (1972b). Genetic studies on flagellum mutants of *Chlamydomonas reinhardtii*. Genet. Res. Camb. 9, 157–164.

Perkins, D. D. (1952). The detection of linkage in tetrad analysis. Genetics 38, 187–197.

Randall, J. and Starling, D. (1972). Genetic determinants of flagellum phenotype in *Chlamydomonas reinhardtii*. In Proceedings of the International Symposium on the Genetics of the Spermatozoon, R. A. Beatty and S. Gluecksohn-Daelsch, eds. (Copenhagen: Bogtrykkeriet), pp. 13–36.

Ringo, D. L. (1967). Flagellar motion and fine structure of the flagellar apparatus in *Chlamydomonas*. J. Cell Biol. 33, 543–571.

Rosenbaum, J., Moulder, J. and Ringo, D. (1969). Flagellar elongation and shortening in *Chlamydomonas*. The use of cycloheximide and colchicine to study the synthesis and assembly of flagellar proteins. J. Cell Biol. 41, 600–619.

Schmidt, J. A. and Eckert, R. (1976). Calcium couples flagellar reversal to photostimulation in *Chlamydomonas reinhardtii*. Nature 262, 713–715.

Walne, P. L. (1967). The effects of colchicine on cellular organization in *Chlamydomonas*. II. Ultrastructure. Amer. J. Bot. 54, 564–577.

Witman, G. B., Plummer, J. and Sander, G. (1978). *Chlamydomonas* flagellar mutants lacking radial spokes and central tubules. J. Cell Biol. 76, 729–747.

Part VII
ORIGIN OF MITOSIS, MEIOSIS, AND SEX

Editors' Comments
on Papers 33 and 34

33 CLEVELAND
Sex Produced in the Protozoa of Cryptocercus by Molting

34 MARGULIS and SAGAN
Excerpt from Evolutionary Origins of Sex

   Eukaryotic cell division (mitosis) differs from that of prokaryotes and constitutes one of the primary characteristics of the eukaryotes. The eukaryotic DNA is coiled with histone proteins to form chromosomes during parts of the cell cycle, unlike prokaryotic DNA, which is a single loop of nucleic acids without histone proteins. The movement of chromosomes in eukaryotic cell division is accomplished by spindle fibers made of tubulin microtubules whereas in prokaryotes the loop of DNA is attached inside the cell membrane, which grows and pulls the replicated genome apart. The evolution of eukaryotic cell division has been reviewed by Pickett-Heaps (1974).
   The tubulin-based spindle fibers represent another unique characteristic of eukaryotes, the presence of this versatile protein thus far not found in any prokaryotes except possibly spirochetes, spiroplasmas, and *Azotobacter*. The significance of the exceptions is discussed in Part VI. Tubulin is also used for motility of substances within the cell and for support of the cell architecture.
   Eukaryotic cells with the ability to divide by pulling apart chromosomes with spindle fibers were also pre-adapted to a type of cell division reducing a diploid chromosome number to haploid. This type of division (meiosis) is primarily used to create gametic genomes, which may then fuse and recombine with other gametic genomes. This division and recombination fits the biological definition of sex. Sex is one of the many mechanisms by which cells establish and maintain variability in their genomes. Sex is not necessarily linked to reproduction, although this is a popular misconception because sex is best studied in mammals where the two processes are connected.

Small groups of protoctists, fungi, and plants studied in which sex is not always linked to reproduction and is not always obligate abound in fascinating sexual variations too numerous to summarize here. A surprisingly modern discussion on the origin of meiosis as an environmentally induced event is presented in a paper by Cleveland that is seldom read (Paper 33). Only by studying sexuality as a diverse phenomenon with a long evolutionary history that began with the protoctists can the origins of sex in eukaryotes be understood (Paper 34).

REFERENCE

Pickett-Heaps, J., 1974, The Evolution of Mitosis and the Eukaryotic Condition, *Biosystems* **6**:37–48.

Sex Produced in the Protozoa of *Cryptocercus* by Molting

L. R. CLEVELAND

Department of Biology, Harvard University[1]

Twenty-five species of flagellate protozoa live in the hind gut of the wood-feeding roach, *Cryptocercus punctulatus* (1). They represent a varied assortment of organisms, ranging from small, fairly simple cells to large, complex ones, and comprise 2 orders, 8 families, and 12 genera. They all exhibit some form of sexual behavior when their host molts, but between molts there is no sexual behavior, they have the haploid number of chromosomes, and division is mitotic. However, under the influence of molting, their chromosome number is doubled, and they remain diploids until two meiotic divisions convert them again to haploids. But this reversible process of mitosis to meiosis does not occur in all of them at the same time: some begin it in the early stages of molting, finishing two or three days before the exoskeleton is shed; others do not begin it until 10–15 hours after the exoskeleton is shed and do not complete it until two to three days later. Also, they do not all employ the same method in changing from haploidy to diploidy.

The sexual process is fertilization in *Trichonympha, Leptospironympha,* and *Eucomonympha,* but in each genus there are interesting fundamental differences in the details of the process. In *Trichonympha* gametogenesis occurs within cysts, each gametocyst producing two gametes. In *Eucomonympha,* no cysts have been seen, and the method by which the gametes are formed is unknown. In *Saccinobaculus, Barbulanympha,* etc., other processes occur.

In *Trichonympha,* both gametes are the same size, but there are very definite, easily recognized differences between them before fertilization begins; and sometimes these differences may be seen before excystation, although they are usually not evident until later. The female or egg has an area of clearly defined, large, dense granules embedded in a jelly-like matrix. This area lies in the posterior end of the body and occupies about one-tenth of the entire cell. There is a clear, open space in the center of the area into which the male or sperm, which has no area of specialized granules, inserts its somewhat pointed, anterior, rostral end when fertilization begins. In a comparatively short time very firm contact is made between the gametes, and from this point on the egg plays the active role by ingesting the entire cytoplasmic and nuclear contents of the sperm. Soon after ingestion is complete, the extranuclear organelles of the sperm begin to disintegrate. First the nuclear sleeve, the outer and inner caps, and the postrostral flagella go, then the rostral flagella and the parabasals, leaving only the rostral tube, the rostral lamella, and the two centrioles (a long one and a short one at interphase). In the meantime the sperm nucleus becomes free to move, because of the disintegration of the extranuclear organelles holding it in position. It moves toward the egg nucleus, which is not free to migrate because those organelles which keep it in place, like the others of the egg, do not disintegrate. The membranes of the two nuclei touch each other and soon join so firmly that it is impossible, in living material, to separate one from the other without destroying both nuclei. Their chromosomes, which can be seen clearly in the living state, come closer and closer together, and an attraction between homologues is plainly evident. Complete nuclear fusion results, and the two groups of chromosomes become one group, all lying in a common, enlarged nuclear membrane. By this time, or shortly thereafter, the remaining extranuclear organelles of the sperm (rostral tube, rostral lamella, and centrioles) begin to disintegrate and soon disappear. Now a duplication of the male and female chromosomes occurs, and they enter the first meiotic prophase. As they shorten, synapsis and tetrad formation occur. At this point, the centrioles of the egg produce an achromatic figure which functions to separate the shortened, rod-shaped, metaphase chromosomes into two groups, the chromosomes going to the poles as dyads. This is the first meiotic division, and the second, which follows quickly—before the chromosomes have time to divide—is as typical as the first. This returns the chromosomes to the haploid condition.

In *Leptospironympha,* no cytoplasmic differentiation in the gametes has been seen, but their nuclei are clearly differentiated into male and female. As in *Trichonympha,* one gamete begins to enter the other, is ingested, and loses its extranuclear organelles. Nuclear fusion and zygotic meiosis are as in *Trichonympha.* However, *Leptospironympha* differs markedly from *Trichonympha,* in that its gametes are quite unlike the ordinary (somatic?) cells from which they arise. They have only the short rostral portion of flagellar bands. The two long spiral portions, which in nongametic cells extend from the anterior to the posterior end, are absent. The lack of these heavy, rather rigid bands obviously greatly facilitates the entrance of the sperm and the cytoplasmic union of the two cells.

In *Eucomonympha* one gamete does not enter the other; the two join in a manner similar to conjugation in ciliates, although the process is not conjugation in any sense of the word. When the gametes first come together, only a small portion of their

---

[1] Some of this work was done at the Mountain Lake Biological Station of the University of Virginia. It was aided by a grant from the Penrose Fund of the American Philosophical Society.

surfaces is joined, and they present a picture like that of very late cleavage; later, after joining more completely, they present a picture closely resembling early cleavage. Finally, the furrow between their joined surfaces disappears, and what was once two separate cells now looks exactly like a larger but single cell except for the presence of two nuclei and two rostra. In the early stages of cytoplasmic fusion, the rostra, each with a nucleus held in a fixed position at its base, point more or less in the same direction; however, as fusion progresses, the rostra move in opposite directions, finally taking up positions at opposite ends of the zygote. If the situation is examined more closely, it will be seen that the male gamete which, like the female, has flagella over its entire body except for a very small area at the posterior end, began to lose its flagella about the time the furrow between the fusing gametes disappeared. This process continues until the axostyles and all the flagella with the exception of those on the small, anterior, rostral portion of the body disappear. At this time, or slightly earlier in some instances, the nucleus of the male gamete becomes free of its fixed position at the base of the rostrum and begins to migrate toward the nucleus of the female gamete, which, since *Eucomonympha* is a very large cell, may lie 200–300 μ away. However great the distance may be, the male nucleus succeeds in making contact with the female nucleus, which never moves from its fixed position at the base of the rostrum of the female gamete. By the time the nuclei begin to fuse, or thereabouts, the rostrum, lamella, and centrioles of the male gamete are extruded from the cell, pinched off with the loss of practically no cytoplasm. As in *Trichonympha*, the female gamete never loses any of its organelles, while the male loses all except the nucleus.

In the large polymastigote *Saccinobaculus*, the process is autogamy and begins by simultaneous division of many of the nuclei. The daughter nuclei may move a considerable distance apart in the cell and return to the interphase condition, yet the cytoplasm makes no effort to divide. When the nucleus divides, the flagella and the very large, broad, heavily staining axostyle are discarded and renewed, one new set being produced for each nucleus. This occurs three to five days before the roach sheds its exoskeleton, and the cells remain in this condition for one to two days. Then the posterior ends of the axostyles begin to move together. This process continues until the axostyles lie side by side, from end to end, so closely that one has to look carefully to see that two are present. Since the nuclei are securely anchored to the axostyles near their anterior ends, this brings them close together, and fusion follows, but not immediately. It usually occurs shortly before the roach sheds its exoskeleton, or about two days after daughter nuclei and axostyles come together. It should be noted that *Saccinobaculus*, unlike the three genera already considered, does not lose its extranuclear organelles when the nuclei fuse, but at a considerably later time. The duplication of chromosomes occurs 10–20 hours after nuclear fusion. This is followed by synapsis, formation of tetrads, first and second meiotic divisions. Since the extranuclear organelles do not begin the process of disintegration until the prophase of the first meiotic division, and a long time is required for the large, heavily staining axostyle to be dissolved, the presence of two of these structures, instead of one as in all other divisions, serves as a perfect label for the first meiotic division.

Since the processes occurring in *Barbulanympha* and related genera supply valuable information on the origin and evolution of meiosis, these will be dealt with in another paper.

The writer does not know whether the effect of the molting fluid of the roach on the protozoa is direct or indirect, although the experiments carried out so far suggest that it is direct. Withholding of food, addition of CO_2, and removal of some O_2, conditions present at molting, fail to produce any of these sexual phenomena. However, irrespective of whether the effect is direct or indirect, the results set forth here indicate that the evolution of sexual and asexual phases in the life cycles of protozoa began as an environmental response.

Reference

1. CLEVELAND, L. R., et al. *Mem. Amer. Acad. Arts Sci.*, 1934, 17, 185.

Copyright © 1985 by Oxford University Press, England
Reprinted from pages 30–47 of *Oxford Surveys in Evolutionary Biology,* R. Dawkins and M. Ridley, eds., Oxford University Press, Oxford, England, 1985

EVOLUTIONARY ORIGINS OF SEX

L. Margulis and D. Sagan

[*Editors' Note:* In the original, material including Figures 1 through 5 precedes this excerpt.]

MEIOTIC SEX

To discuss the origin of an entirely different, nonbacterial sort of sex – the meiotic sex of eukaryotic cells – it is necessary to leap over millions of years of evolutionary time. Meiotic sex, the kind in which each parent makes an approximately equal genetic contribution to each offspring, probably arose in shallow waters about a billion years ago. This date is based on the interpretation of the late Proterozoic fossils as members of the Protoctist Kingdom (Knoll and Vidal 1983) and the well-documented appearance of soft-bodied metazoans in the late Vendian (Ediacaran animal fossils about 680 million years old: Cloud 1983; Kaveski and Margulis 1983). The presence of metazoans is of course *prima facie* evidence for the prior evolution of meiosis.

By the time the first meiotic sex developed in single-celled eukaryotic microbes two fundamental innovations had already occurred. The first was the origin of the eukaryotic cell itself, which probably resulted from the increasingly integrated symbiotic activities of different kinds of bacteria (Margulis 1981). Second was the innovation of mitosis, by which enormous amounts of genetic material, generally about a thousand times more than that in a bacterium, was reliably distributed to offspring cells. In all eukaryotic cells genetic information resides as DNA bounded by nuclear membrane. In nearly all eukaryotes this nuclear DNA is coiled around histone proteins to form eight-part, knob-like structures called nucleosomes. The knobbed string of nucleosomes, making up the material called chromatin, is further coiled into packages called chromosomes. In standard mitosis chromatin condenses into chromosomes in the early stages of mitotic division and then uncoils to form invisible or barely visible strands at the later stages.

In many types of cells, chromosomes, which by weight are mostly protein, are visible with the light microscope only during mitotic cell division (Fig. 6). Yet standard mitotic division involving visible chromosomes is not present in all eukaryotic cells. Even if visible mitosis can not be found in certain protoctists or fungi, the process itself, or some less visible variant of it must be universal in nucleated cells since accurate distribution of at least one copy of each gene is always imperative. Mitotic cell division and its variants are of course asexual processes. The complex mitotic form of cell reproduction which involves chromatin condensation into visible chromosomes must have evolved prior to the equally visible chromosomal condensation that occurs in meiosis. Meiosis in protoctists, fungi, animals and plants is a sort of series of embellishments on a basic mitotic process. Furthermore, all meiotic organisms also have standard mitotic cell divisions whereas organisms made of cells that do not divide by standard mitosis (e.g., trypanosomes, euglenids, amoebae) do not have life cycle stages involving meiosis. Thus it is generally agreed that the evolution of mitosis precede the evolution of meiosis.

In eukaryotes during mitosis, the chromosomes, bounded by the nuclear membrane, contain dot-like structures called kinetochores attached to

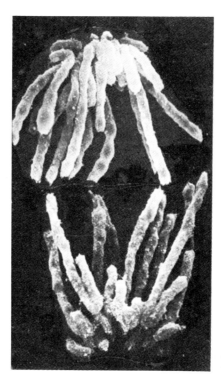

Fig. 6. Anaphase chromosomes. The scanning electron microscope reveals the chromosomes to be complex coiled structures. (Courtesy of Morten Laane, University of Oslo.)

them. The kinetochores, also called spindle fiber attachments or centromeres, are usually located on the chromosomes. Sometimes, however, they may be loose in the nucleoplasm or embedded in the nuclear membrane (Raikov 1981). Kinetochores attach to mitotic spindle microtubules, which in several different ways, depending on species, move the chromosomes. The microtubules of the mitotic spindle are required for the movement of the chromosomes and the segregation of chromosomes to the two poles at either end of a given parent cell. Such a parent cell, containing chromosomes in two discrete groupings at each end of itself, elongates dividing into two offspring cells. The end result of mitosis is the production of two new cells from one, each just like its parent. Protoctists, of which over 60,000 species have been estimated to exist (Corliss 1983), show remarkable variation on the mitotic theme. The variations range from the total absence (e.g., *Pelomyxa palustris*) of mitosis to very strange mitotic patterns involving huge extranuclear cylindrical spindles. For example in some diatoms the chromosomes attach to an amorphous material and move along the surface of an established huge cylinder of microtubules. No direct connection between the chromosomes and the spindle are formed (see Pickett-Heaps *et al.* 1979 and earlier papers in that series). Other protist mitoses also involve large unique associated cell structures, for example the 'attractophores' of hypermastigotes and the centroplast of heliozoans, (see Raikov 1981, for review). It is reasonable then to assume that mitosis arose in protist ancestors to the animals, plants and fungi in the Proterozoic Aeon (earlier than 580 million years ago). Prior to half a billion years ago many kinds of animals have already appeared in the fossil record. The mitotic process must have had a circuitous evolution of its own because it differs in many details in different lineages of eukaryotes (water mold protists and fungi, for example, see Heath 1980 a, b; or for ciliates, see Raikov 1981). No matter how different the details of the process from species to species, mitosis permits large amounts of parent DNA to be distributed equally to offspring cells. The evolution of protists from bacteria, and of multicellular protists (protoctists), fungi, plants and animals from protists fundamentally depended upon the innovations of mitosis. Large amounts of genetic material could be routinely handled in mitosis relative to what was needed for the reproduction of bacterial cells. This ability to replicate and deploy large quantities of nucleoprotein, in the form of chromatin, was a crucial precursor to the sophisticated forms of recombination known as meiotic sex.

MEIOSIS FROM MITOSIS IN PROTISTS

The main bases for the assumption that meiosis evolved from the asexual cell divisions of mitosis are twofold: (a) the wider distribution of mitosis and (b) the identification of the steps required in the conversion of a mitotic cell division to a meiotic one. Vastly different eukaryotes as varied as chlorophyte algae, mushrooms, banana trees and whales always grow by mitosis. In every example fertilization in these different species is either followed by or preceded by mitotic cell divisions. Embryos, spores or any sexual-organism-to-be grows into and maintains its form by mitotic division of somatic cells. The fact that no protoctist, plant, animal or fungus has more than an extreme minority of cells which undergo meiosis at any time argues for the prior evolutionary development of mitosis.

Fertilization, by definition, is the process that restores the chromosome number from the halved value that meiosis produces. Meiosis may be gametic, as it is in most animals. Meiosis may also be zygotic – as it is in many protoctists and most fungi. For example in chlorophytes such as *Spirogyra* (Fig. 7) and *Chlamydomonas*, or in the parasitic apicomplexans, fertilization is followed immediately by meiosis. During most of the life cycle of these organisms cells are haploid. In all the cases of organisms that undergo meiotic reduction of the number of chromosomes and restoration of that number by fertilization, the body of the organism itself, whether haploid or diploid, grows by mitosis.

It is possible to envisage the steps by which meiosis could have evolved from mitosis in protists. Even today these organisms display unique

Spirogyra, a green alga

Mating of haploids

Male on left, on right diploid zygotes

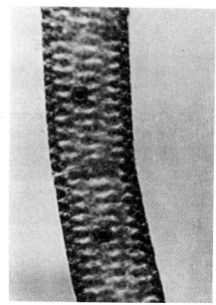
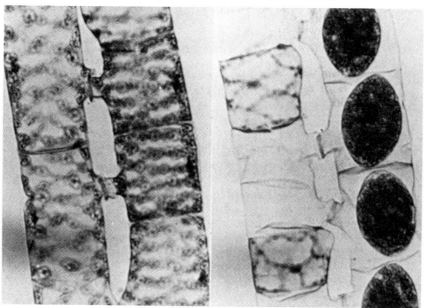

Fig. 7. Conjugating protoctist: *Spirogyra*, a chlorophyte.

variations on the meiotic life cycle, as if they were still experimenting with meiotic options that have been precluded in animals and plants which develop from diploid embryos. L. R. Cleveland (1947) was probably correct in claiming that meiotic sex evolved first in protists.

Ciliates belong to the protist phylum Ciliophora, which includes about eight thousand microbial species. Ciliates undergo meiosis reducing by half their chromosome numbers and then, like animals and plants, they reestablish the number of their chromosomes by fertilization. Yet in none of the several dozen species that have been studied does any sperm fertilize any egg. Two mature ciliates come together and mate. Each mate has at least two nuclei. These nuclei, products of meiosis, contain only half the normal number of chromosomes, and thus only half the quantity of DNA of the original cells. Each ciliate partner has at least one nucleus which it keeps and another which it passes on to its mate. Ciliates thus have sex by the unusual *modus operandi* of trading nuclei. Separate ciliate cells normally do not fuse at all; in many species the mating partners look identical. The trade is an example of organized meiotic sex, of a variation on a common theme. Karyogamy produces in each partner at least one new nucleus which contains half old and half new genetic material.

Chlorophytes like *Clamydomonas* and *Dunaliella* live out their entire lives as sorts of independent 'sperm' or 'egg'. Haploid, they reproduce by mitosis. They show willingness to mate only after external conditions become threatening, for example when the ammonium or light levels in the medium drop below a certain level. Sexual fusion leads to encystment: the doubled organism walls itself off, surviving the drought or famine. These are just a few of many examples of meiotic sex in single cells of the microbial world.

FIRST FERTILIZATION BY CANNIBALISM

The essence of meiotic sexuality if the halving of DNA quantity (chromosome number per cell) and the subsequent reestablishment of normal DNA quantity by the fusion of two cells of separate parentage. Therefore a major step in the evolution of meiotic sex from mitotic ancestors involves the formation of two sets of chromosomes, that is the doubling of the number of chromosomes in a cell. There are at least two known methods by which early protists could have doubled their chromosomes. The first is by cannibalism and the second is normal reproduction of the chromosomes followed by retardation of cytokinesis (Fig. 8).

When hungry organisms live together in crowded quarters circumstances often confront them with the 'choice' of starvation or cannibalism. Eating of conspecifics not only occurs among shipwrecked sailors and alpinists but it has also been observed in populations of protists which have exhausted their food supply. Under severe enough conditions only those microbes which are cannibalistic survive.

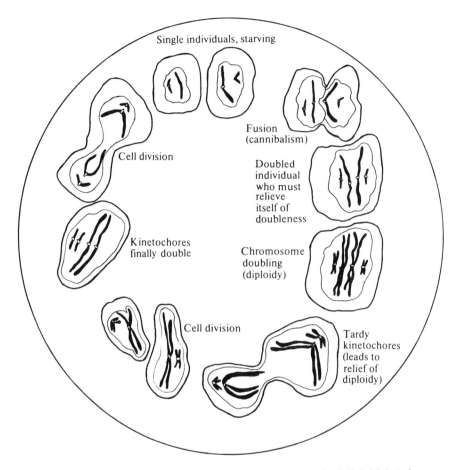

Fig. 8. Premeiosis: the early protist 'cannibalism and relief of diploidy' cycle.

Cannibalistic fusion is not uncommon. *Acanthamoebae* and ciliates including *Stentor* have been observed ingesting their neighbours under threats of starvation. Cannibalism may cause the formation of a fused pair in which cell membranes fuse. The product of fusion is a doubled but still single-celled organism containing two sets of chromosomes, mitochondria and other organelles. This sort of double cell may be considered a first step on the path to the origin of meiotic sex.

Microbes, like all organisms except most vertebrates, lack immunological responses. They tend to accept grafts and fuse easily, at least temporarily. It seems that in the waters of the Proterozoic Aeon protists cannibalistically devoured each other in their quest for food. A minority failed to digest the live food within them. In some cases ingestion without digestion led to a non-destructive co-existence. Perhaps after ingestion of the conspecific, better sources of food for both were found. The combined, cannibalized microbe may have developed an adaptive advantage over its haploid neighbours. Larger size or greater quantity of chromatin may have confered selected advantages. If the nuclei and membranes of the doubled organism were rearranged, a new diploid microbe with twice as many chromosomes would have appeared long before reproduction was coupled in any way to sexuality. The cellular state of this hypothetical ineffective cannibal would then be responsible for the first appearance of a doubled chromosome state, known as diploidy.[3]

The primordial fertilization act apparently emerged from the haphazardly cannibalistic behaviour of hungry haploid protists. This kind of accidental 'fertilization' has been observed in normally haploid microbes. Called 'diploidization', it was documented in the 1940s and 1950s by Cleveland (1935, 1957, 1963). He observed 'hairymen', hypermastigote protists which live in the intestines of wood-eating cockroaches and termites. Occasionally these protists became cannibalistic, ate each other and fused their nuclei (Fig. 9). The partners sometimes lived for a while afterwards but most fusions led to the death of both fused partners. From these studies Cleveland noted that for meiosis to begin as an evolutionary phenomenon it must have been followed by any process which brings the doubled number of chromosomes in the fused or 'fertilized' nucleus back to its original haploid state. Meiosis, Cleveland said, must first have evolved for the 'relief' of diploidy (Cleveland 1947).

Fig. 9. Hypermastigote mating: fertilization cone protrudes from the fertilization ring in the female (the anterior partner). Drawing based on 16 mm film by L. R. Cleveland.

[3] Because most plant and animal cells are normally diploid, we may tend to overestimate the biological importance of diploidy. By way of contrast cells of 'adult' mosses, red algae, apicomplexa and nearly all fungi are not diploid; most of their lives are spent as haploid individuals. Protoctists may be haploid (such as volvocalean green algae), diploid (such as diatoms), polyenergid i.e., with many full sets of chromosomes joined in a single structure (such as radiolarians), polyploid with changing levels of ploidy (such as *Colpoda* and other ciliates), multinucleate haploid or diploid (such as acellular slime moulds) or not even mitotic at all (such as pelobiontid amoebae). They may be enormously polyploid (with multiple sets of chromosomes, such as the spirotrichous ciliates) or polytene, that is, containing the normal number of chromosomes each one with many more than the usual one copy of a double DNA strand (such as the hypotrichous ciliate *Stylonichia* after conjugation). Only in tracheophyte plants and most animals, on the other hand, are diploid cells in the vast majority.

For meiotic sex to evolve and become a regular happening in the life cycle of a species cannibalism is itself insufficient: the fused partners must regularly reduce their doubled set of chromosome back to a single set. A recursive process in which the cannibalistic doubling (fusion) is followed each time by reduction in the number of chromosomes (meiosis) must begin before meiotic sex could have evolved as a mechanism of genetic interchange. In other words, fertilization must be followed, in each and every generation, by some sort of reduction of chromosome number. But, as Cleveland (1947) also pointed out, the relief of early eukaryotes from the burden of diploidy is not too difficult to envisage either.

The major necessary step involves a certain cellular tardiness. In general, the kinetochores, the points of attachment for the movement of the chromosomes, reproduce each time the chromosomal DNA itself reproduces. In mitosis the ratio of kinetochore to DNA replication is one to one. But meiosis involves a delay in kinetochoric reproduction. For, if the chromosomes divide on time, or prematurely, and the kinetochores are tardy in their own reproduction, each kinetochore will be attached to and will pull two chromatids (half chromosomes) to a pole instead of one (Fig. 8). The relief of diploidy was achieved, in the origin of meiosis, when kinetochoric reproduction was delayed and chromatids, instead of segregating from each other, went together to the offspring cell.

It has been suggested elsewhere that kinetochores and mitotic apparatus generally are descendants of an ancient spirochetal apparatus symbiotically co-opted for chromosomal motion (Margulis 1982a). If this analysis of the fundamentally reproductive behaviour of kinetochores is correct, the probability is high that delayed or accelerated replication of kinetochores occurred many times independently of the chromosomes to which they were attached (Margulis 1982b).

Being of independent origin, delayed kinetochores which were slow to divide, provided the mechanism needed for the 'relief of diploidy'. When they fail to reproduce prior to cytokinesis tardy kinetochores permit the reduction division that is the cornerstone of meiotic sex. But tardy kinetochores alone could not ensure the origin of meiosis. Take, for example, a protist cell with delayed kinetochores and only two different chromosomes, such a cell might divide becoming two offspring cells lacking the normal complement of chromosomes. In the first case, the protist might become two protists with two still-doubled chromosomes in one offspring and none in the other. Secondly, it might become two protists one of which has three chromosomes and the other just one. Thirdly it might become two protists one of which has the correct two distinct chromosomes and the other also an appropriate combination of two. All of these protist offspring, except those with one copy each of the chromosome, would die. Only the equal distribution of complete sets of chromosomes to offspring holds any potential either for immediate survival or for the origin of meiotic sex. Selection always must have been very vigorous in maintaining offspring with at least one complete set of chromosomes.

Small events accumulated to trigger the beginnings of meiotic sex. Two complete haploid sets of chromosomes were led into separate cells by the tardy kinetochores attached to spindle microtubules. The original doubled parent probably had been formed by cannibalism. However, a doubled parent also could have been produced by reproduction of the nucleus followed by a failure of the rest of the cell to divide. Such a delay in cytokinesis would lead to dikaryosis, two nuclei in the same cell. In some cases these nuclei must have fused. The dikaryosis and fusion led to autogamy – the fusion of nuclei derived from the same parent. Examples of autogamy are well known. It regularly follows meiosis in some foraminifera, some ciliates and the heliozoan *Actinophrys sol*. In any case diploid cells could have been formed from haploid predecessors either by cannibalism or failure of cytokinesis. Whatever the mistake leading to doubleness was it is doubtful it occurred continually. Diploid, tetraploid and octaploid microbes probably were rare and fatally encumbered. In populations in which haploid protists with single sets of chromosomes were better adapted to growth by mitosis than were diploids, tetraploids and higher ploidy cells, mechanisms for relieving diploidy would have been strongly selected. Eventually the haphazard would become ritualized.

From these inauspicious beginnings animal, plant and fungal meiotic sex emerged. Meiotic processes are not perfect, even as cell biologists know them today they are subject to occasional errors. As long as haploid protists were optimally suited to prevailing conditions, they thrived, reproducing mitotically. Hardships led to coupling, the pooling of cellular resources. Eventually a regular rhythm of nuclear fusion (karyogamy, fertilization) and relief of diploidy (meiosis) emerged. The origin of meiosis is in the 'errors' of the feeding, digesting and mitotic division process. Errors were seen for example by Cleveland, in 1956, who filmed matings of *Leptospironympha* (a hypermastigote). One mating involved an attempted fusion by three instead of two partners. Most likely it, as other fusions by three, led to the lethal state of triploidy. Another kind of evolutionary important error is made by kinetochores. Cleveland observed (1926) and even induced (1956) cells in which kinetochores failed to divide and grow and therefore produced nuclei without chromosomes which of course eventually died. It appears that the microbial cannibalism which still takes place usually results in regurgitation, or death as cells attempt to regurgitate. But in the late Proterozoic Aeon selection presumably honed the products of fusion and relief of diploidy. The prerequisites for the evolution of sexual flower parts, mating birds, and human sweethearts had already appeared a billion years ago or so in protists struggling to survive.

It seems likely that protists, mitotic cells that became diploids, had small chromosome numbers at first. A protist, prior to the evolution of regularized meiosis, with a diploid number of two or four has a far greater chance of returning to haploidy than a cell with a diploid number of twenty or forty. In protists such as certain hypermastigotes that have a few, very large chromosomes, even by chance alone a diploid with a tendency toward lagging kinetochoric divisions will produce a haploid offspring along with a string of aneuploids (cells with incomplete sets of chromosomes or extra chromosomes) doomed to death. In cells doubled by mistake the selection pressure for healthy haploids with single, complete sets of chromosomes rather than unbalanced aneuploids must have been relentless (Margulis 1982).

CROSSING OVER AND SYNAPTONEMAL COMPLEXES

We have seen that meiosis and fertilization could have evolved in protists in a process that involved fusions of whole cells, but not necessarily the recombination of DNA molecules. Crossing over is an exchange of DNA between homologous chromosomes. Many organisms experience meiosis without crossing over, and all organisms may complete meiosis without it. This breakage and reunion of chromosomes themselves is not intrinsic to the mitotic process. DNA breakage and reunion, the excision and insertion of pieces of genetic material was probably first perfected by entities which evolved in the Archean: bacterial viruses and plasmids. Crossing-over presumably used the fundamental DNA recombination process which dates back to the repair of damaged bacterial and virus DNA, but crossing-over involves far more. Genophoric DNA recombination was augmented by the synaptonemal complex and other specifically chromosomal features.

When crossing over occurs during the meiotic cell divisions the special structure formed is called the synaptonemal complex (Fig. 10). This complex protein network ties together lengthwise members of the pairs of chromosomes (homologues). Just before the moment of chromosome separation prior to cell division the synaptonemal complex dissolves. This permits one member of the pair to go to one offspring cell and the other member of the pair to go to the other. The synaptonemal complex is an insurance policy against aneuploidy. The synaptonemal complex ensures an even distribution of already doubled chromosomes. The paired

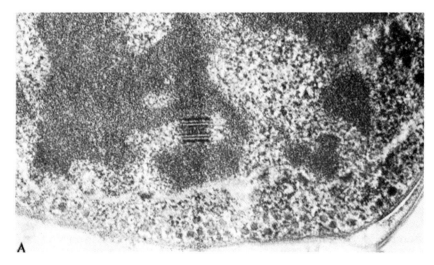

Fig. 10.A . Synaptonemal complex from a marine parasitic protoctist, transmission electron micrograph. (Courtesy of Isabelle Desportes, Paris.)

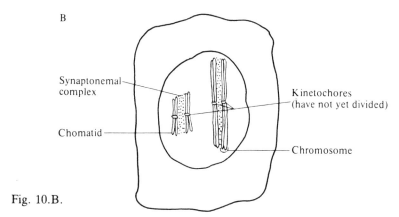

Fig. 10.B.

homologous chromosomes, held together by a reproductively tardy kinetochore, are aligned next to each other by the synthesis of the synaptonemal complex. The synaptonemal complex permits meiosis to proceed normally in organisms with large chromosome numbers. In terms of the origin of sex, the elegant flourish of 'crossing-over', the swapping of stretches of chromatin by paired chromosomes locked in place by the synaptonemal complex is a clever but not really essential refinement of the meiotic masterpiece. It may have evolved under selection pressures for the distribution of increasing quantities of chromatin.

REGULARIZATION OF SEX

Synaptonemal complexes ensuring accurate meiosis, fertilization and embryogeny are prerequisites to the existence of animals and plants. These indispensables have the fortuitous property of forcing sexuality each generation in most animal and plant species. As a consequence of the requirement to mix and match half their genes to form embryos in each generation, sexually reproducing beings are secondarily apt to generate low frequency unique genotypes that survive unpredictable situations (Bell 1982).

Sex takes different forms in plants (which are not always barred, as animals are, from having three or more sets of chromosomes), fungi (which fuse their cytoplasms but not their nuclei; Day 1978), protoctists (whose sexual variations are amazingly diverse) and animals with large eggs and tiny sperm. Therefore meiotic sex very likely evolved from mitosis in protoctist ancestors to animals, fungi and plants more than once. Probably it evolved many times. There is a very important difference in the nature of the problem of the maintenance of sexuality in protoctists, in which meiotic sex developed as an option, and in animals and plants in whose ancestors it became inevitably conjoined to reproduction of the individual.

The maintenance of meiotic sex has both great advantages and great 'costs' (Maynard Smith 1971). Whereas sexual reproducing organisms never have to attract mates, the sexually reproducing organism must attract its mate to an astounding extent. Sexual animal partner's must eventually develop a relationship of enough intimacy to assure the fusion of their bodies (gamontogamy), their single cells (syngamy) and ultimately their nuclei (karyogamy) for the purpose of mingling and even recombining their DNA. Asexually reproducing organisms are usually capable of producing far more offspring at a given moment than sexually reproducing animals and plants. Sexually reproducing populations – when they can escape from the bondage sexuality imposes on them – secondarily revert to asexuality by different routes. This secondary acquisition of asexuality has occurred in nearly every animal taxon (Bell 1982; Buss 1983).

At first, when meiotic sex arose, the eukaryotic partners were nearly identical in appearance as many sexually fusion protists still are today. Eventually mating types or genders evolved. The only rule of mating type or gender is definitional: no organism ever mates successfully and has living offspring with others of the same mating type. Although 'female' is an example of one mating type and 'male' of another, mating types tend in general not to be that fixed. For example, certain protists such as the ciliate *Paramecium multimicronucleatum* have several different mating types. Mating types can change or be lost according to the time of day (Barnett 1967). During the day one clone will be of mating type '1' and will mate only with mating type '2'. At dusk members of the clone will not mate with anyone at all. By nightfall the mating type '1s' will have changed into '2s' and thus will refuse to mate with other mating types '2s'. By changing the lighting conditions, making 'dawn' and 'dusk' come at different times the formation of different sexes or mating types can be controlled in these paramecia. Mating type, it seems, is determined by the presence of proteins on the surface of the cilia. These proteins are such strong attractants that the cilia from organisms of compatible mating types will 'mate' all by themselves. That is, even when sheared from the rest of the protist the cilia will stick to those of the 'opposite sex', as if the rest of the protist cell were in fact still there (Watanabe 1977).

As protists and animals became increasingly long-lived and multicellular the tendency toward sexual dimorphism, for mating partners to diverge into two different forms became increasingly more marked. Sexual dimorphism has evolved millions of times – from the mating type 'plus' in *Chlamydomonas*, to the booming prairie chicken, the bull seal and the malodorous female ginkgo tree. Phenotypically distinct sexual types result from only small genetic differences. At the minimum they involve only the difference of expression of a single gene at a single locus, which determines two opposite mating types (I and II in the group A stocks of *Paramecium aurelia*). In addition, differences in gender, sexual dimorphism, represent intraspecies specializations which are analogous to the selective advantages established by societies which diversify their economies. Since members of the same species generally only mate successfully with each other, each time a new species of animal or plant appeared a new change in gender also evolved. Thus the extent of polyphyly of sex determination is as astounding as it is obvious. It follows that changes in 'gender' occurred at least as many times as sexual species newly evolved. In some cases at least these changes are modifications in the chemistry of secondary compounds and concomitant modifications in reproductive physiology (Swain 1984). What has been treated in the literature as a single problem comprises literally millions of problems with as many solutions.

Many single eukaryotic organisms whose component cells divide asexually by mitosis can produce two or more kinds of gamates capable, later, of fusion with each other. This is the phenomenon of heterothallism so common in many fungi. With time, in many lineages of organisms single individuals could only make one kind – either male or female. The cost of this specialization, like the diminishment of progeny numbers in sexual organisms, was balanced by specific advantages connected with the complexification of large meiotic organisms. Male sex cells, so designated because of their propensity to move on their own, became smaller, more motile and more numerous. Female cells became increasingly large and stationary. The trend from isogamy or sex cells which are equal in appearance to anisogamy, sex cells that are different in appearance, occurred in protoctists, plants and fungi many separate times. Genera of green algae, ciliates, apicomplexa, and even rhodophytes (red algae, organisms in which the smaller 'male' cells are not motile by themselves) have isogamous and anisogamous member species. Indeed the practical definition of 'male' is simply that organism (gamont) itself or single cell (male or microgamete) which produces moving gametes, usually in profusion. A female, by general definition, is usually that gamete which stands still or that gamont which produces gametes that stand still. Relieved of the need to seek out cells of a different mating type, gametes or gamonts in many species were specialized to devote their time to making fewer products thereby retaining more nutrient and expending less energy in motility.

DIFFERENTIATION AND ANIMAL SEXUALITY

All members of the Kingdom Animalia develop from a blastula and hence are intrinsically multicellular. All of the genetic material of at least one mitotic product of the developing zygote must be reserved to insure the continuity of the individual to the next generation. The irreversible use of mitotic microtubule organizing systems and microtubules in cells for physiological purposes accompanies embryological differentiation. In the phyletic history of animals the cells of the germ plasm became those in which the mitotic apparatus remained uncommitted to purposes other than accurate mitotic reproduction. In body cells greater cell specialization followed leading to a progressively greater loss in individual cell reproductive capacity. Cell and tissue differentiation precluded totipotency in all cells but those of the germ line. This has led to an obligatory relationship between sex and differentiation. Organisms such as vertebrates, which relegated somatic cells irreversibly to differentiation, were forced to reserve an undifferentiated germ lineage for reproduction of the individual. Asexual reproduction of individuals, such as budding of hydroids, production of strobili in medusoids or gemma in poriferans is characteristic of animal phyla which show relatively little determinant growth. In animals that have highly differentiated somatic cells asexual reproduction of individuals is a secondary phenomenon. For example, the major mechanism of asexual reproduction in vertebrates is parthenogenesis which, since it involves meiosis and fusion of meiotically produced nuclei, was originally a sexual process. Since sexuality is obligately correlated with differentiation its loss in these animals tends to be lethal. The fact that loss of meiotic sexuality is precluded in animals, especially vertebrates, by the developmental system has not been properly addressed by those modelling the maintenance of sexuality in animal populations.

LOSS OF SEXUALITY

Sexuality, which humans have come to associate not only with reproduction but with the emotions of jealousy and hate, ecstasy and anxiety is never such an ultimate priority as are autopoiesy and reproduction. Autopoiesy and reproduction are imperatives whereas in many species sex never appeared at all or became completely gratuitous: it disappeared. In general it is very difficult to observe the process of loss of sexuality. Since the ancestors of many plants, fungi and animals engaged in well known sexual practices, it is inferred that many species have secondarily lost their sexuality. This is also the case of the ciliate protist *Stentor coeruleus*, a common pond water organism. *Stentor*, which has been under continuous scrutiny since Ehrenberg's pioneer work (c. 1870), seems to be in the process of losing its sexuality.

Each individual *Stentor coeruleus* is capable of a wide range of physiological and behavioural responses. No mutants, altered forms that breed true asexually, have ever been reported. All isolates can 'regulate', that is form a whole repertoire of normal stentors (Tartar 1961). Stentors will occasionally indulge in meiotic sex ... or try to (Fig. 11). But both the mating partners die within about four days of starting the act of conjugation. Since *Stentor* sex always ends in double demise we assume that sex, in this species, is on its way to being lost. Once useful, the orderly process of meiosis followed by nucleus-swapping and then fertilization, has apparently outlived its purpose in this ciliate.

Such plants as seedless grapes, oranges and bananas can be selected for the loss of sexuality quite easily as long as reproduction is insured by other means, such as farmers. Asexual reproduction works faster and more securely. We human beings provide many such food plants, through agricultural grafting and growing from cuttings, with environments in which they can reproduce asexually.

In certain organisms sexuality is reserved for only some members of the species. Many social insects have sacrificed the ability to reproduce independently via sex. Most individual termites no longer are able to propagate themselves. Instead they depend on special reproductives. The workers forage, feed and protect the 'queens' which mate with the male

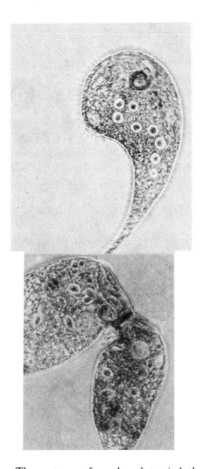

Fig. 11. Mating *Stentor*. The outcome four days later is lethal for both partners.

'kings'. Together they bring more sterile soldiers and workers into the world. Obviously in these societies the individual is selected for only as part of the collective. Sterile casts demonstrate how easily asexuality can be selected for secondarily, when reproduction is usurped by other means such as that of the family unit. Likewise, futurology suggests the possibility of human societies which are asexual through a technological procedure of cloning. *Brave New World* (Huxley 1946), a novel based on the projection of the totalitarian biology of social insects onto human affairs, puts the reproduction of the individual under complete control of the state.

From the broad biological viewpoint of the reproductive imperative, meiotic sexuality is merely an epiphenomenon. It is intimately tied to embryology, tissue and organ differentiation and development in general. The phenomenon itself is the mixing and matching of DNA from different sources in the biosphere to create new 'recombinant' organisms. Sexuality, the production of any given recombinant from a variety of sources, is an absolute precursor to reproduction only in certain animals and plants. Sexuality involves complex, originally separate phenomenon. Recombining bacterial DNA, cannibalistic protists with tardy kinetochores, embryological patterns and development became hooked up in different species in different ways at different times. Reproductive sex has no single origin. It is a set of ancient, embedded accidents endured to accommodate the autopoiesy of organisms whose lives could not be dissociated from sex as easily as a factor is added to an equation. Meiotic sex and fertilization, because they became tied to embryogeny, cell and tissue differentiation and reproduction in animals and plants were difficult to lose. Many lineages secondarily dispensed with sexuality nevertheless.

Sex seems to be a legacy of bacterial DNA patching and protist cannibalism. Meiotic sex depends on cell and nuclear fusion and on the intrinsic property of kinetochores to replicate a bit out of synchrony relative to the chromatin. These are the common denominators of molecular and cellular sexuality. Aside from these conserved products of history there is a multitude of variations on the theme of meiotic sexuality. Bacterial sexuality is fundamentally different and was probably far more important evolutionarily as a generator of genetic novelty than meiotic sex. As one studies living organisms, it becomes clear that the term 'sex' represents separate phenomena (i.e., bacterial and meiotic sex) with distinct evolutionary origins and extraordinary diversity.

Acknowledgements

The authors are grateful to Pamela Hall and Betsey Dexter Dyer for helpful suggestions. We acknowledge the support of this work by the Boston University Graduate School and NASA Life Services Division (NGR-004-025 to L.M.).

References

Anker, P., Stroun, M., Gahan, P., Rossier, A., Greppin,. H. (1971). Natural release of bacterial nucleic acids in plant cells and crown gall induction. In *Informative Molecules in Biological Systems*, (ed. L. G. H. Ledoux) North-Holland, Amsterdam.
Barnett, A. (1966). A circadian rhythm of mating type reversals in *Paramecium multimicronucleatum. J. Cell Physiol.* **67**: 239–270.
Bell, G. (1982). *The Masterpiece of Nature*: The Evolution and Genetics of Sexuality, University of California Press, Berkeley and Los Angeles.
Bradbury, E. M., Maclean, N., Mathews, H. R. (1981). *DNA, Chromatin and Chromosomes*, John Wiley & Sons N.Y.
Bawa, K. (1983). (in press).
Buss, L. (1983). Evolution, development and the units of selection. *Proc. Natl. Acad. Sci.* **80**, 1387–1391.
Cleveland, L. R., (1935). The cell and its role in mitosis as seen in living cells. *Science* **81**, 597–600.
Cleveland, L. R., (1947). The origin and evolution of meiosis. *Science* **105**, 287–288.

Cleveland, L. R., (1956). Cell division without chromatin in *Trichonympha and Barbulanympha. J. Protozoology* **3**, 78–83.
Cleveland, L. R., (1957). Types and life cycles of centrioles of flagellates, *J. Protozoology* **4**, 230–240.
Cleveland, L. R., (1963). Function of flagellate and other centrioles in cell replication. In *The Cell in Mitosis*, (ed. L. Levin), Academic Press, N.Y.
Cloud, P. E. Jr. (1983). The Biosphere. Scientific American **249**, 176–189.
Corliss, J. O. (1979). *The Ciliated Protozoa* 2nd ed. Pergamon Press, N.Y. & London.
Corliss, J. O. (1983). *Composition of the Kingdom Protista* (ms in preparation).
Day, P. R. (1974). *The Genetics of Host-Parasite Interaction*, W. H. Freeman, San Francisco.
Heath, I. B. (1980a). Variant mitosis in lower eukaryotes: indication of the evolution of mitosis? *International Reviews of Cytology* **64**, 1–80.
Heath, I. B. (1980). Mechanisms of nuclear division in fungi. In *The Fungal Nucleus*, (eds. K. Gull and S. Oliver), Cambridge University Press, Cambridge, England.
Huxley, A. L. (1946). *Brave New World*, Harper and Brothers, N.Y.
Kaveski, S., and Margulis, L. (1983). The 'sudden explosion' of animals about 600 million years ago: Why? *Amer. Biol. Teacher* **45**, 76–82.
Knoll, A. and Vidal, G. (1983). Proterozoic evolution. *Bull. Geol. Soc. Amer.* (in press).
Knoll, A. and Barghoorn, E. S. (1977). Archean microfossils showing cell division from the Swaziland System of South Africa, *Scienc* **198**, 396–398.
Lidstrom, M. E., Engebrecht, J. & Nealson, K. H. (1983). Plasmid-mediated manganese oxidation by a marine bacterium, *Fed. European Microbiol. Soc. Letters* (in press).
Margulis, L., (1982a). *Symbiosis in Cell Evolution*, W. H. Freeman and Co., San Francisco.
Margulis, L., (1982). Microtubules in microorganisms and the origins of sex. In *Microtubules in Microorganisms*, (eds P. Cappucinelli and R. Morris) p. 341–350.
Margulis, L. and Sagan, D. (1985). *The Origins of Sex* Yale University Press, New Haven, CT. (in press).
Margulis, L. and Schwartz, K. V. (1982). *Five Kingdoms*, W. H. Freeman and Co., San Francisco.
Margulis, L., Walker, L., J. C. G. and Rambler, M. B. (1976). A reassessment of the roles of oxygen and ultraviolet light in Precambrian evolution Nature **264**: 620–624.
Markert, C. L. and Ursprung, H. (1971). *Developmental Genetics*. Prentice-Hall, Englewood Cliffs, NJ.
Maynard Smith, J. (1978). *The Evolution of Sex*. Cambridge University Press, Cambridge, England.
Pickett-Heaps, J. D. (1974). Evolution of mitosis and the eukaryote condition. *BioSystems* **6**, 37–48.
Pickett-Heaps, J. D., Tippett, D. H. and Andreozzi, J. A. (1978). Cell division in the pennate diatom *Pinnularia* V. Observations on live cells. *Biologie Cellulaire* **35** 295–304.
Primack, R. (1983) (in press).
Ris, H., (1961). Ultrastructure and molecular organization of genetic systems. *Canadian Journal of Genetics and Cytology* **3**, 95–120.
Raikov, I. B. (1981). *The Protozoan nucleus*. Springer-Verlag, Heidelberg and New York.

Rosson, R. and Nealson, K. H. (1982a). Manganese bacteria and the Marine Manganese Cycle in *The Environment of the Deep Sea*. (eds. J. G. Morin and W. G. Ernst). Prentice-Hall, Englewood Cliffs, N.J. p. 201–216.

Rosson, R. and Nealson, K. H. (1982b) Manganese binding and oxidation by spores of a marine bacillus *J. Bacteriol.* **151**, 1037–1044.

Sagan, D. and Margulis, L. (1985). *The Expanding Microcosm*. Summit Books, NY., (in press).

Sonea, S. and Panisett, P. (1983). *The New Bacteriology*. Jones and Bartlett Publishing Co. Boston, MA.

Sulloway, F. J. (1983). *Freud, Biologist of the Mind*. Basic Books, Inc., Publishers, New York.

Swain, T. (1984). *Biochemical Evolution*. Jones and Bartlett (in press).

Tartar, V. (1961). *The Biology of* Stentor. Pergamon Press, Oxford and New York.

Varela, F. and Maturana, H. R. (1974). Autopoiesis: the organization of living systems, its characterization and a model, *Biosystems* **5**, 187–196.

Watanabe, T. (1977). Chemical properties of mating substances in *Paramecium caudatum*: Effect of various agents on mating reactivity of detached cilia, *Cell Structure and Function* **2**, 241–247.

Wilson, E. O. (1975). *Sociobiology: The New Synthesis*. Harvard University Press, Cambridge, MA.

Williams, G. C. (1966). *Adaptation and Natural Selection*. Princeton University Press, Princeton, NJ.

Witzmann, R. F. (1981). *Steroids: Keys to Life*. Van Nostrand Reinhold Company, New York.

Witkin, E. M. (1969). Ultraviolet-light induced mutation and DNA repair, *Annual Rev. of Microbiol.* **23**, 487–514.

Part VIII
CLASSIFICATION OF EUKARYOTES

Editors' Comments
on Paper 35A through 35E

35A MARGULIS and SCHWARTZ
Monera

35B MARGULIS and SCHWARTZ
Protoctista

35C MARGULIS and SCHWARTZ
Fungi

35D MARGULIS and SCHWARTZ
Animalia

35E MARGULIS and SCHWARTZ
Plantae

The five kingdoms system of classification (four eukaryotic kingdoms, one prokaryotic kingdom) is the most logical arrangement yet devised and is consistent with a symbiotic origin of the eukaryotic organelles. Margulis and Schwartz's book *Five Kingdoms* (excerpted here as Papers 35A, B, C, D, and E) is a full treatment of this scheme that includes detailed criteria for each kingdom.

35A

Copyright © 1982 by W. H. Freeman and Company Publishers. All rights reserved
Reprinted from pages 24-29 of *Five Kingdoms,* W. H. Freeman and Company Publishers, San Francisco, 1982

MONERA

Greek *moneres,* single, solitary

L. Margulis and K. Schwartz

Small though they are, bacteria are crucial to health, agriculture, forestry, and the very existence of the air we breathe. The modern food-processing industry began with the awareness of the nature of bacteria. Canning, preserving, drying, and pasteurization are all sterilizing techniques that prevent contamination by even a single bacterium. The success of these techniques is truly remarkable in view of the ubiquity of bacteria and their unbelievable numbers. Every spoonful of garden soil contains some 10^{10} bacteria; a small scraping of film from your gums might reveal about 10^9 bacteria per square centimeter of film—the total number in your mouth is greater than the number of people who have ever lived. Bacteria make up a significant percentage of the dry weight of all animals. They cover the skin; they line nasal and mouth passages, and live in the gums and between the teeth; they pack the digestive tract, especially the colon.

Germs, or pathogens, are simply bacteria (occasionally, fungi) capable of causing infectious diseases in animals or plants. The word *germ,* like the word *microbe,* has no definite technical meaning. A germ is simply a small living organism capable of growth at the expense of another organism; a microbe is simply a small organism of interest to someone. Bacteria cure as well as cause disease. Many of our most useful antibiotics—streptomycin, erythromycin, chloromycetin, and kanamycin—come from bacteria (penicillin comes from fungi).

Although much of the nomenclature is in dispute, more than 5000 species have been named and described in the bacteriological literature. Many more are still unidentified. Few bacteriologists would deny that the vast majority of bacterial species have never been carefully studied and described.

Bacteria are morphologically rather simple. The most complex do undergo developmental changes in form: simple bacteria may metamorphose into stalked structures, grow long branched filaments, or form tall fruiting bodies that release resistant sporelike microcysts. Some produce highly motile colonies. However, detailed knowledge of bacterial structure seldom affords insight into function. In this respect, bacteria are very different from animals and plants. Because their differences lie chiefly in their metabolism, or internal chemistry, many species of bacteria can be distinguished only by the chemical transformations they cause. A universally applied

Monera

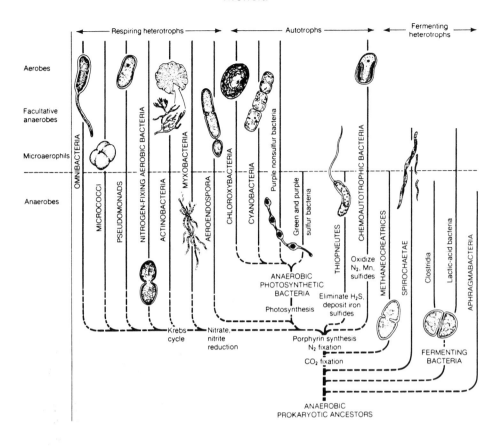

diagnostic test of bacteria is whether they stain purple with the *Gram test,* a staining method developed by the Danish physician Hans Christian Gram (1853–1938). *Gram-positive* organisms (which stain deep purple) differ from *Gram-negative* ones (which stain light pink) in the chemistry of their cell walls.

Bacteria can effect a large number of different chemical transformations and are metabolically far more diverse than all of the eukaryotes. Although some very complicated molecules are made only by certain plants and fungi, the biosynthetic and degradative patterns—the chemistry of reproduction and energy generation—in all plant and fungal cells is remarkably similar. Animals and protoctists exhibit even less variation in their chemical repertoires. In short, the metabolism of eukaryotes is rather uniform; it follows the same fundamental patterns of photosynthesis, respiration, glucose breakdown, and synthesis of nucleic acids and proteins. Certain plants and fungi add flourishes of metabolic virtuosity, but the basic patterns are not altered. Bacteria, on the other hand, are not only very different from eukaryotes but also from each other.

The work of most microbiologists is closely concerned with the role bacteria play in health and disease. The role of bacteria in the environment has been studied much less, but it is equally significant. Bacteria produce and remove all the major reactive gases in the Earth's atmosphere: nitrogen, nitrous oxide, oxygen, carbon dioxide, carbon monoxide, several sulfur-containing gases, hydrogen, methane, and ammonia, among others. Protoctists and plants also make substantial contributions to atmospheric gases, but none that is different from those of prokaryotes. On the other hand, many important reactions are catalyzed only by bacteria.

There is a profound difference between the soil of the Earth and the regolith—loose particles of rock—on the surface of Mars and of the Moon. Of course, Mars and the Moon are very dry and have much less atmosphere than the Earth; but soil and regolith differ in much more than just moisture content. The surface of the Earth—its soil and sediments—is rich in complex organic compounds; less tractable ones, such as tannic acids, lignin, and cellulose, abound as well as much more actively metabolized organics, such as sugars, starches, organic phosphorus compounds, and proteins. All of these organic compounds are the products of chemosynthesis or of photosynthesis, processes that use chemical energy or sunlight to convert the carbon dioxide of the air to the organic compounds of the living biosphere and, ultimately, of organic-rich sediments from which we obtain oil and coal. In fact, the soil and rocks of the Earth contain about one hundred thousand times as much organic matter as living forms do.

Chemosynthesis, the production of organic matter from inorganic by means of chemical energy, is limited to certain groups of bacteria. Photosynthesis is often attributed only to plants, but many bacteria also photosynthesize. Chemosynthesis and photosynthesis are forms of autotrophic nutrition: deriving food and energy from inorganic sources. Heterotrophy, the alternative mode of nutrition, is deriving food and energy from preformed organic compounds—from either live or dead sources. Like plants, most photosynthetic bacteria convert atmospheric carbon dioxide and water to organic matter and oxygen; unlike plants, other bacteria are capable of very different modes of photosynthesis—for example, using hydrogen sulfide instead of water. Bacterial photosynthesis and chemosynthesis are absolutely necessary for cycling the elements and compounds upon which the entire biosphere depends, including, of course, ourselves.

The notion that the food chain starts with the plants, followed by the herbivores, and ends with carnivorous animals is very shortsighted. Zooplankton of the seas feed on protists, which feed on bacteria; bacteria break down the carcasses of animals and algae, releasing back into solution such elements as nitrogen and phosphorus required by the phytoplankton. Bacteria facilitate entire food chains, or rather food webs, by transforming inorganic materials into complex organic compounds and being eaten by other organisms. Life on Earth would die out far faster if bacteria became extinct than if the animals, plants, and fungi disappeared. In fact, we have good reason to believe that life on our planet thrived long before the three more complex kingdoms appeared. Some of the ways we depend on bacteria will be explained in the descriptions of moneran phyla.

Bacteria have an ancient and noble history. They were probably the first living organisms and, in respect to everything but size, have dominated life on Earth throughout the ages. The oldest fossil evidence for bacteria goes back some 3400 million years, whereas the oldest evidence for animals dates back about 700 million years and the oldest fungi and land plants date back about 470 million years. The earliest evidence for protoctists is more equivocal, but they are probably not more than 1000 million years old. It is widely believed that 2000 million years ago the cyanobacteria—oxygen-eliminating photosynthetic prokaryotes that used to be called blue-green algae (Phylum M-7)—effected one of the greatest changes this planet has ever known: the increase in concentration of atmospheric oxygen from far less than 1 percent to about 20 percent. Without this concentration of oxygen, people and other animals never would have evolved.

As a group, bacteria are the most hardy of living beings. They can survive very low temperatures, even total freezing, for years. Some species thrive in boiling hot springs and others survive even in very hot acids. By forming spores, particles of life containing at least one copy of all the genes of a bacterium, they can tolerate total desiccation. Bacteria are the first to invade and populate new habitats: land that has been burned or newly emerged islands. They are capable of survival at great oceanic depths and atmospheric heights, although no organism—not even the hardiest of bacteria—is known to complete its life cycle suspended in the atmosphere.

Some activities of bacteria are still only dimly perceived. The incorporation of metals such as manganese and iron into nodules on lake and ocean floors may possibly be accelerated by bacterial action. Layered chalk deposits called stromatolites are thought to have been produced by the trapping and binding of calcium-carbonate-rich sediment by growing communities of bacteria, especially by cyanobacteria. Even the gold in South African mines is found with rocks rich in organic carbon. In Witwatersrand, the miners find the gold, deposited apparently more than 2500 million years ago, by following the "carbon leader." The carbon is probably of microbial origin. Copper, zinc, lead, iron, silver, manganese, and sulfur seem to have been concentrated into ore deposits by biogeochemical processes involving bacterial growth and metabolism.

All bacteria reproduce asexually. Although a few bacteria are known to be products of unequal sexual donations from two "parents," the extent of bacterial sexuality in nature is not really known. This is partly because the extent of bacterial diversity is not well known, and partly because, with bacteria as with most organisms, sex life is a most elusive object of study. Genes may also be transferred between bacteria by viruses. Some bacteria even excrete DNA. Whether this behavior is a part of bacterial sexuality in nature is not known; in the laboratory, however, DNA excreted by one bacterium is taken up and used by another in the formation of genetically new individuals.

No communities of living organisms anywhere on Earth lack bacteria, but few places are dominated by them. Some exclusive bacterial habitats, most often found in intemperate climates, are the bare rocks of cliffs, the interior of certain carbonate rocks, and muds lacking oxygen. Perhaps the most spectacular are the hot boiling muds of Yellowstone Park, in Wyoming, and the salt flats and shallow embayments of tropical and subtropical areas. Many such flats and bays are dominated by microbial mats, cohesive, more or less flat structures on soil or in shallow water that are caused by the growth and metabolism of microbes, primarily filamentous cyanobacteria. By entrapping bits of sand, carbonate, and other sediment, such growing communities of microbes can grow to be quite conspicuous manifestations of biological activity.

Except for the rather extreme environments where microbial mats or thermal springs abound, eukaryotes dominate our landscape, or so it seems to the naked eye. However, a microscopic examination of any forest, tide pool, riverbed, chaparral, or other habitat apparently dominated by eukaryotes will reveal prokaryotes in abundance. In activity and potential for rapid unchecked growth, they are unexcelled among living organisms. When environmentalists mourn the destruction of habitats by pollution, they are usually thinking of the loss of fish, fowl, and fellow mammals. If their sympathies were with the cyanobacteria and other bacteria instead, they would perceive eutrophication of lakes, for example, as a sign that life is flourishing.

Bacteriologists have not conformed their nomenclatural and taxonomic practices to those of other biologists. Thus, inevitably, several of our groupings differ from those found in the standard reference works such as *Bergey's Manual of Determinative Bacteriology*. Our innovations aim to make the taxonomic level of *phylum* conceptually comparable throughout the five kingdoms. To our knowledge, this has never before been attempted. We recognize sixteen phyla—fewer than in the animals or protoctists but more than in the plants or fungi. These phyla group the bacteria by clearly distinguishable traits, both morphological and metabolic. Where evolutionary information is available, we have tried to keep natural groups together. However, so little is known about the past history of bacteria that our phyla are for the most part frankly pragmatic. We have retained familiar English names, where they were concise and appropriate, for some groupings here given phylum status. New names include Aphragmabacteria (M-1), Thiopneutes (M-4), Methaneocreatrices (M-5), Chloroxybacteria (M-8), Aeroendospora (M-11), and Omnibacteria (M-14).

The habitat scenes are notably arbitrary in this chapter because so many bacteria can be found in both animal and plant hosts, and in soil and water samples, from vastly different habitats and locations. The natural history and ecology of bacteria have been so little explored that little can be said about the distribution and quantity of bacteria in the world's environments.

35B

Copyright © 1982 by W. H. Freeman and Company Publishers. All rights reserved
Reprinted from pages 68-71 of *Five Kingdoms,* W. H. Freeman and Company Publishers, San Francisco, 1982

PROTOCTISTA

Greek *protos,* very first; *ktistos,* to establish

L. Margulis and K. Schwartz

Kingdom Protoctista is defined by exclusion: its members are neither animals (which develop from a blastula), plants (which develop from an embryo), fungi (which lack undulipodia and develop from spores), nor prokaryotes. They comprise the eukaryotic microorganisms and their immediate descendants: all nucleated algae (including the seaweeds), undulipodiated (flagellated) water molds, the slime molds and slime nets, and the protozoa. Protoctist cells have all the characteristically eukaryotic properties, such as aerobiosis and respiration in mitochondria, and most have 9 + 2 undulipodia at some stage of the life cycle.

Why *protoctist* rather than *protist?* Since the nineteenth century, the word *protist,* whether used informally or formally, has come to connote a single-celled organism. In the last two decades, however, the basis for classifying single-celled organisms separately from multicellular ones has weakened. It has become evident that multicellularity evolved many times from unicellular forms—many multicellular organisms are far more closely related to certain unicells than they are to any other multicellular organisms. For example, the ciliates (Phylum Pr-18, Ciliophora), which are unicellular microbes, include at least one species that forms a sorocarp, a multicellular spore-bearing structure; euglenoids (Phylum Pr-6), chrysophytes (Phylum Pr-4), and diatoms (Phylum Pr-11) also have multicellular derivatives.

We have adopted the concept of protoctist propounded in modern times by H. F. Copeland in 1956 (see the bibliography at the end of this chapter). The word had been introduced by John Hogg in 1861 to designate "all the lower creatures, or the primary organic beings;—both *Protophyta,* . . . having more the nature of plants; and *Protozoa* . . . having rather the nature of animals." Copeland appreciated, as several scholars in the nineteenth century had, the absurdity of calling giant kelp by a word, *protist,* that had come to imply unicellularity and, thus, smallness. He proposed an amply defined Kingdom Protoctista to accommodate certain multicellular organisms as well as the unicells that may resemble their ancestors—for example, kelp as well as the tiny brownish cryptophyte alga *Nephroselmis.* The

Protoctista

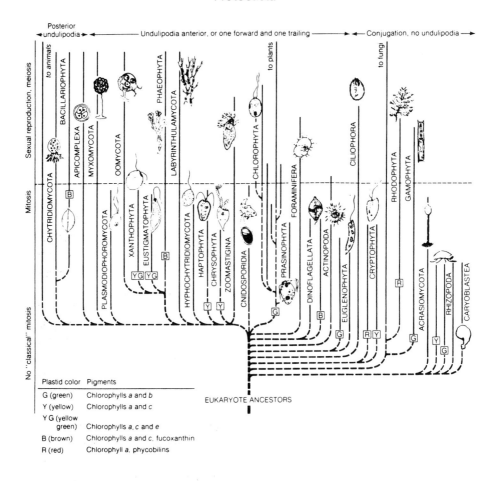

kingdom thus defined also solves the problem of blurred boundaries that arises if the unicellular organisms are assigned to the various multicellular kingdoms.

We propose 27 protoctist phyla. This number is a matter of taste rather than tradition, because there are no established rules for defining protoctist phyla. Our groupings are debatable; for example, some argue that the Caryoblastea (Phylum Pr-1), comprising a single species of nonmitotic amoeba, ought to be placed in Phylum Rhizopoda (Pr-3) with the other amoebas, or that the cellular and plasmodial slime molds (Phyla Pr-22 and Pr-23) should be placed together. Some believe that the chytrids, hypho-chytrids, and oomycotes belong to the fungi and that the chlorophytes belong to the plants. Some insist that chaetophorales and prasinophytes, which here are considered chlorophytes (Phylum Pr-15), ought to be raised to phylum status. There are arguments for and against these views.* Our system has the advantage of defining the three multicellular kingdoms precisely, but the disadvantage of grouping together as protoctists amoebae, kelps, water molds, and other eukaryotes that have little in common with each other.

Protoctists are aquatic: some primarily marine, some primarily freshwater, and some in watery tissues of other organisms. Nearly every animal, fungus, and plant—perhaps every species—has protoctist associates. Some protoctist phyla, such as Apicomplexa (Pr-19) and Cnidosporidia (Pr-20), include hundreds of species all of which are parasitic on other organisms.

No one knows really how many species of protoctists there are; thousands have been described in the biological literature.† Water molds and plant parasites have traditionally been dealt with by the mycological literature, parasitic protozoa by the medical literature, algae by the botanical

---

*For details of alternative kingdom systems, see R. H. Whittaker and L. Margulis, "Protist classification and the kingdoms of organisms." *BioSystems* 10:3–18; 1978.

†Georges Merinfeld estimates that there are more than 65,000 species (personal communication).

literature, free-living protozoa by the zoological literature, and so forth. Inconsistent practices of describing, naming, and defining species have led to a confusion that this book attempts to dispel. Another reason for ignorance is that the group of eukaryotic microbes is large, with much diversity in tropical regions, whereas protozoologists and phycologists are scarce and concentrated in the north temperate zones. Furthermore, distinguishing species of free-living protoctists often requires time-consuming genetic and ultrastructural studies. Funding for such studies is limited because most protoctists are not sources of food and cause no diseases, thus they are of no direct economic importance.

The protoctists show remarkable variation in cell organization, patterns of cell division, and life cycle. Some are oxygen-eliminating photoautotrophs; others are ingesting or absorbing heterotrophs. In many species, the type of nutrition depends on conditions: when light is plentiful, they photosynthesize; in the dark, they feed. However, although protoctists are far more diverse in life style and nutrition than animals, fungi, or plants are, they are far less diverse metabolically than the bacteria.

Increasing knowledge about the ultrastructure, genetics, life cycle, developmental patterns, chromosomal organization, physiology, metabolism, and protein amino-acid sequences of the protoctists has revealed many differences between them and the animals, fungi, and plants. It has even been suggested that the major protoctist groups, here called phyla, are so distinct from each other as to deserve kingdom status, and that nearly 20 kingdoms ought to be created to accommodate them.* With due respect for their differences, a recognition of their common eukaryotic heritage, and a sense of humility toward both their complexity and our ignorance, we present our 27 protoctist phyla.

---

*G. F. Leedale, "How many are the kingdoms of organisms?" *Taxon* 23:261–270; 1974.

35c

Copyright © 1982 by W. H. Freeman and Company Publishers. All rights reserved
Reprinted from pages 144–147 of *Five Kingdoms*, W. H. Freeman and Company Publishers, San Francisco, 1982

FUNGI

Latin *fungus*, probably from Greek *sp(h)ongos*, sponge

L. Margulis and K. Schwartz

Kingdom Fungi, as defined in this book, is limited to eukaryotes that form spores and are amastigote (lack undulipodia) at all stages of their life cycle.* It is estimated that there are 100,000 species of fungi, mostly terrestrial, although a few truly marine species are known. Because fungi often differ only in subtle characteristics, such as the pigments and complex organic compounds they produce, it is likely that many yet unknown species exist.

Fungal spores can germinate to grow slender tubes called *hyphae* (singular, *hypha*), which are divided into "cells" by cross walls called *septa* (singular, *septum*). Each such cell may contain more than one nucleus—the exact number depends on the species of fungus. The septa seldom separate the cells completely; thus, cytoplasm can flow more or less freely through the hyphae. In fact, the hyphae of some fungi have no septa at all.

A large mass of hyphae is called a *mycelium* (plural, *mycelia*), which is the vegetative form of most fungi. From time to time, reproductive structures are formed, also made of hyphae. Such structures are commonly noticed as molds, morels, and mushrooms. The largest and most complex are the large mushrooms and shelf fungi, some of which arise from mycelia meters in diameter. Many others are microscopic.

Fungi are sexual by conjugation, in which hyphae of different mating types come together and fuse (see the photographs illustrating Phylum F-3, Basidiomycota). Even after conjugation, the hyphal nuclei, which are always haploid, do not immediately fuse. Instead, each parental nucleus grows and divides within the hyphae, often for long periods of time. The offspring nuclei remain in pairs, one nucleus descended from each parent. A hypha containing paired haploid nuclei, whether or not they have been shown to come from separate parents, is called *dikaryotic*. A mycelium of such

---

* Traditional classifications often include funguslike microbes in the fungi (such as chytrids and oomycotes) all in the Plantae as Subkingdom Fungi. We have classified these that have undulipodia as protoctists (Phyla Pr-25, 26, and 27).

Fungi

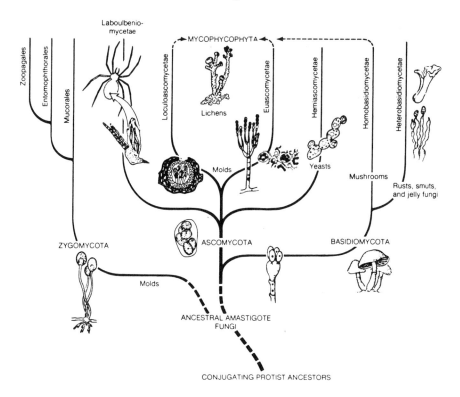

hyphae is called a *dikaryon*. If the nuclei of each pair are demonstrably different, the mycelium is called a *heterokaryon*. If the hyphae contain only single, unpaired nuclei, the mycelium is called a *monokaryon*.

In most fungal phyla, the dikaryotic state is eventually followed by fusion of nuclei to form diploid zygotes, but diploidy is transient. The zygote immediately undergoes meiosis to reestablish the haploid state. Meiotic cell division results in the formation of spores. All fungi form some sort of spores. The deuteromycotes, which lack sexual stages, produce only asexual spores. In all cases, the spores are haploid and are capable of germinating into haploid hyphae. In most fungi, hyphae of different mating types fuse later in the life cycle, and the dikaryotic or sexual stage follows. In asexual species, the hyphae are permanently haploid.

Fungi lack embryological development: the spores develop directly into hyphae or, in some cases, single vegetative cells. In many species, spores are formed in special structures called sporangia, asci, or basidia. Most fungi, even those that have sexual stages, can form spores vegetatively, without sex. In fact, most reproduce asexually more often than they do sexually. Vegetative spores called conidia form at the tips of hyphae. Most conidia are dispersed by the wind and can endure conditions of heat, cold, and desiccation unfavorable to the growth of fungi. Under favorable conditions, the conidia grow into hyphae and form mycelia. This is the general reproductive pattern, although there are many variations.

Nearly all fungi are aerobes, and all of them are heterotrophs. Fungi do not ingest their food but absorb it. They excrete powerful enzymes that break down food into molecules outside the fungus; these molecules are then transported in through the fungal membrane. The various fungal strategies for survival include the production of complex organic compounds, such as the ergot and amanita alkaloids, which can induce hallucinations or even death in mammals.

Fungi are tenacious, resisting severe desiccation and other insults. Their cell walls, composed of the nitrogenous polysaccharide chitin, are hard and stiff and resist the loss of water. Some grow in acid, others survive in environments that contain nearly no nitrogen. They are the most resilient of the eukaryotes.

Many fungi cause diseases, especially in plants. However, many more form important and constructive associations with plants. These associations can be quite intimate. Most orchid seeds, for example, require specific fungal partners in order to germinate, and fungi inhabiting the roots of forest trees are apparently responsible for transporting nutrients from the soil to the plants. Some fungi are important sources of antibiotics: *Penicillium chrysogenum* (Phylum F-4, Deuteromycota) produces penicillin. Molds and yeasts are used in the production of cheese and beer.

The ancestry of fungi is not well understood. The fungal way of life has evolved many times and in many groups—slime molds and nets, chytrids, and oomycotes (Phyla Pr-21 to Pr-27). It is possible that the true fungi descended from conjugating protists and thus share an ancestor with the gamophytes (Phylum Pr-14) or the rhodophytes (Phylum Pr-15). The ascomycotes and basidiomycotes seem related to each other and probably did descend from a zygomycote ancestor; the deuteromycotes clearly descended from either the ascomycotes or the basidiomycotes by loss of sexual stages. At any rate, fungi differ from animals and plants in life cycle, in mode of nutrition, in pattern of development, and in many other ways. Thus, many mycologists feel that it is high time fungi were raised to kingdom status.

The oldest fossil fungi, dating from the Devonian, are intimately associated with fossil plant tissue. It has been suggested that plant-fungus associations made it possible for plants to become truly terrestrial: the fungi could have transported nutrients to plants and prevented them from drying out. In any case, the association between plants and fungi has persisted for at least 300 million years.

35D

Copyright © 1982 by W. H. Freeman and Company Publishers. All rights reserved
Reprinted from pages 160-165 of *Five Kingdoms,* W. H. Freeman and Company Publishers, San Francisco, 1982

ANIMALIA

Latin *anima*, breath, soul

L. Margulis and K. Schwartz

In traditional two-kingdom systems, the multicellular animals were referred to broadly as *metazoa* to distinguish them from one-celled "animals," the *protozoa*. In our classification system, the traditional protozoa belong to Kingdom Protoctista; in our system, animals may be defined as multicellular, heterotrophic, diploid organisms that develop anisogamously—from two different haploid gametes, a large egg and a smaller sperm. The product of fertilization of the egg by the sperm is a diploid zygote that develops by a sequence of mitotic cell divisions. These mitoses result in first a solid ball of cells and then a hollow ball of cells called a *blastula*. All animals (as defined in this book) develop from a blastula. In most animals, the blastula invaginates, folds inward at a point, to form a *gastrula,* a hollow sac having an opening at one end. Further growth and movement of cells produce a hollow digestive system called an *enteron* if it is open at only one end, and a *gut* or *intestine* if it has developed a second opening.

The details of further embryonic development differ widely from phylum to phylum but are fairly constant within each phylum. Such developmental details provide very important criteria for determining relationships between the phyla. In many phyla, developmental details are known for very few species; in some, not for any. In all cases they cannot be summarized in a few words. For this reason, concise and precise definitions of the phyla cannot always be given here; our descriptions are rather more informal.

Although multicellularity is found in all the kingdoms, it has developed most impressively in the animals—their cells are joined by complex junctions into tissues. Such elaborate joints—desmosomes, gap junctions, and septate junctions, for example—ensure and control communication and the flow of materials between cells. These junctions—and there are more kinds

Animalia

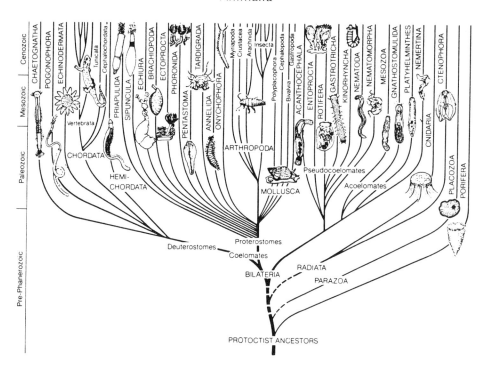

than those listed here—can be seen with an electron microscope. Indeed, the study of cells in tissues, and of tissues in organs, is a science in itself—histology.

Most animals have ingestive nutrition: they take food into their bodies and then either engulf particles or droplets of it into digestive cells by the process of phagocytosis ("cell eating") or pinocytosis ("cell drinking") or absorb food molecules through cell membranes. Although behavior of various kinds (attraction to light and avoidance of noxious chemicals, sensing of dissolved gases, and so forth) can be found in members of all five kingdoms, the animals have elaborated upon this theme too, far more than members of the other kingdoms. Mammalian behavior is perhaps the most complex.

The animals are the most diverse in form of all the kingdoms. The smallest are microscopic—smaller than many protists—and the largest today are whales, sea mammals in our own class (Mammalia) and phylum (Chordata). The members of most phyla are found in shallow waters. Truly land-dwelling forms are found only in two phyla, Arthropoda (Phylum A-27) and Chordata (Phylum A-32). Several phyla contain species that live on land in the soil (for example, earthworms), but these require constant moisture and have not really freed themselves from an aqueous environment throughout their life cycle. In fact, most animal phyla are aquatic worms of one kind or another. Most species of animals are extinct—these form the subject of another science: paleontology. Only living forms have been included in this book.

Of all organisms on Earth, only the animals have succeeded in actively invading the atmosphere. Representatives of all five kingdoms that spend significant fractions of their life cycle in the atmosphere can be found (for example, spores of bacteria, fungi, and plants). However, none in any kingdom spend their entire life cycle in the atmosphere, and only animals fly. Active locomotion of animals through the air has been independently achieved several but not many times, and only in two phyla: Arthropoda, class Insecta, and Chordata, classes Aves (birds), Mammalia (bats and *Homo sapiens* only), and Reptilia (several extinct flying dinosaurs).

For many years (and even today), biologists divided the animals, protozoans and metazoans together, into two large groups: the invertebrates, those without backbones, and the vertebrates, those with. In fact, all animals except the Craniata, a subphylum of Phylum Chordata, belong to the invertebrate group. This invertebrate/vertebrate dichotomy amply represents the skewed perspective we have as members of Phylum Chordata. Our pets, beasts of burden, sources of food, leather, and bone—that is, the animals closest to our size and best known to us—are members of our own phylum. We now realize that, from a less species-centered point of view, characteristics other than backbones are more basic and reflect much earlier evolutionary divergences.

The animal phyla are described here in approximate order of increasingly morphological complexity. Two phyla of animals, set apart as Subkingdom Parazoa, lack tissues organized into organs and have an indeterminate shape. These are the newly discovered Placozoa (Phylum A-1) and the well-known sponges (Phylum A-2, Porifera). The other thirty phyla, constituting Subkingdom Eumetazoa (true metazoans), have tissues organized into organs and organ systems.

There are two branches of the Eumetazoa. One consists of radially symmetrical organisms, the coelenterates (Phylum A-3) and the comb jellies (Phylum A-4, Ctenophora). These animals are planktonic and thus face a uniform environment on all sides; their radial symmetry is both internal and external. All the rest of the 28 phyla show bilateral symmetry, at least internally.

The bilaterally symmetrical phyla may be divided into three groups, or *grades:* those that lack a coelom (Acoelomata, Phyla A-5 through A-8), those that have a body cavity but lack a true coelom (Pseudocoelomata, Phyla A-9 through A-15), and those that develop a true coelom (Coelomata, Phyla A-16 through A-32). What is the coelom? The process of gastrulation leads to the development of three tissue layers in all animals more complex than the coelenterates and ctenophores. These tissue layers, called the endoderm, mesoderm, and ectoderm (listed from the inside out), are the masses of cells from which the organ systems of animals develop. In general, the intestine and other digestive organs develop from endoderm, the muscle and skeletal materials from mesoderm, and the nervous tissue and outer integument from the ectoderm. In the coelomates, the mesodermal tissues open to contain a space that widens and eventually forms a body cavity in which digestive and reproductive organs, among others, develop and are suspended. This true body cavity is called the coelom. A pseudocoelom is an internal space that does not develop from a space surrounded by mesoderm.

Two groups, called *series,* of coelomate animals are distinguished according to the fate of an early developmental feature called the *blastopore.* The invagination of the blastula, the hollow ball of cells into which the animal zygote develops, is the blastopore. This embryonic structure enlarges as cells divide, grow, and move over each other. In animals of Series Protostoma (Phyla A-16 through A-27), the blastopore eventually becomes the mouth of the adult. In Series Deuterostoma (Phyla A-28 through A-32), the blastopore becomes the anus, the rear end of the intestine; the mouth forms as a secondary opening at the end of the animal opposite from the anus. The five deuterostome phyla are thought to have common ancestors more recent than the ones they have with any protostome phyla. However, this divergence probably occurred more than 680 million years ago, judging from the putative presence of both protostomes and deuterostomes in the Ediacaran fauna.

Virtually all biologists agree that animals evolved from protoctists. However, which protoctists, when, and in what sort of environments are questions that are still actively debated. E. D. Hanson has amassed a great deal of information on the protoctist-animal connection but admits the problem

Animalia

has not been solved for the Eumetazoa.* The Parazoa, or at least the Porifera (Phylum A-2), are thought to have evolved from the choanoflagellates (Phylum Pr-8, Zoomastigina). This is deduced from the details of fine structure of the cells. It is possible, in fact likely, that the other animal phyla, especially the eumetazoans, had different ancestors among the protoctists.

---

*E. D. Hanson, *Origin and early evolution of animals.* Wesleyan University Press; Middletown, Connecticut; 1977.

35E

Copyright © 1982 by W. H. Freeman and Company Publishers. All rights reserved
Reprinted from pages 248-251 of *Five Kingdoms*, W. H. Freeman and Company Publishers, San Francisco, 1982

PLANTAE

Latin *planta*, plant

L. Margulis and K. Schwartz

Members of the plant kingdom are multicellular, sexually reproducing eukaryotes. Their cells contain green plastids, chloroplasts, which contain such pigments as chlorophylls *a* and *b*, xanthophylls, and other yellow and red carotenoid pigments. Today, they are the major mechanism for transforming solar energy into food, fiber, coal, oil, and other usable forms. Photosynthesis by plants sustains the biosphere, not only converting solar energy into food but producing oxygen as well.

Plants are adapted primarily for life on land, although many live in water during part of their life cycle. Some half a million species are known, and it is believed that there are many still undiscovered plants, especially in the tropics. Furthermore, because many plants resemble each other in form but are chemically different, it is highly likely that this estimate is a low one. The vast majority of plants living today belong to Phylum Angiospermophyta, or flowering plants (Phylum Pl-9).

Within the plant kingdom there are two basic groups: the bryophytes, or nonvascular plants, and the tracheophytes, or vascular plants. The latter are distinguished by conducting tissues called xylem and phloem. Xylem transports water and ions from the roots upward through the plant, and phloem transports photosynthate, sugar, and other products of the leaves throughout the plant. These rigid but metabolically active tissues are absent in the bryophytes.

Unlike the origin of fungi and animals, the origin of plants is generally agreed upon: green land plants descended from green algae (Chlorophyta, Phylum Pr-15). This hypothesis is based on such properties as similar pigmentation (including the presence of chlorophyll *a* and *b* in the chloroplasts), sperm having two undulipodia, and intercellular connections called plasmodesmata. Furthermore, some chlorophytes (such as *Klebsormidium*) have cellulosic walls and patterns of mitotic cell division identical to those of plants. In both, a cell-wall structure called a cell plate or phragmoplast develops perpendicular to the mitotic spindle.

Plantae

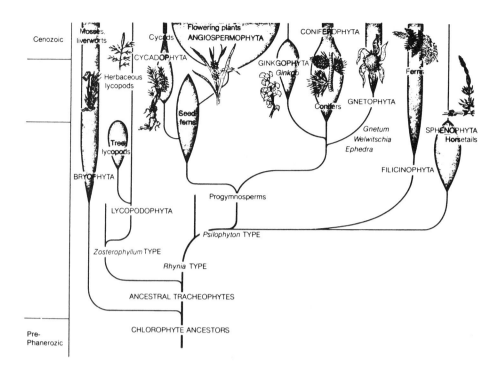

Land plants first appeared in the Devonian Period, as rootless, leafless, and stemless but upright seaweedlike organisms. The earliest for which there is a good fossil record were ancestral tracheophytes of two sorts, represented by the extinct genera *Zosterophyllum* and *Rhynia*. The bryophytes are presumed to have evolved before the appearance and stabilization of vascular tissue—that is, before the appearance of these tracheophytes—although there is no early bryophyte fossil record.

The *Zosterophyllum* types gave rise to lycopods (Phylum Pl-2), a group that speciated extensively at the end of the Paleozoic Era but is now reduced to a few small herbaceous genera. The *Rhynia* types were the ancestors of all the other vascular land plants. Many groups, such as the seed ferns (Cycadofilicales—all extinct) and the horsetails (Sphenophyta, Phylum Pl-3), were far larger and more important in the past than they are now.

Although the flowering plants (Angiospermophyta, Phylum Pl-9) are an enormous group, they are relatively young, having appeared on the scene about 100 million years ago. They are apparently descended from seed ferns. The actual steps that led to the origin of seeds and fruits are not known, but that evolutionary innovation changed the living world by producing an environment in which man and other mammals could survive.

If it seems that there are far fewer plant than animal groups, it is partly because plant taxa are defined by morphological rather than chemical criteria. The differences between many plants are invisible—they produce different chemical compounds called secondary metabolites. Such compounds are not directly required for survival and reproduction, but they play a role in the plant's defenses against fungi, animals, and other plants. They include feeding deterrents, toxins, psychoactive compounds such as the marijuana alkaloids, and nerve poisons such as cyanide. Some plants leak compounds into the soil to prevent plants of other species from growing around them. These poisons and other secondary metabolites, even gaseous compounds, are important in determining the distribution, growth rate, and abundance of plants in natural communities. Thousands of secondary metabolites are known; many are starting materials for the manufacture of drugs.

Plants are distinguished from organisms belonging to the other eukaryotic kingdoms by their life cycles. They develop from embryos, diploid multicellular young organisms supported by sterile or nondeveloping tissue. Unlike animals, most of whose cells are diploid, and fungi, which are mostly haploid or dikaryotic, plants alternate haploid and diploid generations in an orderly fashion. Haploid plants are called gametophytes; diploid plants are called sporophytes. In the bryophytes, the more conspicuous green plant is the gametophyte; the sporophyte is small and brown and different looking. In the tracheophytes, on the other hand, the sporophyte is green and larger and more conspicuous than the gametophyte. The sporophyte generation dominates the life cycle in the most recently evolved phyla: in the flowering plants, the gametophyte is so reduced that, instead of being a separate plant, it is only a small group of cells entirely dependent on the sporophyte.

AUTHOR CITATION INDEX

Aaronson, S., 124
Abelson, P., 65
Abernethy, J. L., 122
Abram, D., 219
Abu-Erreish, G., 123
Ackrell, B. A. C., 144
Adams, D. K., 241
Adams, G. M. W., 175, 238, 283
Adelberg, E., 67, 86, 221
Adler, J., 219
Afinogenova, A. V., 220
Agsteribbe, E., 155, 197
Ahmadjian, M. E., 219
Ahmadjian, V., 219
Ahn, T. I., 96
Akin, D. E., 197
Akroyd, P., 94
Albergoni, V., 123
Alberte, R. S., 166
Alexander, M., 220
Alexander, N. J., 175
Alexander-Johnson, E., 124
Al-Hafidh, M., 66
Allen, C., 264
Allen, M. B., 66, 82, 166
Allen, M. M., 166
Allen, R. D., 222
Almassy, R. J., 76, 86
Althauser, M., 219
Altmann, R., 16
Ames, B. N., 264
Ammer, D., 124
Anastasi, A., 123
Anderson, C., 219
Anderson, D. R., 176, 182
Anderson, K., 272
Anderson, R. G. W., 264
Anderson, R. S., 197
Anderson, T. W., 124
Andreozzi, J. A., 307
Andrews, P., 123
Anita, N. J., 182
Anker, P., 306
Appel, P. W. U., 86
Archer, D., 239
Argetsinger, J., 264
Asada, K., 122, 123

Asano, A., 144
Atchison, B. A., 197
Attardi, G., 140
Auffret, A. D., 122
Avers, C. J., 136, 140

Bailey, C. R., 76
Baker, J. R. J., 182
Balasingam, M., 127
Balch, R. S., 123
Balch, W. E., 86, 192
Baldensperger, J. B., 123
Baldwin, I. L., 94
Banerjee, S., 265, 272
Bang, S. S., 123
Bannister, J. V., 123, 124, 127
Bannister, L. H., 110
Baptist, J. N., 221
Barath, Z., 140, 232
Bardele, C., 272
Barghoorn, E. S., 307, 65, 86
Barikowiak, A., 123
Barile, M. F., 182
Barker, W. C., 76
Barnabas, J., 76
Barnett, A., 306
Barnett, W. E., 140
Barra, D., 124
Barrell, P. G., 192
Barth, L. G., 264
Barth, L. J., 264
Bartley, W., 144
Bassot, J. M., 123
Baugh, L. C., 265
Baumann, P., 123, 222
Bawa, K., 306
Bayley, S. T., 86
Beale, G. H., 82, 182
Beatrice, M. C., 144
Beauchamp, C. O., 123, 127
Beckman, G., 123
Begel, O., 197
Belar, K., 65
Belcour, L., 197
Bell, G., 123, 306
Ben-Shaul, Y., 67, 183
Bendich, A. J., 197

Beringer, J. E., 110
Berliner, J., 265
Bernal, J. D., 65
Bernheim, B. C., 182
Berns, M. W., 264
Bibby, B. T., 182
Bibring, T., 272
Binder, L. I., 264, 265
Birky, C. W., Jr., 176
Bisalputra, B. R., 272
Bisalputra, T., 182
Bjorn, G. S., 202
Black, J. A., 123
Blacklock, J. S. W., 241
Blakemore, R., 123
Blakesly, R. W., 123
Blankenship, M. L., 182
Blaschke, G., 144
Block, K., 65
Bloodgood, R. A., 232, 272
Blum, J., 124
Boeshore, M. L., 197
Boezi, J. A., 123
Bogomolni, R. A., 202
Bogorad, L., 202
Boisvert, H., 123
Bolen, P. L., 175
Bonaventura, C., 123
Bonaventura, J., 123
Bonen, L., 86, 123, 152, 182, 192, 233
Borders, C. L., 123
Borisy, G. G., 284, 264, 272
Borst, P., 232
Bottomley, W., 197
Bouck, G. B., 182
Boughter, J. M., 272
Boulter, D., 76
Boyer, P. D., 155
Boyle, J. E., 110, 111, 123
Boynton, J. E., 175, 176
Boyum, R., 123
Brachet, J., 65
Bracker, C. E., 110
Bradbeer, J. W., 197
Bradbury, E. M., 305
Bradfield, J. Y., 197

Author Citation Index

Brady, F. O., 123
Braverman, G., 66
Brenner, R. M., 264
Breuninger, E., 284
Brewin, N., 110
Breyer, E. P., 144
Breznak, J., 272
Briand, J., 197
Bridges, S. M., 123
Briggs, W., 202
Brimacombe, R., 152
Brinkley, B. R., 264, 265, 272
Britton, L., 123
Brock, C. J., 122
Brock, T. D., 232
Broda, E., 76, 192
Brodie, A. F., 144
Brokaw, C. J., 283
Brosius, J., 152
Brothers, A. J., 264, 265
Brown, J. S., 166
Brown, R. D., 155
Brown, R. H., 197
Brown, R. M., 182
Brownlee, G. G., 192
Bruff, B., 221
Brunori, M., 76
Bruschi, M., 76
Bryson, V., 76
Buchanan, R. E., 76
Bücher, T., 140, 232
Buchner, P., 65, 219
Buck, C. A., 82
Buetow, D. E., 183, 202
Bullough, W. S., 115
Burger, A., 219
Burnet, M., 219
Burnham, J. C., 219
Buss, L., 306
Butow, R. A., 140, 197

Cabioch, L., 220
Cachon, J. G., 272
Cachon, M., 272
Calhoon, R. E., 124
Calkins, G. N., 66
Calvayrac, R., 197
Cammack, R., 124
Canale-Parola, E., 272
Cande, Z., 272
Cantor, M. C., 183
Capano, M., 144
Carbon, P., 152
Carlson, K., 265
Carr, N. G., 192
Carrico, R. J., 123
Carroll, T. W., 221
Cassini, A., 123
Castro E Melo, J., 219
Cavalier-Smith, J., 86
Cavalier-Smith, R. A., 284
Cavlier, T., 182

Cerami, A., 127
Cernichiari, E., 111
Chance, B., 144
Chao, S., 197
Chapman, D. J., 82, 166
Chapman-Anderson, C. A., 144
Chappell, J. B., 144
Chase, D., 111
Chatelain, R., 123
Chatterjee, A. K., 221
Chatton, E., 251
Chen, K. N., 123
Chen, M. W., 152
Cheng, J. Y., 182
Cheng, L., 166
Cheung, A. T. W., 272
Child, F., 251
Child, T. M., 264
Chilton, M. D., 124
Chin, D., 272
Chou, D., 219
Christensen, T., 182
Clark-Walker, G., 136, 232
Claus, G., 82
Clayton, R., 202
Cleveland, D. W., 265
Cleveland, L. B., 66
Cleveland, L. R., 305, 306, 307, 238
Cloud, P. E., Jr., 66, 76, 307
Cochran, W. G., 124
Coen, D., 176
Cohen, H., 124
Cohen-Bazire, G., 111, 192
Cole, W. J., 82
Collins, C. R., 110
Conde, M. F., 175
Conti, S. F., 219, 222, 232
Cooper, G., 272
Copeland, H. F., 66
Corliss, J. O., 307
Cornish-Bowden, A., 123
Cosbey, E., 175
Cottrell, S. T., 140
Council, K. A., 123
Cowdry, E. V., 16
Cox, E. R., 183
Crane, F. L., 144
Crapo, J. D., 123
Crawford, N., 152
Crawford, R. M., 182
Creech, H. J., 284
Crichton, R. R., 123
Crick, F., 86
Crompton, M., 144
Cronquist, A., 66
Cross, N., 233
Crothers, S. F., 219
Crotty, W. J., 197
Cullis, K. F., 197
Cummins, J. E., 232
Cunningham, R. S., 86, 151, 152

Cunningham, W. P., 265
Cutting, J. A., 94

Dagert, M., 197
Dahlberg, M. D., 123
Daily, W. A., 166
Dalgarno, L., 152
Daniels, E. W., 144
Darland, G., 232
Darley, W. M., 182
Darlington, C. D., 66
Darnell, J. E., 86, 140
Dartnall, H. J. A., 202
Davey, P. J., 140
Davidson, E. H., 115
Davies, B. H., 166
Davies, J. M., 182
Davis, B. J., 94, 123, 127
Davis, B. K., 219
Day, E. D., Jr., 123
Day, P. R., 307
Dayhoff, M. O., 76, 122, 124
Dearlove, G. E., 275
De Bary, A., 219
DeDuve, C., 232
Delbrück, M., 176
Dembowska, W. A., 275
Denny, C. F., 219
Dentler, W. L., 264, 265
Desikachary, T. V., 76, 192
Deutsch, H. F., 123
Deutsch, J., 176
DeVay, J. E., 221
De Vries, H., 155
De Wachter, R., 152
Dickbuch, S., 222
Dickerson, R. E., 76, 86, 144
Diedrich, D. L., 219
Diehn, B., 202
Dillon, L. S., 66
Dilworth, M. J., 94
Dippell, R. V., 251, 264, 272
Disbrey, C., 265
Dixon, W. J., 124
Dobson, G. V., 192
Dodge, J. D., 182, 183
Doelle, H. W., 144
Doflein, F., 66
Doolittle, P. R., 192
Doolittle, W. F., 86, 122, 151, 152, 182, 232
Dorey, A. E., 110
Doudorff, M., 67, 221
Dougherty, E., 66
Doutt, R. L., 219
Downs, A. J., 144
Drews, G., 219
Droop, M., 82
Drouet, F., 166
Drucker, I., 222
Drum, R. W., 182
Drummond, M. H., 124

Drysdale, J. W., 123
Duboscq, O., 272
DuBuy, H., 66
Duclaux, G. N., 110
Dujon, B., 176, 232
Dunn, J. E., 219
DuPraw, E. J., 82
Durasay, P., 123
Dutton, P. L., 144
Dyer, T. A., 123

Earl, A. J., 197
Eaton, M., 127
Ebel, J. P., 152
Ebel, R. A., 192
Ebersold, W. T., 176, 284
Echlin, P., 66, 82
Eck, R. V., 76
Eckert, R., 284
Edelhoch, H., 123
Edelman, M., 66, 82, 192
Edwards, H., 197
Edwards, M. E., 202
Ehrenberg, L., 265
Ehresmann, C., 152, 192
Ehrlich, M. A., 219
Eichmann, R., 86
Eisenstadt, J., 66
Ellis, R. J., 82
Engebrecht, J., 307
Engel, A. E. J., 82
Engelking, H. M., 219
Engelland, A., 112
Enmanji, K., 123
Ephrussi, B., 232
Epler, J. L., 140
Epstein, H. T., 66, 67, 192
Erickson, S. K., 144
Erlich, S. D., 197
Etkin, N. L., 127
Evans, E. H., 192
Evans, E. L., 166
Evans, L. V., 182
Evans, T. E., 232

Fackrell, H. B., 219
Falconet, D., 152
Falk, H., 182
Fan, H., 140
Farineau, J., 197
Farkas, G. L., 265
Farr, A. L., 123, 140
Farr, N. J., 123
Farrar, J. F., 110
Farrelly, F., 197
Farris, J. S., 76
Fauron, C. M. R., 197
Fedorcsak, I., 265
Fedorova, A. M., 220
Feeley, J., 82
Feinlieb, M., 202
Fellner, P., 192

Ferguson, S. J., 144
Ferrier, N. C., 182
Fewson, C. A., 66, 144
Fichte, B. A., 220
Fischer, A. G., 66
Fischer, M., 144
Fitch, W. M., 76
Fitzharris, T. P., 232
Flavell, R., 182, 264
Fonesca, J. R., 272
Ford, P. J., 155
Forget, P., 144
Fork, D. C., 82
Foster, K. W., 284
Foulds, I., 192
Fox, G. E., 76, 86, 122, 152, 192, 233
Fox, S. B., 175, 176
Francis, G. W., 166
Francis, S., 272
Franke, W. E., 182
Fred, E. B., 94
French, C. S., 166
Fricke, B., 183
Fridovich, I., 122, 123, 124, 127
Fried, R., 123
Friedlander, M., 264
Fritsch, F. E., 66
Fujita, Y., 202
Fujiwara, K., 272
Fuller, G. M., 272
Fulton, C., 264

Gabay-Laughnan, S., 197
Gabriel, M., 66
Gahan, P., 306
Gallop, A., 110
Galmiche, J. M., 66
Galper, J. B., 140
Gawlik, S. R., 183
Gellissen, G., 197
Gerhart, T., 264
Gerling, D. A., 232
Getz, G. S., 144
Gharagozlou, I., 272
Ghei, O. K., 144
Ghosh, A., 232
Gibbons, I. R., 66
Gibbons, N. E., 76
Gibbs, M., 66
Gibbs, S. P., 182, 183
Gibor, A., 66, 232
Gibson, J., 123
Gilbert, W., 86, 155, 197
Gillham, N. W., 175, 176, 232
Gilmore, E. B., 140
Gilvarg, C., 265
Girard, A. E., 144
Gloor, L., 220
Glotz, C., 152
Goldschmidt, R., 66
Goldsmith, T. H., 166

Goldstein, L., 96
Gonatas, N. K., 265
Gooday, G. W., 111
Goodman, M., 76
Goodwin, T. W., 166
Gordon, M. P., 124
Goscin, S. A., 123
Gott, C., 94
Gould, R. R., 284, 264
Gould, S. J., 124
Gowans, C. S., 284
Graham, J. R., 166
Grandi, M., 140
Granick, S., 66, 232
Grassé, P., 272
Gray, M. W., 86, 151, 152, 159, 197
Green, G. R., 207, 233
Green, T. G. A., 111
Greenwood, A. D., 182
Greenwood, C., 76
Greenwood, P. H., 124
Greppin, H., 306
Griffiths, H. B., 182
Grimes, G. W., 275
Grimstone, A. V., 238
Grisham, L., 272
Grivell, L. A., 155
Gromov, B. V., 220
Gross, S. R., 140
Grossman, I. W., 232
Grossman, L. I., 82
Gruber, H. E., 284
Grula, J. W., 197
Grunstein, M., 197
Guélin, A., 220
Guilley, H., 197
Gupta, R., 123, 152
Gurney, T., Jr., 264
Gutell, R., 152
Guttes, E. W., 82
Guttes, S., 82

Haarhoff, K. N., 144
Haddock, B. A., 144
Hagenbüchle, O., 86
Haimo, L. T., 265
Haldane, J. B. S., 66
Hale, M. E., 219
Halfen, L. N., 166
Hall, D. O., 122, 123, 124
Hall, R., 220
Hall, W. T., 82
Halvorson, H. O., 140
Hamilton, W. A., 144
Hammersmith, R. L., 275
Hanawalt, P. C., 82
Haneda, Y., 123
Haniu, M., 76
Hansen, K. L., 219
Hanson, E. D., 330
Hanson, M. R., 197

Author Citation Index

Harington, A., 159
Harkins, R. N., 123
Harris, C. E., 123
Harris, C. H., 275
Harris, E. H., 175, 176
Harris, J. I., 122, 123
Hartman, H., 264
Hartmann, M., 66
Hase, T., 76
Hasemann, V., 122
Hashimoto, T., 219
Haskins, F. A., 140
Haslam, J. M., 140
Haslett, B., 76
Hastings, J. W., 123, 124
Hastings, P. J., 175
Hatchikian, C. E., 123
Hattori, A., 202
Haupt, W., 202
Haury, J., 202
Hawley, E. S., 140
Haxo, F. T., 82, 166
Heath, I. B., 307, 182, 272
Hebbel, R. P., 127
Hedrick, J. L., 123
Heidemann, S. R., 264
Heitkamp, D. H., 232
Helms, A., 140
Helwig, J. T., 123
Henry, L., 122
Henry, Y. A., 123
Hensgens, L. A. M., 155
Heocha, C. O., 82
Hernault, S. W. L., 275
Hespell, R. B., 123, 220
Heywood, P., 182
Hibberd, D. J., 182
Highfield, D. P., 272
Hill, R. L., 122
Hinde, R., 111
Hirth, L., 197
Hoare, D. S., 192
Hodge, T. P., 197
Hoffman, E. J., 66, 264
Hogan, J. J., 152
Hogness, D. S., 197
Holdeman, L. W., 272
Hollande, A., 272
Holligan, P. M., 111
Holowinsky, A. W., 166
Holzer, G., 86
Honda, S. I., 197
Hongladarom, T., 197
Hopkins, J. M., 265
Hoppe, J., 155
Hopps, H. E., 182
Hopwood, A., 197
Horio, T., 144
Horowitz, A. T., 220
Hovind-Hougen, K., 239, 272
Howe, C. J., 197
Howland, G. P., 202

Huang, B., 283, 284
Huang, J. C. C., 219, 220, 221
Huberman, J. A., 115
Hudock, M. O., 175
Huh, T. Y., 152
Hunt, L. T., 76
Huxley, A. L., 307
Hyams, J. S., 284

Ichihara, K., 123, 124
Imai, K., 144
Ingraham, J., 86
Ingram, L. O., 192
Inman, O. L., 166
Inoué, S., 264, 272
Ishiguro, E. E., 220
Iwamatsu, T., 264
Izhar, S., 197

Jackl, G., 155
Jacobs, H. T., 197
Jahn, J. R., 272
Jakob, H., 136, 233
Janekovic, D., 86
Jaskunas, S. R., 176
Jearnpipatkul, Y., 127
Jeffrey, S. W., 166, 182
Jeon, K. W., 96, 111
Jeon, M. S., 96, 111
Jergensen, E. G., 197
Jinks, J. L., 66, 232
Johansen, J. T., 122
John, P., 144, 232
Johnson, D., 144
Johnson, R. T., 124
Johnson, S. C., 123
Johnson, U. G., 284
Johnston, A. W. B., 110, 197
Johnston, C. S., 182
Jonard, G., 197
Jones, C. W., 144
Jope, C., 197
Jorgensen, R. A., 192
Junge, C. E., 86
Jurand, A., 82, 111, 182

Kahane, I., 124
Kahn, W., 183
Kalf, G. F., 82
Kalley, J. P., 182
Kalnins, V., 272
Kamen, M. D., 144
Kamen, P. L., 144
Kanematsu, S., 122, 123
Karakashian, M. W., 111, 182
Karakashian, S. J., 66, 111, 182
Katz, J., 166
Kay, W. W., 144
Keele, B. B., Jr., 123
Keiding, J., 140
Kelleher, J. K., 265, 272
Keller, A., 197

Kelly, M. R., 238
Kemble, R. J., 197
Kessell, M., 220
Keya, S. O., 220
Kidd, P., 264
Kirby, T., 124
Kirk, J. T. O., 82
Kirschner, M. W., 264, 265
Kissil, M. S., 183
Kitchener, K., 127
Kivic, P. A., 182
Kleinig, H., 182
Klingenberg, M., 140, 144
Klubek, B., 220
Knoll, A., 86, 307
Ko, C., 96
Kocur, M., 144
Kolodner, R., 197
Kono, Y., 122
Konovalova, S. M., 220
Kop, J., 52
Kozak, M., 86
Kretsinger, R. H., 233
Kroon, A. M., 232, 233
Krungkrai, J., 127
Kudo, R. R., 144
Kuhn, P., 241
Kunisawa, R., 192
Küntzel, H., 140, 197, 232
Kusel, J. P., 233
Kusunose, E., 124
Kusunose, M., 123, 134, 124
Kuznicki, L., 272

Labelle, J. W., 115
Lackenbruch, P., 124
Lacks, S. A., 124
Ladwig, R., 219
Laemmli, U. K., 272
Laetsch, W. M., 111
Lam, Y., 144
Lambie, W., 123
Lambina, V. A., 220
Lang, N. J., 182
Lardy, H. A., 144
Lascelles, J., 66
Lassen, W. H. K., 124
Laughnan, J. R., 197
Laval-Martin, D., 197
Lavelle, F., 123
Lawford, H. G., 144
Lazaroff, N., 192, 202
Lazarus, C. M., 197
Leaver, C. J., 152, 197
Ledbetter, M. C., 197
Lederberg, J., 66
Lee, R., 176, 182, 197
Leedale, G. F., 183, 202, 320
Lefebvre, P. A., 284
Leff, J., 66
Le Gall, J., 76, 123
Lehninger, A. L., 144

Leine, M., 144
Lemieux, C., 197
Lengyel, P., 76, 176
Lerch, K., 124
Levine, E. E., 175, 284
Levine, R. P., 175, 284
Levings, C. S., 197
Levisohn, S., 67
Lewin, R. A., 76, 155, 166, 183, 202
Lewis, B. J., 123, 152, 192
Lewis, D., 82, 220
Li, T. -K., 123
Lidstrom, M. E., 307
Lifshitz, I., 197
Linnane, A. W., 136, 140, 232, 233
Littlefield, L. J., 110
Liu, T. -Y., 123
Livingston, A., 124
Livingston, V., 124
Lloyd, W. J., 144
Lockhart, C. M., 111
Loeblich, A. R., 183
Logan, M., 112
Lonsdale, D. M., 192, 197
Loppes, R., 175
Lorch, I. J., 96
Lowry, O. H., 123, 140
Lozier, R. H., 202
Lpyko, W., 123
Luchrsen, L. J., 86
Luck, D. J. L., 140, 233, 265, 283, 284
Luduena, R. F., 272
Luehrsen, K. R., 123
Lukins, H. B., 136
Lumsden, J., 122, 123, 124
Luttrell, E. S., 220
Lwoff, A., 251, 264
Lyman, H., 66
Lyttleton, J. W., 66, 82

McAuley, P. J., 111
McCord, J. M., 123, 127
McCoy, E., 94
MacCoy, M. T., 140
McElvany, K. D., 123
McGill, M., 264
Machlieidt, W., 155
McIntosh, J. R., 265, 272
MacKay, R. M., 152
Mackie, P., 192
McLain, G., 144
McLaughlin, P. J., 76
Maclean, N., 306
M'Cluskie, J. A. W., 241
MacVittie, A., 284
Maeda, Y., 123
Magnes, L., 123
Magrum, L. J., 86, 152, 192
Mahler, H. R., 76, 144

Makhlin, E. E., 96
Maliga, P., 197
Malinowski, D. P., 123
Maller, T., 264
Mamkaeva, K. A., 220
Mandel, M., 66, 67, 192, 221
Maniloff, J., 123, 232
Mans, R. J., 86, 197
Marchalonis, J. J., 123
Marcker, K. A., 82, 140
Marcus, A., 82
Margoliash, E., 76
Margrum, R. S., 123
Margulis (Sagan), L., 76, 82, 86, 96, 111, 115, 122, 144, 159, 183, 202, 232, 233, 265, 272, 307, 308
Markert, C. L., 307
Marklund, S., 123
Markowitz, H., 124
Marmur, J., 67, 82
Maroudas, N. G., 140
Marsh, H. V., 66
Marshall, C. E., 16
Martin, B., 122
Martin, J., 124
Martinec, T., 144
Martini, F., 124
Marvin, D. A., 115
Mason, T. R., 76
Massinger, P., 140
Masui, Y., 264
Mathews, H. R., 306
Matin, A., 220
Matsubara, H., 76
Matsuda, G., 76
Mattern, C. F. T., 66
Mattox, K. R., 183
Maturana, H. R., 308
Maxam, A. M., 155, 197
Maynard Smith, J., 307
Mazanec, K., 144
Mazia, D., 251
Menke, W., 183
Merechowsky, M., 66
Mereschkowsky, C., 82
Meshnick, S. R., 127
Messer, G., 183
Messing, J., 197
Michel-Wolwertz, M. R., 82
Michelson, A. M., 123
Mickey, R. M., 124
Mignot, J. P., 183
Miki-Nomura, T., 264
Minchin, E. A., 66
Misra, H. P., 123, 124
Mitchell, A., 140
Mitchell, G., 123
Mitchell, H. K., 140
Mitchell, M. B., 140
Mitchell, P., 144
Monier, R., 76

Montes, G., 197
Moore, J., 183
Moore, K. H., 233
Moore, W., 76, 272
Mori, T., 124
Morman, M. R., 144
Morris, I., 66, 82
Morse, M. L., 115
Moulder, J., 111, 284
Mounolou, J., 136, 233
Moustafa, E., 94
Moyle, J., 144
Munkres, K. D., 82
Murphy, M., 155
Murphy, P. S., 144
Muscatine, L., 82, 111
Myers, G., 124
Myers, J., 166

Nagy, B., 82
Nagy, F., 197
Naik, V. R., 122
Nakayaza, K., 123
Nalbandyan, R. M., 123
Nass, M. M. K., 82, 233
Nass, S., 82
Nealson, K. H., 111, 124, 307, 308
Needleman, S. B., 76
Nelson, J. S., 124
Nester, E. W., 124
Netter, P., 176
Neuberger, M. R., 123
Neupert, W., 140, 232
Newcomb, E. H., 166
Newton, N., 144
Nicholas, D. J. D., 144
Nienhaus, A. W., 155
Nisbet, E. G., 86
Nishikawa, K., 76
Nishioka, D., 264
Noda, Y., 123, 124
Noll, H., 82, 140
Noller, H. F., 152
Nordstrom, S. A., 284
Normura, M., 76
Northrop, F. D., 122
Notton, B. M., 144
Novick, R. P., 115
Nultsch, W., 202

Oakley, B. R., 183, 272
Oakley, D. C., 272
Obar, R., 239
Ochs, S., 272
Oehler, D. Z., 76
Ohta, T., 264
Ohyuama, T., 123
Okunuki, K., 144
Olitsky, P. K., 16
Olmsted, J. B., 264
Olmsted, M. A., 284
Opanrin, A. I., 86

Author Citation Index

Orian, J. M., 155
Ormerod, W., 272
Ornstein, L., 123
Oro, J., 86
Osborn, H. F., 16
Osborn, M., 123, 264
Oschman, J. L., 111
Osumi, M., 140
Ouchterlony, O., 123
Outka, D. E., 183
Overballe-Peterson, C., 122
Owen, J. A., 94

Page, L. A., 222
Palmer, J. D., 197
Palmieri, F., 144
Panijpan, B., 127
Panisett, P., 308
Pankratz, H. S., 272
Park, C. M., 76
Parsons, D. F., 144, 233
Pechman, B., 233
Pechman, K., 152, 192
Pechnikov, N. V., 220
Penman, S., 140
Pennell, L., 123
Perkins, D. D., 284
Perrot, B., 197
Petrochilo, E., 176
Petzold, H., 222
Pica-Mattoccia, L., 140
Pichnoty, F., 144
Picken, L., 66
Pickett-Heaps, J., 264, 272, 287, 307
Piperno, G., 265, 284
Pirie, N. M., 66
Plaskitt, K., 239
Plaut, W., 67
Plummer, J., 284
Poccia, D., 264
Pollack, R., 272
Polsky, R. F., 284
Pool, R. R., 111, 112
Popoff, M., 66
Porter, K. R., 265, 284
Posakony, J. W., 197
Postma, P. W., 144
Poulsen, C., 197
Preer, J. R., 82, 111, 182
Preer, L. B., 111
Prescott, D. M., 67
Prescott, G. W., 82
Pribula, C. D., 76
Primack, R., 307
Pring, D. R., 197
Puget, K., 123
Pugh, T. D., 166
Puma, J. D., 264

Quagliaricello, E., 144

Rabinowitz, M., 233
Raff, E. C., 265
Raff, R. A., 76, 144, 265
Raftery, M., 82
Raikov, I. B., 307
Rajagopalan, K. V., 123
Ramanis, Z., 176, 284
Rambler, M., 76, 307
Ramshaw, J. A. M., 76
Randall, J., 265, 284
Randall, R. J., 123, 140
Rao, C. R., 124
Rao, K. K., 76
Ratner, J. B., 264
Raven, P. H., 76
Ravindranath, S. D., 123
Ray, D. S., 82
Reich, E., 233
Reichelt, J. L., 124, 222
Reichenow, E., 66
Reiner, A. M., 220
Renaud, F. H., 67
Renshaw, A., 144
Rhoades, M. M., 202
Richards, K. E., 197
Richardson, D. C., 124
Richmond, M. H., 115
Richter, D., 140
Rifkin, M., 284
Riggs, A. D., 115
Righetti, P., 123
Rigsby, L. L., 197
Riley, F., 66
Ringo, D., 284
Riordan, J. F., 123
Ris, H., 67, 307
Rittenberg, S. C., 220, 221
Robbins, E., 265
Roberts, J. W., 197
Robinow, C. F., 221
Robinson, J., 219, 221
Rohlf, F. J., 123
Romer, A. S., 124
Roodyn, D. B., 136, 233
Rosario, B., 284
Rosebrough, N. J., 123, 140
Rosen, D. E., 124
Rosen, W. G., 183
Rosenbaum, J. L., 264, 265, 284
Rosenfeld, A. C., 265
Ross, E. J., 221
Rossier, C., 306
Rosson, R., 220, 308
Rotilio, M. E., 124
Round, F. E., 183
Rowell, P., 111
Rubin, M. S., 144
Rubin, R. W., 265
Ruby, E. G., 111
Rudda, G. K., 144
Rudzinska, M. A., 182

Rusch, H. P., 232
Rutten, M. G., 67
Ruttenberg, C. M., 232

Sebald, W., 140
Sagan, C., 67
Sagan, D., 307, 308
Sager, R., 175, 176
Sala, F., 140
Salin, M. L., 123, 127
Samallo, J., 155, 197
Saman, G. R., 265
Samsonoff, W. A., 210, 232
Sander, G., 284
Sando, N., 140
Sanger, F., 192
Santer, M., 86
Santore, U. J., 182
Sato, R., 144
Sato, S., 123
Saunders, G., 136
Sauvageau, C., 166
Sawada, Y., 123
Scawen, M. D., 76
Schäfer, K. P., 140
Scheibe, J., 202
Scherff, R. H., 221
Schidlowski, M., 86
Schiff, J. A., 66, 67, 192, 202
Schildkraut, C., 67
Schinina, M. E., 124
Schmidt, J. A., 284
Schnare, M. N., 152
Schnepf, E., 82, 183
Scholes, P. B., 144
Schonbohm, E., 202
Schopf, J. W., 65, 76, 82, 86, 192
Schulman, H. M., 94
Schwab, A. J., 140
Schwartz, K. V., 307
Schwartz, R. M., 76, 86, 122
Schweikhardt, F., 144
Schweizer, E., 140
Scott, N., 197
Seaman, G. R., 67
Searcy, D. G., 207, 233
Sebald, W., 152, 155, 232
Sederoff, R. R., 197
Seely, G. R., 82
Seidler, R. J., 219, 220, 221
Senger, H., 202
Sepsenwol, S., 183
Shalijain, A. A., 123
Shapiro, L., 82
Sharoyan, S. G., 123
Sharp, D. G., 265
Shatz, G., 82
Sheeler, P., 183
Sherman, F., 144
Sherman, I. W., 127
Shields, C. R., 197

Shields, G. S., 124
Shields, M., 272
Shilo, M., 220, 221, 222
Shine, J., 152
Shipp, W. S., 144
Shropshire, W., 202
Shugart, L. R., 140
Siegal, S., 123
Siegel, A., 197
Siegel, R. B., 152
Siegel, R. W., 67
Siegelman, H. W., 82
Siegesmund, K. A., 183
Sierra, M. F., 144
Silberman, H. J., 94
Silva, P. C., 82, 183
Simpson, F. J., 221
Simpson, R. J., 123
Sinclair, J. H., 233
Singer, B., 176
Singh, R. N., 192
Sitte, P., 182
Skryabin, G. K., 220
Slankis, T., 183
Sleigh, M., 67
Slonimski, P. P., 136, 176, 232, 233
Smillie, R., 67
Smith, A. E., 82, 140
Smith, A. J., 123
Smith, D., 82, 110, 111, 140
Smith, L., 144
Smith, M. J., 123
Smith, T., 221
Smith-Sonneborn, J., 67
Smyth, R. D., 284
Sneath, P. H. A., 192
Snedecor, G. W., 124
Snell, W. J., 265
Snellen, J. E., 221
Snyder, J., 265, 272
Sogin, M., 192
Sokal, R. R., 123, 192
Sokatch, J. R., 233
Solymosy, F., 265
Sonea, S., 308, 115
Sonneborn, T. M., 251, 275
Sorokin, S. P., 265
Soyer, M. O., 272
Spencer, D. F., 152
Spolsky, C., 86, 144, 183
Stackebrandt, E., 123
Stahl, D., 123, 152, 192
Stamatoyannopoulos, G., 155
Stanier, R. Y., 17, 67, 86, 111, 115, 192, 221
Starling, D., 284
Starlinger, P., 197
Starr, M. P., 220, 221, 222, 233
Stein, D. B., 207, 233
Steinberg, H., 127

Steiner, G., 241
Steinman, H., 122
Steitz, J. A., 86
Stephens, R., 272
Stern, D. B., 197
Stetter, K. O., 86
Stevens, B. J., 233
Stewart, J. W., 144
Stewart, K. D., 183
Stewart, W. D. P., 111
Stiegler, P., 152, 192
Stigbrand, T., 123
Stirpe, F., 144
Stockenius, W., 202
Stolp, H., 222
Storz, J., 222
Strain, H. H., 166
Straley, S. C., 222
Strampp, A., 124
Strand, R., 111
Stroun, M., 306
Stubblefield, E., 265
Stutz, E., 82
Sueoka, N., 67
Sullivan, J. B., 123
Sulloway, F. J., 308
Surzycki, S. J., 175
Suthipark, U., 127
Suyama, Y., 82
Svendsen, I. B., 122
Swain, T., 308
Swieb, C., 152
Swift, H., 67
Swinton, D., 192

Tait, G. H., 144
Takahashi, M., 122, 123
Takemura, S., 76
Tamir, H., 265
Tanaka, M., 76
Tanner, W. E., 123
Tartar, V., 251, 308
Tauro, P., 140
Taylor, D. L., 111, 183
Taylor, E. W., 272
Taylor, F. J. R., 86, 111, 144, 183, 272
Taylor, V. I., 222
Teller, D. C., 123
Terasima, T., 140
Tewari, K. K., 197
Tham, S. H., 136
Thomas, D. Y., 136
Thomas, M. M., 166
Thomas, P. S., 155
Thomashow, M. F., 220
Thompson, G. A., Jr., 265
Thompson, W., 144, 197
Thornber, J. P., 166
Thornley, A. L., 159
Thurston, E. H., 192

Tilney, L. H., 272
Timkovich, R., 76, 144
Timmis, J. N., 197
Tingle, C. L., 175, 176
Tippett, D. H., 307
Tissieres, A., 76, 140
To, L., 111, 272
Tolmach, L. J., 140
Tomas, R. N., 183
Tornabene, T. G., 86
Török, I., 197
Townsend, R., 239
Trager, W., 67, 272
Trang, N. L., 127
Trench, R. K., 112
Tsuchihashi, M., 124
Turner, G., 197
Tzagoloff, A., 144

Uchida, K., 122
Uchida, T., 152, 233
Uematsu, T., 222
Urano, M., 123
Urey, H. C., 67
Ursprung, H., 307
Uzzell, T., 86, 144, 183

Van Baalen, C., 192
Van Dam, K., 144
Van Winkle-Swift, K. P., 175, 176
Van Winkle-Swift, N. W., 175
Vance, P. G., 123
Vandenberghe, A., 152
Van den Boogaart, P., 155, 197
Vanderslice, K., 272
Vanopdenbosch, B., 123
Varela, F., 308
Varon, M., 220, 222
Vernon, L. P., 82, 144
Vesk, M., 182
Vidal, G., 307
Vieira, J., 197
Vierny, C., 197
Virgin, H., 202
Visconti, N., 176
Visser, A. S., 144
Vogel, H., 76
Vogelman, T. C., 202
Vogt, M., 144
Von Brunn, V., 76
Von Jagow, G., 140

Wachter, E., 155
Wada, K., 76
Wagner, R. P., 82, 140
Wahrman, J., 264
Wakimoto, S., 222
Walker, J. C. G., 76
Walker, J. E., 122
Walker, L., 265, 307
Walker, S., 272

Author Citation Index

Walkup, R., 192
Wallace, P. G., 136
Wallin, I. E., 16
Wallin, J. E., 67
Walne, P. L., 284
Walter, P., 144
Wang, W. L., 175
Ward, B. L., 197
Watanabe, T., 308
Watson, M. R., 265
Waud, W. R., 123
Weber, K., 123, 264, 272
Weber, M. M., 233
Weill, L., 176
Weingarten, M. D., 265
Weinrich, D. H., 67
Weisenberg, R., 265
Weisiger, R. A., 122
Weislogel, P. O., 140
Weiss, H., 140
Weiss, P., 251
Weitzman, S. H., 124
Wellburn, A. R., 197
Wellburn, F. A. M., 197
Weltman, J. K., 123
Wenyon, C. M., 67
Werner, S., 232
Westergaard, O., 140
Whatley, F. R., 144, 232, 233
White, B. N., 197
White, D. C., 144

Whitfield, P., 197
Whittaker, R. H., 82
Wilbur, K. M., 182
Wildman, S. G., 197
Wiley, R. D., 123
Wilkie, D., 136, 140, 233
Wilkinson, B. J., 144
Wilkinson, C. R., 112
Williams, D. E., 183
Williams, G. C., 308
Williams, G. R., 144
Williamson, D. H., 140
Wilson, D., 144
Wilson, E. B., 67, 233
Wilson, E. O., 308
Wilson, L., 272
Wilson, M. L., 284
Windom, G. E., 219
Winston, M., 265
Wintersberger, E., 82
Wiseman, A., 175, 176
Withers, N. W., 183
Witkin, E. M., 308
Witman, G. B., 265
Witzmann, R. F., 308
Woese, C. R., 76, 86, 123, 152, 183, 192, 233
Wolbach, S. B., 16
Wolfe, J., 265
Wolfe, R. S., 86, 192
Wolken, J. J., 82

Wood, E., 127
Woodruff, D., 124
Woodward, D. O., 82, 272
Woolhouse, H. W., 197
Woolkalis, M. J., 123
Wunsch, C. D., 76
Wyatt, G. R., 197

Yamakura, F., 123
Yamanaka, T., 144
Yamin, M. A., 233
Yasunobu, K. T., 76
Yoshikawa, K., 123
Yost, F. J., Jr., 123
Younger, K. B., 265
Yow, F. W., 67
Yu, R., 140
Yudin, A. L., 96
Yuthavong, Y., 127

Zablen, L., 76, 123, 152, 183, 192, 233
Zackroff, R., 265
Zeldin, B., 192
Zillig, W., 86
Zimmermann, R. A., 192
Zuelzer, M., 241
Zuker, M., 152
Zurawski, G., 197

342

SUBJECT INDEX

Amitosis, 34
Amoebae, 2, 28, 95, 109, 227, 238
Animalia. *See* Animals
Animals, 1, 86, 325, 330, 334
   mitochondria of, 225-227
Archaebacteria, 83, 84, 86, 228
Autogenous origin theory, 85. *See also* Direct filiation theory

Bacillus, 184
Bacteria, 1, 11, 12. *See also* Monera; Prokaryotes
   age of, 2
   definition of, 11, 13, 15
   relationship to other forms of life, 13, 15
Bacteriocuprein, 116-122
Beta-carotene, 165, 166
Bioluminescent symbionts, 109, 117, 122
Blue-green bacteria, 17, 21, 40, 42, 44, 46, 48, 49, 51, 53-55, 58, 62-64, 68-76, 78, 80, 83, 84, 86, 101, 164, 166, 184, 314, 315, 317
Brown algae, 2, 64, 198. *See also* Phaeophytes

Cannibalism, 295, 300
Carbon dioxide, 52, 54
Carotenoids, 331
Centrioles, 24, 26, 35-37, 48, 58, 60, 63, 228, 252, 265
Centromeres, 24, 34, 36, 59, 60, 291, 292, 296
Chlamydomonas, 2, 43, 65, 167-175, 252, 276, 284, 302, 395
   basal bodies of, 252-265
Chloramphenicol, 137-140
Chlorophylls, 202
   chlorophyll a, 80, 81, 164-166, 177, 318, 331
   chlorophyll b, 80, 81, 165, 166, 177, 318, 331
   chlorophyll c, 80, 81, 165, 318
Chlorophyta. *See* Chlorophytes
Chlorophytes, 81, 162, 165, 177, 319, 331, 332. *See also* Green algae
Chloroplast DNA, 78, 154, 165-175, 182, 193, 198
Chloroplasts, 2, 46, 48, 57, 61, 62, 68-72, 74, 75, 80, 81, 84, 101, 105, 108, 144, 145, 150, 162, 177, 198, 262, 331
Chromosomes, 1
Chrysophyta. *See* Chrysophytes
Chrysophytes, 65, 81, 181, 317. *See also* Golden-brown algae

Chrysoplasts, 64, 162
Ciliate cortex, 244, 251, 273-275
Ciliates, 32, 64, 227, 273-275, 295
Closed chromosomes, 114
Compartmentalization, 84, 85
Criteria for symbiotic origin, 6, 47, 48, 237, 238
Cryptocercus, 288, 289
Cryptomonads, 42, 80, 81, 181
Cryptophyceae. *See* Cryptomonads
Cyanobacteria. *See* Blue-green bacteria
Cycloheximide, 153, 154, 199, 225
Cytochrome c, 72-75, 141, 224
Cytochromes, evolution of, 142-144, 224
Cytoplasm, 71
Cytoplasmic DNA, 110
Cytoplasmic inheritance, 49, 57, 62, 65, 132, 136, 231, 244, 251

DCCD-binding protein, 153, 154
Diatoms, 2, 64, 65, 80, 165, 317
   symbionts of, 100, 101
Dinoflagellates (dinomastigotes), 2, 64, 65, 177, 178, 181, 236
Direct filiation theory, 5, 6, 16, 17, 72, 110, 266
DNA
   distribution of, 22, 24
   loss of, 81, 110, 157-159, 225
DNA base ratios, 64, 65

Elongation factors, 137-146
Embryos, 2
Endoplasmic reticulum, 177, 182, 199, 228, 229
Endosymbiotic theory, 4, 6, 8, 9, 71, 72, 74, 84, 85, 141-144, 149-151, 157-159, 177, 198
Erythrocytes, mouse, 125, 126
Ethidium bromide, 137-140
Eubacteria, 84, 85
Euglena, 2, 43, 46, 61, 78, 81, 165, 177, 198
Euglenoids, 317
Euglenophyta. *See Euglena*
Euglenoplasts, 162
Eukaryotes, 1, 3, 4, 69
Eukaryotic motility organelles (undulipodia), 1, 2, 24, 26, 28, 30, 32, 34, 37, 47, 49, 55, 57-65, 71, 79, 99, 236, 266, 276, 284, 317, 321, 331
Eumitosis, 24, 30, 32, 34, 40, 64
Evolution, rates of, 69, 109, 190, 191
Extrusomes, 6

343

Subject Index

Ferns, 2
Ferredoxins, 70-72, 74, 75
Flagella, bacterial, 1, 60, 64, 238
Flagellates, 36. See also Mastigotes
Flagellin, 1
Flavodoxins, 75
Fossil record, 4, 68, 69, 293
Fungi, 1, 72, 198, 266, 321, 324, 333, 334
    mitochondria of, 225-227

Genetic code, 131
Genetic elements, 157-159
Gene transfer, 69, 72, 75, 98, 110, 116-122, 131, 153, 154, 157-159, 162, 163, 193
Golden-brown algae, 198. See also Chrysophytes
Green algae, 42, 76, 80, 164-166, 177, 178, 198, 303, 331. See also Chlorophytes

Hydrogen (molecular), 50, 51
Hydrogenosomes, 6
Hydrogen sulfide, 54

Intervening sequence, 85

Kinetochores, 296-303. See also Centromeres
Kinetosomes, 237, 250

Leghemoglobin, 90-94
Lichens, 72, 98, 99, 105-107

Mastigotes, 2. See also Flagellates
Meiosis, 49, 285, 308
Metazoa, 325
Microfossils, 2, 51, 53, 69, 83, 84, 184, 291
Microtubule-organizing centers, 252, 265, 271
Microtubules, 1, 237, 238, 252, 272, 286, 292
Mitochondria, 1, 2, 10-13, 15, 16, 22, 47-49, 57-59, 62, 63, 68, 71-79, 84, 86, 106, 109, 130-140, 163, 198, 206, 207, 223
Mitochondrial ATPase, 153, 154
Mitochondrial DNA, 64, 78, 85, 130-134, 136, 137, 140, 145-151, 153, 154, 167, 169, 193, 197, 198, 224
Mitosis, 17, 271
Mitotic apparatus, 59
Mitotic figures, 24-27, 29, 31, 38, 39, 41
Molds, 2
Monera, 1, 311, 316. See also Bacteria; Prokaryotes
Moss, 2
Motility organelles, 62, 289
Mushrooms, 2
Mutualism 97, 98, 103, 108
Mycoplasm, 206
Myxomcetes, 2

Neurospora, 137-140, 153, 154
Nine-plus-two organelles. See Eukaryotic motility organelles (undulipodia)
Nucleocytoplasm, 84-86, 206, 234
Nucleus, 1, 49, 50, 58
    "changed," 96
Nutrient transfer, 98, 104

Oligonucleotide catalogue, 184
Organelles, retention of, 56
Origin of life, 50, 63
Oxidative respiration, 1
Oxygen (molecular), 4, 21, 50-53, 69, 78, 125, 224, 315

Paracoccus, 130, 141-144, 224, 228
Paramecium, 46, 65, 79, 180, 244, 251, 302
Parasites, 2, 102
Parasitism, 97, 98, 103, 104, 108, 125, 127, 319
Phaeophytes, 81, 162. See also Brown algae
Phaeoplasts, 162
Phagocytosis, 102-104, 179, 206, 327
Photosynthesis, 1, 2, 20, 21, 52, 53, 55, 57, 74-76, 78, 197-202
Photosynthetic bacteria, 72
Phycobilins, 69, 70, 186, 318
Phylogenetic trees, 68, 69, 72, 73, 75, 76, 116-122, 186
Pinocytosis, 327
Plantae. See Plants
Plants, 1, 331, 334
    mitochondria of, 225-227
Plasmodium, 125, 126
Plastid DNA, 61, 64, 65, 85
Plastids, 40, 45, 49, 58, 63-65, 77-79, 86, 131, 141, 162, 163, 198, 206, 207, 331
    secondary loss of, 45
Porphyrins, 20, 51, 52, 80
Positional information, 276, 284
Prasinophyceae, 180
Prebiotic evolution, 84
Primitive atmosphere, 50
Prokaryotes, 1, 68, 76, 83, 86, 184, 192, 198, 202. See also Bacteria; Monera
Protein synthesis, 137-140
Protists, 83, 293, 303, 322. See also Protoctists
Protoctista. See Protoctists
Protoctists. 1, 317, 320, 325, 326, 329, 330
Pseudomonads, 209. See also Protists
Pyrrophyta, 81

Recombination, 167
Red algae, 2, 44, 64, 80, 178, 198. See also Rhodophytes
Rhizobium, 90-94, 103, 104
Rhodophyta. See Rhodophytes
Rhodophytes, 65, 81, 162, 303. See also Red algae
Rhodoplasts, 64, 66, 162
Ribosomal RNA, 71-73, 78, 84, 85, 137, 184, 225, 237, 262
    genes for, 145-151, 225
Ribulose bisphosphate carboxylase, 193
RNA, role in centriole, 252, 265

Saccharomyces, 132-136, 153, 154
Satellite band DNA, 57, 61, 64
Selection pressure, 2
Sequence evolution, 68, 76, 190, 191
Sex, 28, 69, 285, 308

Subject Index

Sperm, 2
Sperm tails, 1
Spindle fibers, 1
Spirochaetes, 60, 61, 71, 104, 237, 238, 240, 243, 266, 272, 286, 298
Spirochetes. *See* Spirochaetes
Spiroplasms, 286
Sporozoa, 2
Stentor, 249, 304, 305
Stromatolites, 3, 69, 75
Superoxide dismutase, 116–122, 125, 126
Symbiont integration, 96, 99
Symbiont specificity, 211, 212
Symbiosis, 2, 6, 8, 13, 15, 17, 19, 21, 37, 40, 45, 55, 57, 59, 64, 70, 73, 88, 90, 91, 94, 116, 130, 177, 209
  maintenance of, 88, 89
Synaptonemal complex, 300, 301
Syncyanoms, 46
Synechocystis, 164–166

Taxonomic criteria, 44, 63
Taxonomy, 185
Terpenoids, 230
Tetrahymena
  basal bodies of, 252, 265
  tissue, 2
Transfer RNA, 137, 225, 237
Transfer RNA genes, 225
Tubulin, 1, 2
Tubulin-like protein, 1

Undulipodia. *See* Eukaryotic motility organelles (undulipodia)
Urkaryote, 83

Volvox, 2

Xanthophylls, 116, 181, 318, 331
Xanthophyta. *See* Xanthophytes
Xanthophytes, 81

About the Editors

BETSEY DEXTER DYER received the Ph.D. in microbiology from Boston University and teaches at Wheaton College in Norton, Massachusetts. Her research interests include metal cycling, the Gaia hypothesis, and the origins of eukaryotes.

ROBERT OBAR received the Ph.D. in biochemistry from Boston University and is a research associate at the Worcester Foundation for Experimental Biology. His research interests include the origins and evolution of the eukaryotic cytoskeleton.